Catalytic Ab's

▲ Janda, K.D. et.al.
- Science 241:1188, 1988
- " " 437, 1989
- J. Am. Chem. Soc. 112:1274
* - Biotechnol. Prog., in press

▲ Iverson, B. et.al.
* - Science 243:1184, 1989
- " 249:659, 1990

▲ Huse, W.D. et.al.
* - Science 246:1275, 1989

▲ Lerner, R.A et al
* - Sci Am. 256:58 1988

▲ Tramontano, A et al
* - J. Am. Chem. Soc. 110:22282, 1988
* - Meth. Enzymol. 178:531, 1989

▲ Janjic, N. et.al.
- J. Am. Chem. Soc. 111:6374, 1989

Topics in Medicinal Chemistry
4th SCI–RSC Medicinal Chemistry Symposium

Special Publication No. 65

Topics in Medicinal Chemistry

The Proceedings of the 4th SCI–RSC Medicinal Chemistry
Symposium, organised by the Fine Chemicals Group of the
Society of Chemical Industry and the Fine Chemicals and
Medicinals Group of the Industrial Division of the Royal
Society of Chemistry

Cambridge, England, 6th–9th September 1987

Edited by
P. R. Leeming
Pfizer Central Research,
Pfizer Limited, Sandwich, Kent

ROYAL
SOCIETY OF
CHEMISTRY

CHEM

Sep/ae

British Library Cataloguing in Publication Data

SCI-RSC Medicinal Chemistry Symposium
(4th : 1987 : Cambridge, England)
Topics in medicinal chemistry.
1. Medicine. Biochemistry
I. Title   II. Leeming, P. R.
III. Series
612′.015

ISBN 0-85186-726-X

The cover illustration shows the structure of the antifungal agent fluconazole (see pp. 255–266)

Published by the Royal Society of Chemistry, Burlington House, London, W1V 0BN

Printed in Great Britain by Henry Ling Ltd., at the Dorset Press, Dorchester, Dorset

SD 88
81251 HK

Introduction

The Cambridge biennial series of Medicinal Chemistry Symposia are organised jointly by the Royal Society of Chemistry (Fine Chemicals and Medicinals Section of the Industrial Division) and the Fine Chemicals Group of the Society of Chemical Industry. Since the first Symposium in 1981, the series has become established as a major international forum for presentation and discussion of recent scientific advances relevant to drug discovery.

The 1987 Symposium commenced with a session on Neuropeptides. The 1981 Symposium included lectures on opiate receptors and regulatory peptides at a time when these were relatively new topics. The rapid strides in understanding the structure-activity relationships of various neuropeptides since then can be judged from the first five papers of this book.

The second topic was Control of Enzymatic Processes. After a review of the different approaches to enzyme inhibition, there are papers on renin inhibitors, the use of the chemistry of small rings to moderate enzyme activity, and the synthesis of amino-sugar inhibitors of glycosidases.

The third topic was Antiviral Agents - the first time this subject has featured in the series. After many years of slow progress in this field - the importance of which has been greatly increased by the emergence of the Human Immunodeficiency Virus - there have been many striking achievements of which acyclovir, for the treatment of herpes, is the most well-established. Appropriately, this is discussed in a plenary paper. One of the other papers covers zidovudine (AZT), the first drug of clinical value in combatting the HIV virus, and there are two papers which describe more recent laboratory discoveries of compounds with antiviral activity. The session as a whole illustrates the considerable advance made in this therapeutic area in recent years.

The fourth topic includes a paper on the avermectins, a complex series which has engendered enormous interest in veterinary medicine for the treatment of endo- and ecto-parasite infections, and the last paper of this section describes total synthesis of these substances which have provided organic chemists with such excellent targets for demonstration of ingenuity in the control of stereochemistry in molecules possessing numerous chiral centres. There are also papers on the exciting new antifungal agent fluconazole and one on recent work to find drugs to combat the growing threat of malaria.

The fifth and final session contained four very different papers illustrating the use (and abuse!) of computational chemistry. This topic has featured in all three previous Symposia which illustrates the degree of fascination that the subject has for medicinal chemists and for Symposia organisers! The power of the computer to calculate and display conformations and structural intricacies of complex molecules is not in question. The debate continues, however, on the prospective value of computational techniques in the search for new bio-active molecules - or perhaps on how best to apply computational power to that end.

We are grateful to all speakers for the very high standard of both the papers and the slides at this Symposium. Both are so important for such an occasion. The superb organisational ability of Mrs. R.A. Potter, Mr. H.L. Bennister and of the Conference Manager at Churchill College, Miss Hammerton, and her staff are gratefully acknowledged.

The Symposium Organisers also acknowledge the generosity of Hoechst U.K. Ltd., ICI plc, Pfizer Central Research and Wyeth Research U.K. who supported the Symposium in various ways.

The Fifth Symposium will be held at Churchill College, Cambridge on September 10-13, 1989.

Organising Committee

Dr. P.H. Bentley (Hoechst U.K. Ltd.)

Professor M.M. Campbell (University of Bath, U.K.)

Dr. R.W. Lambert (Roche Products Ltd., U.K.)

Dr. P.R. Leeming, Hon. Sec., (Pfizer Central Research, U.K.)

Dr. A.C. White (Wyeth Research, U.K.)

Contents

Neuropeptides

Control of Enzymatic Processes in Medicinal Chemistry

Antiviral Agents

Antibacterial, Antifungal, and Antiparasitic Agents

Application of Computing to Medicinal Chemistry

Multiple Receptors for Substance P and Related Tachykinins

L. L. Iversen,* K. J. Watling, A. T. McKnight, B. J. Williams, and
C. M. Lee

MERCK SHARP & DOHME RESEARCH LABORATORIES,
NEUROSCIENCE CENTRE, HARLOW, ESSEX CM20 2QR, UK

1 INTRODUCTION

Substance P is an undecapeptide that has been known for more
than 50 years to occur in mammalian peripheral tissues and in
brain [1,2]. It belongs to a family of related peptides, the
tachykinins, which share a common C-terminal sequence:-

$\cdots$-Phe-X-Gly-Leu-MetNH$_2$

The other mammalian tachykinins are neurokinin A and neurokinin
B (Figure 1). These peptides exert a variety of potent actions
on smooth muscle, glandular tissue and in the central nervous
system. In CNS particular interest has focussed on the
possibility that substance P, released as a neurotransmitter or
neuromodulator from sensory nerve endings in spinal cord may act
as a mediator of pain information as it enters CNS. Thus,
suitable antagonists of this peptide might offer a novel means
of controlling pain, at the CNS level, but not involving an
opioid mechanism. The pharmacology of the tachykinins, like
that of a number of other mammalian peptides, is complicated and
still only poorly understood. The tachykinins seem to act on
several different receptor sub-types, and this will form the
substance of the present review.

SUBSTANCE P	Arg-Pro-Lys-Pro-Gln-Gln-Phe-Phe-Gly-Leu-Met-NH$_2$
PHYSALAEMIN	pGlu-Ala-Asp-Pro-Asn-Lys-Phe-Tyr-Gly-Leu-Met-NH$_2$
ELESOISIN	pGlu-Pro-Ser-Lys-Asp-Ala-Phe-Ile-Gly-Leu-Met-NH$_2$
KASSININ	Asp-Val-Pro-Lys-Ser-Asp-Gln-Phe-Val-Gly-Leu-Met-NH$_2$
NEUROKININ A	His-Lys-Thr-Asp-Ser-Phe-Val-Gly-Leu-Met-NH$_2$
NEUROKININ B	Asp-Met-His-Asp-Phe-Phe-Val-Gly-Leu-Met-NH$_2$

FIGURE 1:   TACHYKININS

The evidence for the existence of multiple classes of receptor for the naturally occurring tachykinins has been reviewed previously [3,4]. Our original proposal for a subdivision of SP-P (NK-1) and SP-E (NK-2) receptors [3] was based on pharmacological data, showing differences in the relative potencies of tachykinins in different peripheral tissue bioassays. These differences have since been confirmed in other laboratories [5,6]. More recently, several different radio-labelled tachykinin derivatives have been used to characterise tachykinin receptors in peripheral tissues and in CNS, using binding studies [7-10]. Such studies have suggested that there may be three or more types of tachykinin receptor binding sites in mammalian tissues.

2 MULTIPLE TACHYKININ BINDING SITES

Most laboratories agree that the ^{125}Iodine-labelled-Bolton-Hunter conjugate of substance P (^{125}I-BHSP) labels sites in peripheral tissues and in brain which bear a close similarity to one another. This is indicated by the pattern of specificity revealed by a range of different displacing ligands. The pattern of specificity of these sites also closely resembles that previously described for the NK-1 ("SP-P") receptor*; SP-labelled sites of this type are seen in many regions of CNS, and in peripheral tissues including guinea pig ileum, guinea pig urinary bladder and rat salivary gland. These sites can also be labelled selectively in peripheral tissues and in brain by using ^{125}I or ^{3}H-labelled physalaemin [9,11].

Studies in which ^{125}I-BH-eledoisin (^{125}I-BHE) was used as a radioligand have revealed a different type of receptor binding site in brain membranes, quite distinct from that labelled by ^{125}I-BHSP [7,10]. The finding that these sites preferred eledoisin, kassinin and neurokinins to substance P suggested at first that they might represent the NK-2 ("SP-E")* receptor type previously identified in peripheral tissues. This now seems incorrect; the sites labelled by eledoisin in brain appear to possess their own distinct profile of specificity – the most notable feature being that neurokinin B is by far the most potent displacing ligand (Table 1) as first pointed out by Torrens et al. [12]. In this respect the ^{125}I-BHE-labelled sites in brain resemble the NK-3 ("SP-N") sites proposed by Laufer et al. [13]. The sites in brain labelled by ^{125}I- BHE are thus referred to by us [14] (Table 1) as NK-3, to

---

*The receptor nomenclature used here was agreed to by the participants at a symposium on "Substance P and Neurokinins" held in Montreal, Canada in July 1986. (ed. J. Henry, Springer-Verlag. in press).

TABLE 1 : MULTIPLE TACHYKININ BINDING SITES - RELATIVE POTENCIES OF TACHYKININS AND ANALOGUES AS DISPLACERS

	NK-1 ^{125}I-BHSP in rat salivary gl	NK-2 ^{125}I-BHE in hamster urin. bl.	NK-3 ^{125}I-BHE in rat cerebr. cortex
Substance P	100	100	100
Physalaemin	24.6	6.2	163
Eledoisin	2.5	763	912
Kassinin	1.5	1010	912
Neurokinin A	4.6	2845	154
Neurokinin B	1.6	1252	6933
[Glp6,D-Pro9]-SP(6-11)	0.003	111	6.9
[Glp6,L-Pro9]-SP(6-11)	0.34	0.3	2.5
SP-0-methyl	5.9	0.3	0.1
"D-Pro/L-Pro ratio" *	124.0	0.003	0.36
"SPOMe/I-BHSPOMe ratio" **	0.3	164.0	1085.0
Preferred tachykinin	SP	NK-A	NK-B

Results are expressed relative to substance P = 100, large numbers indicate high potency; calculated from data of Lee et al 14. *D-Pro/L-Pro ratio is ratio of IC50's for [Glp6,D-Pro9]-SP(6-11) to [Glp6,L-Pro9]-SP(6-11); ** SPOMe/ I-BHSPOMe ratio is ratio of IC50's for SP-0-methyl ester and its iodinated Bolton-Hunter conjugate.

distinguish them from sites which can also be labelled by
[125]BHE in certain peripheral tissues, which possess the
pharmacological specificity of NK-2 (SP-E) receptors. The NK-3
sites in brain can also be labelled selectively by using
^{125}I-BH conjugated neurokinin A [15] or ^{3}H-neurokinin B
[Bergstrom et al., this symposium] which yield very similar
data. In peripheral tissues also, neurokinin A can be used as
an alternative radioligand for NK-2 sites [16] which we believe
are equivalent to the "SP-K" sites identified by Buck et al.
[8]. Some of the differences between these three proposed
categories of binding site are summarised in Table 1. Our own
recent work [14] has shown that certain selective ligands can be
useful too in distinguishing the NK-1, NK-2 and NK-3 sites.
Thus, substance P-O-methyl ester was previously described as a
selective agonist for NK-1 receptors [17] and it has proved
useful as a discriminating ligand in binding assays, showing
considerable preference for the NK-1 sites versus the other
two. In attempts to use this as a more selective radioligand
for NK-1 sites we prepared the non-radioactive iodinated-Bolton
Hunter conjugate. To our surprise this derivative showed an
unexpectedly high affinity for NK-2 and NK-3 sites, losing most
of the NK-1 selectivity seen in the parent compound. The ratio
of potencies of the SP-O-methyl ester versus its I-BH conjugate
is a useful discriminator of NK-1 versus the other two sites
(Table 1). Even more valuable are the pair of analogues in
which D-Pro or L-Pro replace Gly in position 9 of substance
P[18]. (Figure 2). Here the analogue [Glp[6],L-Pro[9]]SP-(6-11)
proved to be more than 100 times more potent than [Glp[6],D-
Pro[9]]SP-(6-11) as a displacer at NK-1 sites, whereas the order
of specificity was reversed at NK-2 sites, with [Glp[6],D-
Pro[9]] SP-(6-11) being more than 300 times more potent than the
L-Pro[9] analogue (Table 1).

Choice of radioligands for binding studies

Unfortunately the radioligands currently available leave
much to be desired, since none of them shows complete
selectivity for a particular receptor type. ^{125}I-BHSP is
perhaps the best, with a preferential selectivity of 50-100 fold
in its affinity for NK-1 versus the other sites. ^{125}I-BHE, on
the other hand, possesses only about 5-10 fold selectivity for
NK-2 versus NK-1 sites, and fails to distinguish between NK-2
and NK-3 sites. In tissues that contain significant numbers of
NK-1 sites ^{125}I-BHE may, thus, yield confusing results. In
this respect ^{125}I-BHNKA may be preferable as a ligand for NK-2
sites [14], and ^{3}H-NKB may prove to be the ligand of choice
for NK-3 sites in brain and elsewhere.

DiMeC−7

Glp−Phe−(N−Me)Phe−(N−Me)Gly−Leu−Met−NH$_2$

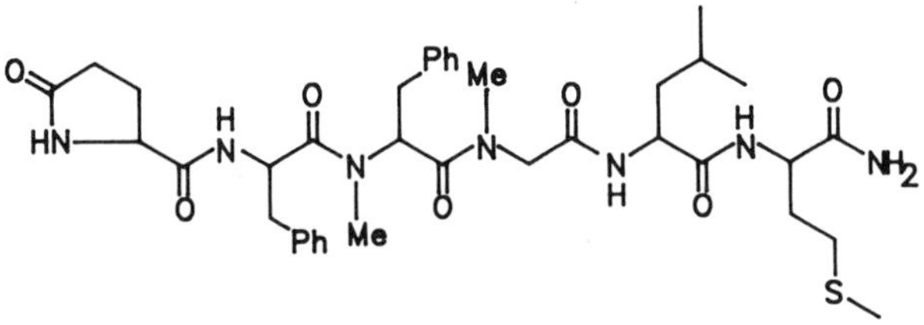

SENKTIDE

Suc−Asp−Phe−(N−Me)Phe−Gly−Leu−Met−NH$_2$

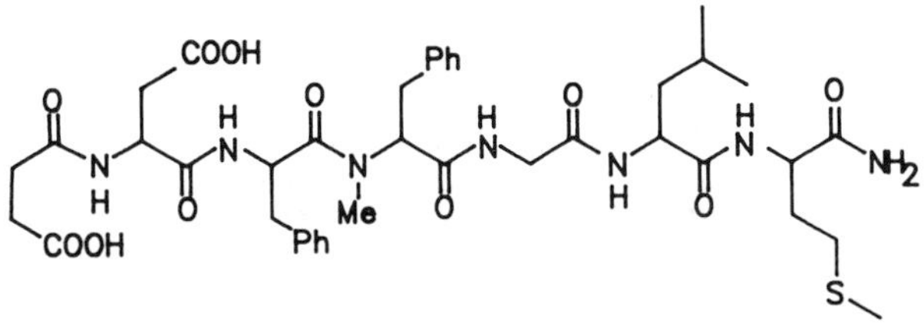

L−363,851

Glp−Phe−Phe−(R)Gly[ANC−2]Leu−Met−NH$_2$

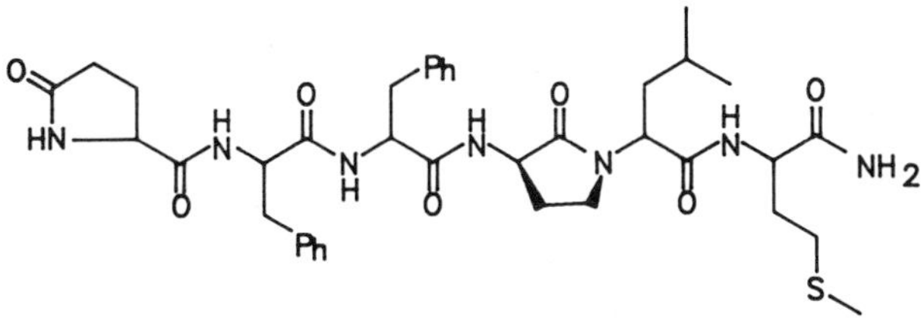

<u>Figure 2</u> − SOME SYNTHETIC TACHYKININ AGONIST ANALOGUES

3 FUNCTIONAL CORRELATES OF MULTIPLE BINDING SITES

PI Breakdown

We have found in previous studies that the ability of tachykinins to induce a breakdown of inositol phospholipids (PI response) offers a convenient biochemical model for comparing the potencies of a range of agonists. In rat salivary gland for example,[19] the specificity of agonists for the PI response is the same as that for eliciting a secretory response, and this has the characteristic features of an NK-1 receptor. However, Watson [20] also found that tachykinins could elicit a PI response in rat ileum, with a pattern of specificity resembling that of an NK-2 site. More recent data from Bristow et al. in this laboratory [21] clearly indicate that both NK-1 and NK-2 receptors in peripheral tissues may be coupled to PI turnover. Thus, in hamster urinary bladder the relative potencies of various tachykinins and analogues in eliciting PI breakdown correlated well (r = 0.94) with their potencies in displacing ^{125}I-BHE in the same tissue. Furthermore, the analogues [Glp^6,D-Pro^9] SP-(6-11) and [Glp^6,-L-Pro^9] SP-(6-11) exhibited the predicted order of potencies in NK-1 (rat salivary gland) versus NK-2 (hamster urinary bladder) tissues.

Pharmacological Bioassays

A number of studies have compared tachykinin potencies in a wide range of peripheral tissue models [3,5,6]. Our own recent findings indicate that it is possible to identify bioassay preparations that display the characteristic patterns of agonist specificity predicted for the NK-1, NK-2 and NK-3 receptor types described above. The guinea pig ileum (in presence of atropine) has long been regarded as a typical NK-1 preparation, and our own bioassay data shows an excellent agreement between EC50 values for a variety of agonists, and the IC50's of the same compounds in displacing ^{125}I-BHSP binding in rat cerebral cortex (r = 0.87; Table 2). The rat vas deferens likewise serves as a typical model for the NK-2 receptor type, and again a significant correlation was observed between bioassay results and ^{125}I-BHE binding data (hamster urinary bladder; r = 0.87).

Dr. McKnight has also found a bioassay preparation whose specificity mirrors that of the NK-3 site, hitherto described only as a neuronal site in brain membranes (binding data) or in neurones of the myenteric plexus of guinea pig ileum [13]. The ability of tachykinins to increase the frequency and amplitude of contractions in rat portal vein seems to represent a novel and useful model for the NK-3 site. Neurokinin B is the most potent agonist (EC50 = 2 nM) and substance P-O-methyl ester is

TABLE 2 : <u>MULTIPLE TACHYKININ RECEPTORS - RELATIVE POTENCIES OF TACHYKININS IN</u> <u>BIOASSAYS</u>

	NK-1 guinea-pig ileum	NK-2 rat vas deferens	NK-3 rat portal vein	NK-4 guinea-pig trachea
Substance P	100	100	100	100
Physalaemin	95	38	330	430
Eledoisin	57	2726	2400	2340
Kassinin	68	5342	195	260
Neurokinin A	8.5	9086	940	2847
Neurokinin B	61	2198	43000	611
[Glp6,D-Pro9]SP-(6-11)	6.5	1969	50	9110
[Glp6,L-Pro9]SP(6-11)	33	<10	2	30400
SP-0-methyl	65	<10	<0.01	106
"D-Pro/L-Pro ratio"	5.14	<0.007	0.04	3.3

Results are expressed as for Table 1 [A. McKnight - unpublished]. EC_{50} of substance P was NK-1=0.003µM; NK-2=13.6µM; NK-3=1.0µM; NK-4=2.81µM.

virtually inactive. The EC50's determined in this preparation correlated significantly with IC50's for ^{125}I-BHE binding in rat cerebral cortex (r = 0.98; Table 2).

More recently Dr. A. McKnight and colleagues have described what appears to be yet another pattern of specificity for the peptide receptors mediating the contractile actions of tachykinins in smooth muscle of guinea-pig trachea [22]. In this tissue the order of potency for mammalian tachykinins, NKA>NKB >>SP is that for the NK-2 receptor, but findings with other peptides, and in particular the very high potency of [Glp^6,L-Pro^9]SP(6-11), make the conclusion that NK-2 receptors are involved untenable. Furthermore, the low potency of the NK-3 selective agonist senktide (Suc-[Asp^6,MePhe^8]-SP(6-11) [23] make it unlikely that NK-3 sites are involved. Although a definitive classification of these receptors is not possible as yet, McKnight et al. [22] proposed the existence of NK-4 receptors in guinea-pig trachea to explain these results. There is as yet no radioligand binding data to suggest what may be the most suitable biochemical model for these sites.

Selective Ligands

The four receptor sub-types proposed here can only be classified tentatively at this stage. A definitive classification must await the development of selective antagonists for each site. Meanwhile, the identification of selective peptide analogues, which act as selective agonists are the most useful tools currently available. For the four sites the most selective agonist ligands currently available are:-

    NK-1 = substance P-O-methyl ester
    NK-2 = neurokinin A
    NK-3 = senktide (Figure 2)
    NK-4 = GlpPhePhe(R)Gly[ANC-2]Leu MetNH$_2$ [22] (L-363,851)
          (Figure 2)

REFERENCES

1. R. Porter & M. O'Connor (eds) 'Substance P in the Nervous System', 1982, CIBA Foundation Symposium No. 91, Pitman, London.
2. T.M. Jessell, In 'Handbook of Psychopharmacology'. Eds. L.L. Iversen, S.D. Iversen & S.H. Snyder, 1983, Plenum, New York. Vol 16, 1.
3. C.M. Lee, L.L. Iversen, M.R. Hanley & B.E.B. Sandberg, N.S.Archiv. Pharmacol., 1982, 318, 281.

4. L.L. Iversen, (1985) In 'Tachykinin Antagonists'. eds. R. Hakanson & F. Sundler. Elsevier, Amsterdam p. 291-304.
5. J. Mizrahi, S. Dion, P. D'Orleans-Juste , E. Escher, G. Drapeau & D. Regoli, Eur. J. Pharmacol., 1985, 118, 25.
6. S.J. Bailey, R.L. Featherstone, C.C. Jordan, I.K.M. Morgan, Br. J. Pharmacol, 1986, 87, 79.
7. M.A. Cascieri, G.G. Chicchi, & T. Liang, J. Biol. Chem., 1985, 260, 1501.
8. S.H. Buck, E. Burcher, C.W. Shults, W. Lovenberg & T.L. O'Donohue, Science, 1985, 226, 987.
9. S.W. Bahouth, D.M. Lazaro, D.E. Brundich, & J.M. Musacchio, Mol. Pharmacol., 1985, 27, 38.
10. J.C. Beaujouan, Y. Torrens, A. Viger & J. Glowinski, Mol. Pharmacol., 1984, 26 248.
11. P. Mohini, S.W. Bahouth, D.E. Brundish, J.M. Musacchio, J. Neuroscience, 1985, 5, 2078.
12. Y. Torrens, S. Lavielle, G. Chassaing, A. Marquet, J. Glowinski, & J.C. Beaujouan, Eur. J. Pharmacol., 1984, 102, 381.
13. R. Laufer, U. Wormser, Z.Y. Friedman, C. Gilon, M. Chorev & Z. Selinger, Proc. Nat. Acad. Sci. US., 1985, 82, 7444.
14. C.M. Lee, N.J. Campbell, B.J. Williams & L.L. Iversen, Eur. J. Pharmacol., 1986, 130, 209.
15. A.C. Foster & R. Tridgett, Br. J. Pharmacol., 1986, 89, 774P.
16. E. Burcher, S.H. Buck, W. Lovenberg & T.L. O'Donohue, J. Pharmacol. Exp. Ther., 1986, 236, 819.
17. S.P. Watson, B.E.B. Sandberg, M.R. Hanley & L.L. Iversen, Eur. J. Pharmacol., 1983, 87, 77.
18 M.F. Piercey, P.J.K. Dorby-Schreur, N. Masiques & L.A. Schroeder, Life Sciences, 1985, 36, 777.
19. M.R. Hanley, C.M. Lee, L.M. Jones & R.H. Mitchell, Mol. Pharmacol., 1980, 18, 78.
20. S.P. Watson, Biochem. Pharmacol., 1984, 33, 3733.
21. D.R. Bristow, N.R. Curtis, N. Suman-Chauhan & K.J. Watling & B.J. Williams, Br. J. Pharmacol., 1987, 90, 211
22. A.T. McKnight, J.J. Maguire & M.A. Varney. Br. J. Pharmacol. 91, 360P.
23. U. Wormser, R. Laufer, Y. Hart, M. Chorev, C. Gilon & L. Selinger, EMBO Journal, 1986, 5, 2805.

Design of Novel Antagonists of Cholecystokynin

R. M. Freidinger,* M. G. Bock, R. S. L. Chang, R. M. DiPardo,
B. E. Evans, V. M. Garsky, V. J. Lotti, K. E. Rittle, D. F. Veber, and
W. L. Whitter

MERCK SHARP & DOHME RESEARCH LABORATORIES, WEST POINT,
PENNSYLVANIA 19486, USA

1. INTRODUCTION

About 100 neuropeptides are now known to have
physiological roles in various body organs. These
peptides occur both peripherally and in the central
nervous system serving as neurotransmitters, neuromodulators, and/or hormones. The diverse biological activity
represented by these substances suggests potential
applications for drugs which either mimic or block their
actions. The development of useful and specific peptide
receptor ligands represents a significant challenge for
the medicinal chemist. Neuropeptides and their direct
analogs are usually not suitable due to their rapid
degradation by proteases, their lack of specificity, and
their poor transport properties across biological membranes.

An approach to useful peptide receptor ligands which
is currently attracting considerable attention is the
design of a partially or totally nonpeptidal agent
(peptide mimetic).[1] Mimetics investigated to date
generally incorporate modifications in the amide backbone
of the parent biologically active peptide.[2] The goal is
to obtain a more pharmacologically useful agent by
eliminating undesirable properties (e.g., cleavage by
proteases) while retaining affinity for the peptide's
receptor. An effective, totally nonpeptidal molecule is
an attractive outcome of this exercise since it is well
established that such agents can have appropriate pharmacokinetic properties to be suitable as drugs. Such a
compound would represent a unique tool for studying in
more depth the role of the peptide of interest in normal
and pathophysiological processes.

Two general approaches to such nonpeptide ligands can be considered. A stepwise, rational design process has received the greatest attention to date. Considerable progress has been made in developing partially peptidal and/or conformationally modified agents with improved properties.[3] Extension of this process to the ultimate objective of an orally active, nonpeptide structure, however, has not yet succeeded. An alternative approach is to search for nonpeptide receptor ligand leads using receptor-based technology for screening. Such a lead could serve as a starting point from which medicinal chemists could design an optimal structure. The application of both approaches to the design of antagonists of cholecystokinin (CCK) is described here.

CCK is a neuropeptide originally isolated from the gastrointestinal tract and is now known to occur widely in the central nervous system as well. CCK is found in a number of molecular sizes, a major form being the carboxy terminal octapeptide, CCK-8(H-Asp-Tyr(SO_3H)-Met-Gly-Trp-Met-Asp-Phe-NH_2).[4] Structure-function studies have shown CCK-8 to be the minimum fully potent endogenously occurring sequence.[5] The interaction of CCK with its receptors in the periphery resulting in the stimulation of pancreatic and biliary exocrine secretion, gall bladder contraction, and gut motility is now well recognized. The role of CCK in the brain is less well understood, although it is thought to function as a neurotransmitter or neuromodulator. A variety of other functions have been attributed to CCK, including satiety, sedation, and antagonism of the analgesic effects of endogenous opiates.[6] In light of this intriguing biology, the development of useful antagonists which interact competitively and selectively with the CCK receptor is desirable.

2. RESULTS AND DISCUSSION

Nonpeptide CCK Antagonists from a Receptor-based Screening Lead

At the beginning of the studies to be described here, only the weak CCK antagonists proglumide, benzotript, and dibutyryl cyclic GMP were known.[7] Considerable progress has been made since that time in the design and synthesis of potent, receptor selective, nonpeptidal antagonists of CCK. Two years ago, the discovery from fermentation broths of the novel CCK antagonist asperlicin, utilizing a radioligand binding assay, was reported.[8] Asperlicin represented a key breakthrough

to a competitive and selective nonpeptide antagonist
(K_i = 600 nM) at the CCK receptor, and an effort was
initiated to design more potent and orally effective
agents from this lead. Early efforts focused on semisynthetic modifications on the natural product itself.
It did prove possible to prepare both more potent and
more water soluble derivatives;[9] however, these compounds
did not exhibit oral activity.

An alternate and ultimately more successful approach
to the design of improved antagonists involved focusing
on structural features of asperlicin which might be key
to its CCK receptor affinity and which could be incorporated into simpler molecules. Particularly attractive were
the 1,4-benzodiazepine and indolenine moieties. Recent
studies in the antianxiety and opiate areas support the
idea that benzodiazepines may be previously unrecognized
nonpeptide ligands of receptors for which the endogenous
ligand is a peptide.[10,11] The existence of common
features of structure and conformation among certain
peptides suggested that this bicyclic system could have
broader application in the design of other peptide
receptor ligands. The indolenine part structure of
asperlicin was attractive because of its resemblance to
tryptophan which structure-function studies have shown to
be a key side chain of CCK.[12] The basic design hypothesis then involved a fusion of the elements of the
5-phenyl-1,4-benzodiazepine with tryptophan. As is
detailed in Figure 1, to achieve maximum correspondence
of the aromatic moieties of asperlicin and the proposed
structure, the D-configuration of tryptophan was
employed.

The target 3R-3-indolylmethylbenzodiazepin-2-one **1**
was synthesized from the requisite aminobenzophenone and
D-tryptophan ethyl ester in refluxing pyridine and proved
to have CCK receptor affinity comparable to asperlicin
(IC_{50} = 3.4 μM). In accord with the design hypothesis,
the 3S enantiomer **2** was 10-fold less potent. A detailed
structure-activity study based on this simplified antagonist was then performed to identify key features for
optimal interaction with the CCK receptor. In addition
to the stereochemistry at the benzodiazepine 3-position,
the nature of the linking group between benzodiazepine
and indole as well as the point of attachment to indole
were found to be crucial. From a variety of indole-
containing 3-amino-1,4-benzodiazepine derivatives, the
indole-2-carboxamido group was found to be far superior
(Table 1). Optimization of this lead produced the
compound designated L-364,718 in which N-1 is methylated

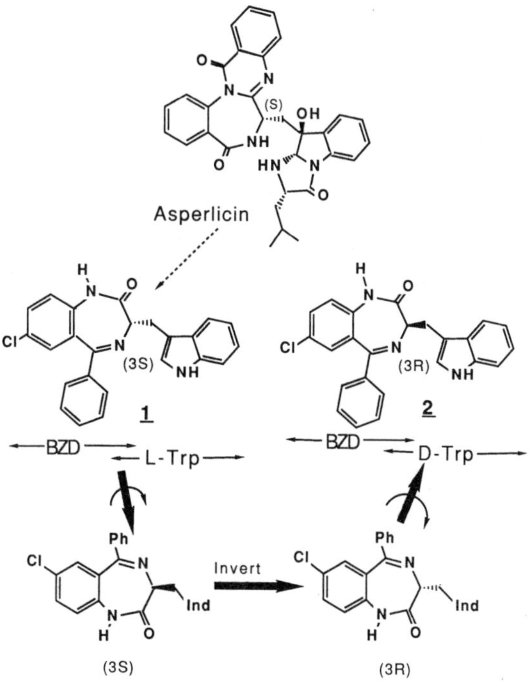

Figure 1. Design of 3-substituted
1,4-benzodiazepine, CCK antagonist
from asperlicin.

and the 3-position configuration is S (same absolute
configuration as 1) (Table 2).[13] This antagonist has
affinity for the peripheral CCK receptor comparable to
CCK itself (IC_{50} = 10^{-10}M) and excellent selectivity with
respect to the central CCK receptor (> 1000-fold) and
other receptors.[14]   In particular, L-364,718 does not
bind to the brain benzodiazepine receptor. This compound
displays no agonist properties and inhibits competitively
the effects of CCK <u>in vivo</u> in a variety of animal models
with long duration and oral activity.[15]   For example,
L-364,718 blocks inhibition of gastric emptying by CCK in
the mouse with an oral ED_{50} of 40 µg/kg.   These studies
suggest that L-364,718 is suitable for evaluation of the
utility of a peripherally selective CCK antagonist in
man.

Table 1. CCK Receptor Affinity
of 3-Acylamino-1,4-benzodiazepine
Analogs

| | IC$_{50}$(μM), ^{125}I-CCK | |
R	Pancreas	Brain
Boc-L-Trp	6	20
Boc-D-Trp	17	>100
	20	>100
	4.8	100
	1.0	11
	0.0047	8.0
	1.1	8.4
CH$_3$C	40	—

Structure-Function Studies in the L-364,718 Series

Additional studies of the benzodiazepine 3-substit-
uent in potent peripherally selective CCK antagonists
have been carried out. It has been found that a substituted phenyl ring is a good replacement for the 2-indole
of L-364,718 (Table 3). Most effective are meta or para
substituted Cl-, Br-, or I-phenyl groups, and disubsti-
tution offers no advantage over mono. For the p-
substituted phenyl analogs, CCK receptor affinity correlates rather well with the Hammett σ-constant (Figure
2). The exceptions to this correlation such as n-pentyl
and phenyl are much larger than the substituents which
correlate. Apparently the receptor binding site for the
benzodiazepine 3-substituent cannot accommodate para
substituents of this size very well. These studies have
produced an additional group of potent and selective
peripheral CCK antagonists which help to further define
the CCK receptor binding site. As examples, the p-Cl and
p-Br analogs also have in vivo properties analogous to
those of L-364,718.

Table 2. Receptor Affinity of Indole-2-carbonylamino-1,4-benzodiazepine analogs

| X | Y | R_1 | R_2 | 3-Stereo | $IC_{50}(\mu M)$ | | ^{3}H-BZD |
| | | | | | ^{125}I-CCK | | |
					Pancreas	Brain	Brain
H	H	H	H	RS	0.0047	8.0	>100
H	H	CH_3	H	RS	0.0011	0.8	>100
H	H	CH_3	CH_3	RS	0.0014	15	—
H	H	CH_2COOH	H	RS	0.0014	6	>100
H	F	CH_3	CH_3	RS	0.0014	0.3	—
H	H	CH_3	H	S	0.00008*	0.27	>100
H	H	CH_3	H	R	0.06	3.7	>100
H	F	CH_3	H	S	0.0006	0.3	} 7.8
H	F	CH_3	H	R	0.019	1.1	

Analogs of Proglumide

Recently, analogs of proglumide have been reported which are substantially more potent and exhibit selective affinity for the peripheral CCK receptor.[16] The most potent of these compounds is 3,4-dichlorobenzoyl-DL-glutamic acid di-n-pentylamide (lorglumide). It was of interest to determine if a relationship could be established between structural features of L-364,718 and lorglumide. A working hypothesis is that the indole-carboxamide and 3,4-dichlorobenzoic amide groups bind to the same receptor site, and that the two benzene rings of L-364,718 and the two n-pentyl chains of lorglumide correspond (Figure 3). As a test, the analogs of proglumide and lorglumide incorporating 2-indole in place of phenyl and 3,4-dichlorophenyl, respectively, were synthesized (Table 4). These derivatives were 68 and 2.5 times more potent than their parent compounds, respectively, and both exhibit peripheral receptor selectivity. These results support the proposed structural correspondence. Furthermore, computer modeling comparisons between L-364,718 and the indole analog of lorglumide show that a good match of benzene rings and pentyl chains can be obtained when the indole-2-carboxamides are superimposed.

Table 3. CCK Receptor Affinity
of Substituted 3-benzoylamino-
1,4-benzodiazepine analogs

R	X	Y	3-Stereo	IC50 (nM), 125-I CCK	
				Pancreas	Brain
H	H	p-Cl	RS	41	>40,000
H	H	3,4-di-Cl	RS	29	>40,000
CH3	H	p-Cl	RS	8.3	40,000
CH3	F	p-CL	S	2.5	2,900
CH3	F	p-Cl	R	49	11,000
CH3	F	p-Br	S	3.5	4,000
CH3	F	m-Br	S	3.5	3,500
CH3	F	o-Br	S	6,200	70,000
CH3	F	p-I	S	3.4	3,000
CH3	F	p-CN	S	43	31,000
H	F	p-NO2	S	94	>50,000
H	F	p-N(CH3)2	RS	270	>100,000
H	F	p-OCH3	RS	96	>40,000

In contrast to results obtained with benzodiazepine
CCK antagonists, there was little difference between the
potencies and selectivities of the individual enantiomers in the proglumide series. These results suggest
that the glutamic acid side chain is not an important
binding element for these compounds. It has proven
possible, however, to identify analogs of this type where
the activities of the enantiomers do differ, and the
peripheral receptor affinity and selectivity are improved. Illustrative are the m-methoxyphenylurea analogs
of D- and L-glutamic acid di-n-pentylamide. These preliminary results indicate significant potential for
development of useful CCK antagonists in this series such
as were found in the substituted benzodiazepine series.

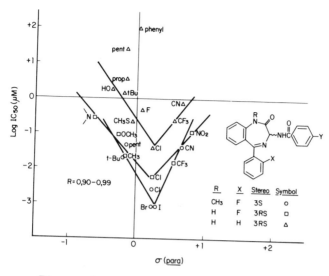

Figure 2. Correlation of pancreas CCK receptor affinity with the Hammett σ-constant of substituent for substituted 3-benzoylamino-1,4-benzodiazepines.

Probing the Relationship Between L-364,718 and CCK Peptide Structure.

Since L-364,718 is the first nonpeptide antagonist acting at a peptide receptor which has comparable receptor affinity to the endogenous ligand, a key question for potential extension of these studies to other peptide systems is how the interactions of antagonist and CCK with the receptor compare in molecular terms. A working hypothesis has been that CCK peptide analogs which are antagonists are more likely than CCK itself to have receptor binding elements corresponding with such features of nonpeptide antagonists. Efforts have, therefore, been focused on developing constrained peptide analogs of CCK which are antagonists. Models of these more conformationally defined peptides should be obtainable using theoretical and spectral techniques, and their comparison with L-364,718 should offer the potential for greater insights.

These studies began with the antagonist Cbz-Tyr (SO₃H)-Met-Gly-Trp-Met-Asp-NH2 (IC$_{50}$ = 3 μM).[17] The possibility of a turn in the receptor bound conformation

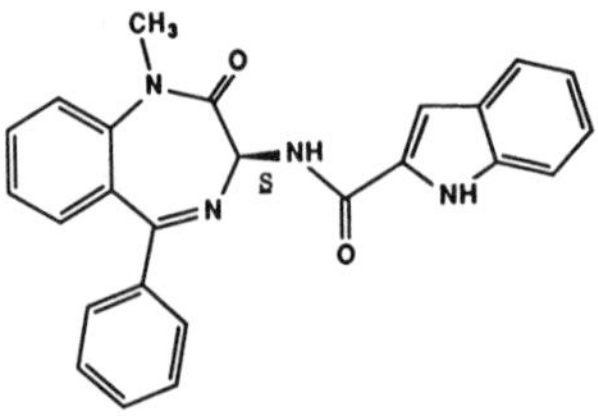

L-364,718

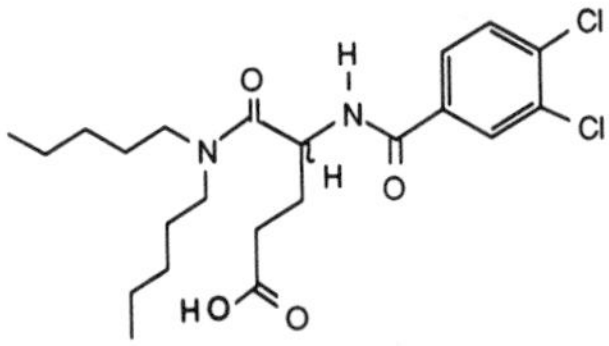

Lorglumide

Figure 3. Comparison of L-364,718
and lorglumide. Structures are
oriented to illustrate proposed
correspondence of CCK receptor
binding elements.

was suggested when substitution of D-Trp for L-Trp gave
only a 4-fold loss in receptor affinity. Furthermore,
removal of the sulfate from Tyr was found to have a
negligible effect on binding, and subsequent analogs have
utilized Phe in this position. To further test for a
turn conformation, cyclic analogs were designed (Table
5). Cyclization of the sequence Phe-Met-Gly-D-Trp-Met
through a proline gave an analog with receptor affinity
comparable to the best linear antagonist (IC_{50} = 5.3 µM).
This result further supports the turn hypothesis. Additional analogs in which Phe and Pro were interchanged
(IC_{50} > 30 µM) and in which Ala replaced Phe (IC_{50} = 13
µM)[5] show, respectively, that backbone conformation is
important in these analogs and the Phe side chain is
contributing little to binding. In an attempt to gain
additional receptor affinity by mimicking the C-terminal
Phe of CCK, Pro was replaced by N-δ-benzyl-D-Asn to give
cyclo-[Phe-Met-Gly-D-Trp-Met-D-Asn(Bzl)]. This analog is
the most potent cyclic peptide antagonist of CCK (IC_{50} =
0.21 µM) to date. In analogy with L-364,718, it also

Table 4. CCK Receptor Affinity of
Proglumide, Lorglumide, and Analogs
Designed from Comparison with L-364,718

			IC 50 (μM)	
R1	**R2**	**Stereo**	**Pancreas**	**Brain**
1. (CH2)2CH3	Phenyl	DL	250	800
Proglumide				
2. (CH2)2CH3	2-Indolyl	L	3.7	88
3. (CH2)2CH3	2-Indolyl	D	1.1	47
4. (CH2)4CH3	3,4-Di-Cl-Phenyl	DL	0.02	2.2
Lorglumide				
5. (CH2)4CH3	2-Indolyl	DL	0.008	0.23
6. (CH2)4CH3		D	0.005	0.71
7. (CH2)4CH3		L	0.5	9.7

selectively binds to the peripheral CCK receptor ($IC_{50} >$
100 μM, brain receptor). It elicits no CCK-like
contractile effects of the guinea pig ileum at concentrations as high as 10^{-5}M. Determination of its solution
conformation and modeling comparisons with L-364,718 are
in progress.

Current goals are to develop cyclic peptide antagonist analogs of CCK with potencies similar to L-364,718
in order to make modeling comparisons more legitimate.
Conformational models of these cyclic peptides will be
constructed utilizing solution data and conformational
search techniques. The ultimate objective is to establish a structural relationship between peptide and
benzodiazepine with basis in experiment which may have
utility in future design studies. The several types of
novel CCK antagonists reported here have considerable
potential for increasing our understanding of the role of
CCK.

Table 5. CCK Receptor Affinity
of Cyclic Hexapeptide CCK Analogs

	IC_{50} (µM) ^{125}I-CCK	
	PANCREAS	BRAIN
Phe–Met–Gly | | Pro–Met–D–Trp	5.3	17
Pro–Met–Gly | | Phe–Met–D–Trp	> 30	> 100
Ala–Met–Gly | | Pro–Met–D–Trp	13	> 100
Phe–Met–Gly	1.2	17
Phe–Met–Gly	0.21	> 100
Phe–Met–Gly W = L | | W–Asp–Met–D–Trp W = D	5.8 8.6	60 100

Reference List

1.   P. S. Farmer, 'Drug Design', E. J. Ariens, Ed.,
Academic Press, New York, 1980, p. 119.
2.   A. F. Spatola, 'Chemistry and Biochemistry of Amino
Acids, Peptides and Proteins', B. Weinstein, Ed., Marcel
Dekker, Inc., Basel, 1983, p. 267.
3.   D. F. Veber and R. M. Freidinger, <u>Trends Neurosci</u>.,
1985, <u>8</u>, 392.
4.   J. A. Williams, <u>Biomed. Res</u>., 1982, <u>3</u>, 107.
5.   M. A. Ondetti, F. Pluscec, E. F. Sabo, J. T.
Sheehan, and N. Williams, <u>J. Am. Chem. Soc</u>., 1970, <u>92</u>,
195.
6.   J. E. Morley, <u>Life Sciences</u>, 1982, <u>30</u>, 479.
7.   W. F. Hahne, R. T. Jensen, G. F. Lemp, and J. D.
Gardner, <u>Proc. Natl. Acad. Sci. USA</u>, 1981, <u>78</u>, 6304.

8.   R. S. L. Chang, V. J. Lotti, R. L. Monaghan, J. Birnbaum, E. O. Stapley, M. A. Goetz, G. Albers-Schonberg, A. A. Patchett, J. M. Liesch, O. D. Hensens and J. P. Springer, Science, 1985, 230, 177.
9.   M. G. Bock, R. M. DiPardo, K. E. Rittle, B. E. Evans, R. M. Freidinger, D. F. Veber, R. S. L. Chang, T. Chen, M. E. Keegan, and V. J. Lotti, J. Med. Chem., 1986, 29, 1941.
10. H. Alho, E. Costa, P. Ferrero, M. Fujimoto, D. Cosenza-Murphy, and A. Guidotti, Science, 1985, 229, 179.
11. D. Romer, H. H. Buscher, R. C. Hill, R. Maurer, T. J. Petcher, H. Zeugner, W. Benson, E. Finner, W. Milkowski, and P. W. Thies, Nature, 1982, 298, 759.
12. H. M. Rajh, E. C. M. Mariman, G. I. Tesser, and R. J. F. Nivard, Int. J. Pept. Prot. Res., 1980, 15, 200.
13. B. E. Evans, M. G. Bock, K. E. Rittle, R. M. DiPardo, W. L. Whitter, D. F. Veber, P. S. Anderson, and R. M. Freidinger, Proc. Natl. Acad. Sci. U.S.A., 1986, 83, 4918.
14. R. S. L. Chang and V. J. Lotti, Proc. Natl. Acad. Sci. U.S.A.. 1986, 83, 4923.
15. V. J. Lotti, R. G. Pendleton, R. J. Gould, H. M. Hanson, R. S. L. Chang, and B. V. Clineschmidt, J. Pharmacol. Exp. Ther., 1987, 241, 103.
16. C. Niederau, M. Niederau, J. A. Williams, and J. H. Grendell, Am. J. Physiol., 1986, 251, G856.
17. M. Spanarkel, J. Martinez, C. Briet, R. T. Jensen, and J. D. Gardner, J. Biol. Chem., 1983, 358, 6746.

Novel Approaches to the Pharmacological Modification of Peptidergic Neurotransmission

B. P. Roques

DÉPARTEMENT DE CHIMIE ORGANIQUE, UER DES SCIENCES
PHARMACEUTIQUES ET BIOLOGIQUES, 4 AVENUE DE
L'OBSERVATOIRE, 75006 PARIS, FRANCE

1 INTRODUCTION

In the central nervous system, neuropeptides such as the enkephalins, CCK, SP, etc... behave both as classical neurotransmitters, interacting with post-synaptic receptors to ensure the transmission of the nerve impulse, and as neuromodulators, acting presynaptically to modulate the release of various effectors (monoamines or peptides) (review in ref. 1). As illustrated by CCK and DA in the mesolimbic pathway, neuropeptides are also able to modify the threshold of the physiological responses induced by the colocalized neurotransmitter (review in ref. 2). Furthermore, the interruption of the responses induced by the interaction of neuropeptides with various receptor types is ensured by more or less specific peptidases which cleave the native peptide into inactive fragments (review in ref. 3).

Analysis of the physiological relevance of a given neuropeptidergic pathway requires the use of molecules (agonists or antagonists) interacting selectively with the different receptor types. Moreover the occurrence of a physiological control of the responses induced by stimulation of the various types of receptors through a tonic (or phasic) release of neuropeptides can be investigated by inhibition of their degrading enzymes. Although the molecular architecture of the different targets belonging to a neuropeptidergic pathway is still unknown, selective agonists or antagonists can be rationally designed by taking into account the conformational properties of the native peptides, while specific peptidase-inhibitors can be prepared from crystallographic data on related enzymes. The results of following this strategy for the enkephalins are summarized in this paper. In addition the main biochemical and pharmacological results obtained with the designed probes are briefly reported and discussed.

2 RECEPTOR-SELECTIVE PROBES : A CRUCIAL REQUIREMENT FOR PHARMACOLOGICAL INVESTIGATIONS

As in the case of monoaminergic effectors, it appears that each brain peptide possesses several classes of binding sites, which must be selectively stimulated in order to determine their physiological functions. Taking into account the relative proportion of two distinct receptors in various tissues, theoretical calculations have shown that investigations on the physiological responses associated with the stimulation of a single receptor require molecules exhibiting a binding affinity at least 100 times higher for one class of binding site (4). There is therefore a very interesting challenge in designing specific effectors which belong to the group of modified peptides or to that of classical synthetic substances. These latter compounds could possess more favourable pharmacokinetic properties and may therefore be more appropriate for possible therapeutic use. Such a search has been exemplified in the case of enkephalins, which have been shown to interact with two opioid binding sites designated μ and δ.

Theoretical support for a rational design of selective ligands for neuropeptide receptors

Receptors and enzymes are proteins, and crystallographic studies have shown that the binding energy of a substrate bound in the active-site of an enzyme is mainly due to Van der Waals and electrostatic interactions of the lateral chains of its constituting amino acids. Such a mechanism of interaction is assumed to occur also for a peptide interacting with two or more distinct receptor-types. In this case different parts of the same molecule are expected to be involved in the recognition of each binding site.

Obviously neither the solvated conformations nor the computed energetically stabilized forms of a short peptide correspond exactly to the biologically active structure at the receptor site. However comparison of data from crystallographic studies of enzyme-inhibitors such as rhizopus chinensis-pepstatin (5) and thermolysin-β-phenylpropionyl-L-phenylalanine (βPPP) (6) and from NMR studies in solution of pepstatin (7) and βPPP (6) have shown that no drastic changes occur at the level of the peptide backbone between the solvated and the bound forms. Moreover a good relationship appears to exist between the immunogenicity of peptides and their tendency to form turns both in proteins and in solution (8,9). Likewise, purification of receptors by means of antiidiotypes is based on the structural

analogies between the solvated form of a potent and selective
ligand recognized by the primary antibody and the peptidic
epitope of the antiidiotype able to fit the receptor binding site
(10,11). In addition peptidase-inhibitors with a structure
mimicking the transition state of a substrate have given rise to
antibodies endowed with very efficient enzymatic potencies, even
higher than those exhibited by chemically designed artificial
enzymes (12). All these results show that the affinity and the
selectivity of a biological effector for its targets is due to a
limited number of interactions, ensured by well-adapted chemical
groups. The determination of these main components and of their
spatial relationships could lead to the design of simple (peptide
or non-peptide) molecules able to fit the appropriate subsites of
the biological targets. These findings have led to proteins
(enzymes and receptors) and their ligands being considered as
dynamic entities interacting through a zipper rather than by a
lock and key mechanism (13,14). This latter binding mode should
be considered only as a particular case occurring with severely
conformationally   restricted   ligands.   Given   this
receptor-recognition process the knowledge of the solvated
conformation of a peptide seems to be an essential pre-requisite
for a rational design of selective ligands.

These principles, which have only recently been proposed,
represent a drastic change in the use of the results of
conformational analysis, considered for a long time as irrelevant
for the rational design of selective effectors. The discovery of
folded conformations in solution for peptides as small as the
enkephalins (15) or CCK_8 (16) has largely contributed to this
renewal of interest since previously, linear sequences of less
than 10 aminoacids were considered too small to exist under
privileged conformations in solution.

3 SYNTHESIS AND SELECTIVITY OF MU AND DELTA OPIOID PROBES

The presence in the brain, as well as in peripheral organs,
of at least two binding sites for enkephalins is now well
established (review in 17). The high-affinity site called δ, or
the enkephalin receptor, exhibits a high preference for peptide
structures. The low-affinity site designated μ, or morphine
receptor interacts preferentially with natural and synthetic
opiates. There is only a ten-fold difference in the binding
affinity of the enkephalins for μ and δ receptors, suggesting the
occurrence of large similarities in both binding sites. NMR (15)
and crystallographic (18) studies have shown that the enkephalins
exist in equilibrium between folded and extended forms of similar
energy. These data combined with the results of structure-activity
studies with modified enkephalins were used to propose structural
requirements for preferential recognition of opioid μ and δ

Table 1. Inhibitory potencies of enkephalin analogues and opiates on the specific binding of 2 nM [³H]DSTBULET at δ-sites and of 1 nM [³H]DAGO at μ-sites in rat brain tissue at 35°C.

[³H]DSTBULET	[³H]DAGO	$\dfrac{K_I[^3H]DAGO}{K_I[^3H]DSTBULET}$
DSLET	3.80+0.63	31.0+5.0	8.2
DTLET	1.61+0.22	25.3+2.5	16
DSTBULET	2.81+0.64	374+35	130
BUBU	1.69+0.45	480+44	280
DPDPE	8.85+1.69	993+151	110
DPLPE	7.08+1.17	873+210	120
ICI 174,864	311+17	29200+4500	94
DAGO	629+13	3.90+0.80	0.0062
TRIMU 5	3130+50	10.0+2.1	0.0050
Naloxone	50.5+4.9	3.43+0.44	0.068
CTOP a)	2.8+0.5	13,500+2,750	0.0002

a)from ref 27 with [³H]Naloxone and [³H]DPDPE as μ and δ ligands.
DSLET = Tyr-D.Ser-Gly-Phe-Leu-Thr
DTLET = Tyr-D.Thr-Gly-Phe- Leu-Thr
DSTBULET = Tyr-D.Ser(OtBu)-Gly-Phe-Leu-Thr
BUBU = Tyr-D.Ser(OtBu)-Gly-Phe-Leu-Thr(OtBu)
DPDPE = Tyr-D.Pen-Gly- Phe-D.Pen
DPLPE = Tyr-D.Pen-Gly-Phe-L.Pen
ICI 174,864 = bis allyl-Tyr-Aib-Aib-Phe-Leu
DAGO = Tyr-D.Ala-Gly-(NMe)Phe-Gly-ol
TRIMU 5 = Tyr-D.Ala-Gly-NH(CH₂)₂-CH(CH₃)₂
CTOP = D.Phe-Cys-Tyr-D.Trp-Orn-Thr-Pen-Thr-NH₂.

receptors (19). This allowed the design of mu agonists such as TRIMU 5, Tyr-D-Ala-Gly-NH(CH₂)₂-CH(CH₃)₂ (20) and several cyclic enkephalins synthesized by Schiller et al. (review in 21). These agonists exhibit a mu-selectivity analogous to that of DAGO, the first described specific mu agonist (22). Similarly, δ-specific agonists belonging to the series of linear hexapeptides, such as DSLET and DTLET, were prepared (4). All these peptides and their tritiated analogues have been extensively used to characterize the μ and δ opioid receptors (review in 23).

Recently two cyclic enkephalins c(D.Pen²-L.Pen⁵) enkephalin (DPLPE) and c(D.Pen²-D.Pen⁵) enkephalin (DPDPE), characterized by the presence of a disulfide bond linking two highly constrained penicillamino residues, have been shown to exhibit an even better δ selectivity than the linear DSLET and DTLET (24). However this δ selectivity is associated with a large decrease in δ affinity (25), suggesting the existence of conformational restrictions in the cyclic peptides which strongly inhibit the interaction with the mu site but induce some unfavourable constraints for optimal δ receptor recognition.

Following this, new compounds belonging to the linear hexapeptide series were designed in order to increase both the selectivity and the affinity for δ sites. Very efficient δ ligands were obtained by the introduction in DSLET of the bulky O-t-butyl group in positions 2 and 6, producing more constrained structures. One of these new compounds [³H]DSTBULET is now proposed as a reference ligand for studies of δ sites (26), Table 1.

Concerning the antagonists, it must be emphasized that naloxone is only partially selective for mu sites in contrast to the cyclic analogue of somatostatin CTOB which has recently been shown to be highly selective for these sites (27). Modified enkephalins bearing two allyl groups on the amino group of Tyr (ICI 154,129 and ICI 174,864) behave as selective δ antagonists but their affinity remains low (28).

Biochemical properties of mu and delta opioid receptors

The binding of [³H]DAGO and [³H]DTLET to rat brain membranes and competition experiments with highly selective ligands have shown that μ and δ sites correspond to independent binding sites (29). Furthermore the mu and delta receptors of rat brain, respectively labelled with [³H]DAGO and [³H]DTLET, were shown to differ in their sensitivity to ions, especially Na$^+$ and in their kinetic properties (29).

The use of [³H]DAGO and [³H]DTLET allowed a precise and unambiguous distribution of μ and δ receptors to be obtained in rat (30) and human (31) brains. Interestingly, μ receptors were found in higher proportion than δ sites in brain regions involved in the control of pain while the reverse situation occurred in the limbic system, especially in structures associated with dopaminergic pathways. No important changes in the density of μ and δ receptors were observed in brain and spinal cord of arthritic rats (32) and in brains of Parkinsonians (31).

4 PEPTIDASES-INDUCED METABOLISM OF ENKEPHALINS

A weak and transient analgesia was obtained only for high doses (≈ 100 μg per mouse) of intracerebroventricularly administered Met⁵-enkephalin (Tyr-Gly-Gly-Phe-Met) or Leu⁵-enkephalin (Tyr-Gly-Gly-Phe-Leu) (33). This suggested that, in line with their neurotransmitter role, these peptides were quickly removed from the synaptic cleft. In vitro incubation of enkephalins with brain tissue has shown that several peptidases are able to cleave the endogenous pentapeptides into inactive fragments. The Tyr-Gly bond is hydrolyzed by membrane-bound aminopeptidases, one of which resembles aminopeptidase N. Furthermore, a dipeptidylaminopeptidase activity releasing the Tyr-Gly fragment is also involved in enkephalin degradation in vitro (3). Finally, the enkephalins are easily metabolized by cleavage of the Gly^3-Phe^4 bond under the action of a peptidase, originally designated enkephalinase (34) but identical to the neutral metalloendopeptidase NEP previously isolated by Kerr and Kenny from rabbit kidney (35). Very interestingly all the enkephalin - inactivating enzymes belong to the group of Zn metallopeptidases, offering therefore the possibility of designing mixed inhibitors.

Structural characteristics of metallopeptidases and rational design of inhibitors

As elegantly shown from the crystallographic analysis of two metallopeptidases : the carboxypeptidase A and the bacterial endopeptidase thermolysin, all the zinc metalloproteases have similarities in their active sites and in their respective mechanisms of action (36). The catalytic process involves the coordination of the oxygen of the scissile bond to the Zn atom, followed by a glutamate-promoted nucleophilic attack of a water molecule on the polarized carbonyl carbon. A simplified model of active site of carboxypeptidase A was used by Ondetti and Cushman to design captopril, the highly potent and clinically used inhibitor of the zinc-containing peptidase angiotensin converting enzyme (ACE). Following the same strategy we have developed highly potent and selective inhibitors of the three enkephalin degrading enzymes (review in ref. 3) Table 2. The specificity of enkephalinase is essentially ensured by the preferential interactions of the S_1' subsite with aromatic or large hydrophobic residues. Taking this into account, two highly potent "enkephalinase" inhibitors have been designed able to recognize the S_1'-S_2' subsites and to interact with the Zn atom present in the catalytic site via a thiol group. These two compounds were designated thiorphan, N-[2(RS)-(mercaptomethyl)-1-oxo-3-phenyl-propyl]-glycine (37) and retro-thiorphan, 3-[1(RS)-(mercapto-methyl)-2-phenylethyl]-amino]-3- oxopropanoic acid] (38). The retro-inversion of the amide bond in thiorphan and derivatives was shown to induce a complete differentiation between enkephalinase and ACE inhibition (38).

The inhibitory potency of the separate enantiomers of thiorphan and retrothiorphan has evidenced several similarities between the active site of thermolysin and the neutral

Table 2. Inhibitory potencies and selectivity of the most commonly used inhibitors of enkephalin degrading enzymes. NEP : neutral endopeptidase E.C. 24.11 ; Amino N : aminopeptidase N ; DAP : dipeptidylaminopeptidase.

	K_I (nM)		
	NEP	Amino M	DAP
AMINOPEPTIDASE			
Bestatin (R.S)	>10,000	500	>10,000
Leucine-thiol (S)	>10,000	16	>10,000
DIPEPTIDYLAMINOPEPTIDASE			
Tyr-Phe-NHOH (S.S)	>10,000	>10,000	50.
ENDOPEPTIDASE			
Thiorphan (R+S)	2.5	>10,000	>10,000
Retro-thiorphan (R)	2.3	>10,000	>10,000
Kelatorphan (R.S)	1.7	380	0.9
RB 38 (R.S)	0.8	110	0.7

Figure 1. Computer graphic representations of the energetically-minimized conformation of (S)thiorphan (left) and (R)retro-thiorphan (right) into the active site of thermolysin.

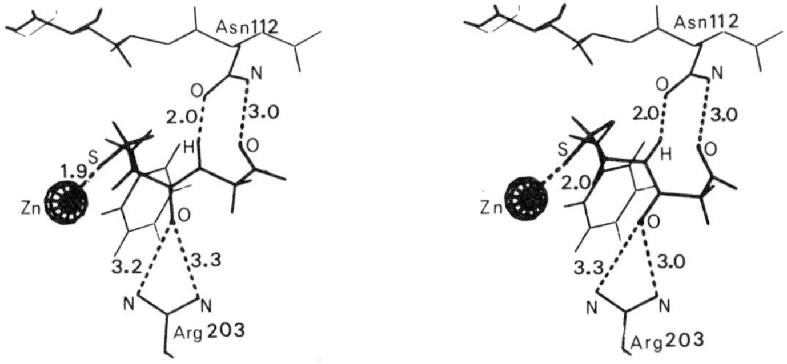

a) Despite the retro-inversion of the peptide bond, all the important hydrogen bonds can be formed.

endopeptidase 24.11, such as their abilities : i) to recognize a retroamide bond as well as a standard amide bond ; ii) to interact similarly with residues in the P_1' position of either the R or S configurations in the thiorphan series but contrastingly to discriminate between the R and S isomers in the retrothiorphan series. These four inhibitors were modelled in the thermolysin active site and their spatial arrangement compared to that of a thiol inhibitor co-crystallized with thermolysin. In all cases, the essential interactions involved in the stabilization of the bound inhibitor were conserved (Fig.1). However, the bound (R)retrothiorphan presented unfavorable intramolecular contacts, accounting for its lower inhibitory potency for the two metallopeptidases (39).

The relevance of this approach was recently reinforced by the determination of the sequence of NEP from rabbit kidney (40). Although the enzyme consists of 749 aminoacids and thermolysin only 315, two highly conserved sequences, including most of the residues present in the active site of the bacterial enzyme, are also present in NEP (41). This is in accordance with biochemical experiments which have shown the occurrence of essential histidine and arginine residues in the active site of both NEP and thermolysin (42).

Taking into account these results and the computed homologies in the primary structures of both peptidases, the modelling of the three-dimensional structure of the active-site of NEP is now possible (41)

Bidentate peptides as highly potent and mixed inhibitors of enkephalin-degrading enzymes

As previously noted, the enkephalins are cleaved in vitro by enkephalinase, a dipeptidylaminopeptidase, and a membrane-bound aminopeptidase. All these enkephalin-degrading enzymes belong to the group of metalloproteases characterized by a wide specificity. It was, therefore, theoretically possible to design a compound able to inhibit the three peptidases, provided that the expected loss of binding affinity, due to a relative inability of the lateral chains of the inhibitor to fit adequately the respective subsites of the three different enzymes, could be counterbalanced by the strength of coordination to the Zn atom. This was indeed obtained with bidentate-containing inhibitors (43).

Among these compounds kelatorphan, $(R)HONH-CO-CH_2-CH(CH_2\phi)-CONH-CH(CH_3)-COOH$, exhibits, as expected, high inhibitory potencies versus : enkephalinase (Ki~1 nM), DAP (Ki ~ 2 nM), aminopeptidase N (Ki ~ 0.4 µM). Bestatin, a currently used aminopeptidase inhibitor, is equipotent on aminopeptidase N and on the various aminopeptidases present in brain tissue. In contrast, kelatorphan displays an approximately 200 fold higher affinity (Ki = 0.4 µM) for aminopeptidase N than for the total membrane-bound aminopeptidases (IC_{50} ~ 90 µM) (44).

In vitro and in vivo protection of enkephalins from degrading enzymes.

The use of incubated or superfused slices of brain or spinal cord tissue allows enkephalin metabolism to be studied in conditions not too far from those existing in vivo. Under conditions where the recovery of intact exogenous [^{3}H]Met-enk, from superfused rat striatal or spinal cord slices amounted to 53% and 45% respectively, these percentages rose to about 80% in the presence of 1 µM thiorphan and 20 µM bestatin but to 91% (striatum) (44) and 98% (spinal cord) with kelatorphan (45). Even more relevant biologically, kelatorphan increased the recovery of endogenous Met-Enk released either by spontaneous outflow or by K^+-evoked overflow in both in vitro (rat striatal slices) and in vivo (superfused spinal cord of anesthetized rats) models (44-45) (Fig. 2). Under similar conditions the association of bestatin and thiorphan were found to be inactive on the basal release of endogenous enkephalins and less potent than kelatorphan in all situations.

Figure 2. A. Release of endogenous Met-enkephalin-like-material (MELM) from superfused spinal cord of anesthetized rats under nociceptive stimuli in the presence or in the absence of kelatorphan. B. Inhibition (%) of the firing of nociceptive neurons of rat dorsal horn after intrathecal administration of enkephalin-degrading enzyme inhibitors.

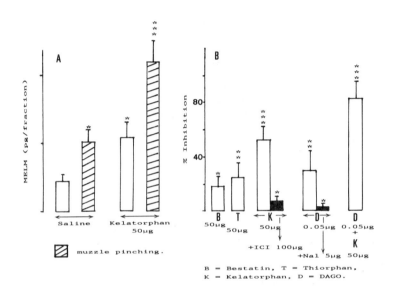

B = Bestatin. T = Thiorphan.
K = Kelatorphan. D = DAGO.

Distribution of enkephalinase in brain tissue by autoradiography

As for many metallopeptidases including ACE, the specificity of NEP is wide since this enzyme is able to cleave a large number of peptides (enkephalins, S.P., bradykinin, neurotensin, CCK_8, etc...) *in vitro*. The localization of the enzyme in brain structures was therefore studied by autoradiography using a tritiated inhibitor [^{3}H]HACBO-Gly (46), which has a high affinity (KD~0.5 nM) for NEP and by immunohistochemistry with a polyclonal antibody (47). By both techniques NEP was found to be mainly localized in brain regions enriched in opioid receptors. Nevertheless it cannot be completely excluded that in particular structures NEP might cleave other peptides, such as SP in the substantia nigra. Furthermore the neuronal localization of NEP in rat brain was demonstrated by lesion experiments followed by quantitative autoradiographic measurements of [^{3}H]HACBO-Gly binding (48).

Spinal and supraspinal analgesic effects of kelatorphan

According to their ability to completely inhibit enkephalin metabolism in vivo, kelatorphan and its derivatives exhibit analgesic potency higher than that produced by thiorphan, bestatin or their association after icv administration (49).

Interestingly kelatorphan was shown to induce the same type of analgesic response as δ agonists on the nociceptive neurons of the rat dorsal horn. This effect was blocked by δ-selective antagonists, showing that endogenous enkephalins produce δ opiate receptor-mediated neuronal inhibitions at the spinal level (50). Furthermore the naloxone-reversible analgesic effect of intrathecally administered DAGO was not altered in the presence of kelatorphan, so demonstrating an additive effect of the μ agonist and the endogenous enkephalins acting via the δ-opioid receptor (51) (Fig. 2). Moreover, kelatorphan was shown to induce potent antinociceptive effects in normal and especially in arthritic rats, these latter being considered as an experimental model of chronic pain. At doses as low as 2.5 mg/kg iv in normal rats kelatorphan produced a naloxone-reversible increase in the vocalization threshold to pain pressure, comparable to that observed in these animals with morphine 1 mg/kg iv (52).

5 PHARMACOLOGICAL RESPONSES INDUCED BY SELECTIVE ACTIVATION OF MU AND DELTA OPIOID RECEPTORS

The development of highly selective ligands for μ and δ opioid binding sites has resulted in extensive investigations on the physiological functions associated with each receptor type (review in 23). Thus, most of the pharmacological responses elicited by morphine-administration have been related either to the stimulation of a single class of opioid receptors or to the activation of both the μ and the δ types. Nevertheless, owing to the clinical interest of morphine and surrogates, the most interesting results concern the role for μ and δ receptors in analgesia, control of respiration and behavioural responses.

Respiratory depression is one of the most prominent side effects which occur after the systemic administration of opiates. This effect is naloxone-reversible and susceptible to tolerance-dependence mechanisms. Both μ and δ receptors located in the bulbar areas were shown to be involved in the control of central respiratory-related neurons. The μ and δ receptors could act independently to control the tidal volume and the respiratory frequency respectively (53). Interestingly, no significant effects on respiration were observed after local or systemic administration of kelatorphan, suggesting the absence of a pharmacologically significant tonic release of endogenous

TABLE 3. Effects of DAGO (µ), DTLET (δ) and kelatorphan :
  A : on the basal release of newly synthesized [³H]DA
  from rat striatal slices ;
  B : on behavioral activity of rats after intracaudate
  administration.

A

Compound	% [³H]DA released a)
Saline	100+8
DAGO 10⁻⁶M	102+5
DTLET 10⁻⁶M	190+10*
DTLET (10⁻⁶M) + ICI 10⁻⁵M	116+12
Kelatorphan 50 µg	168+9*
Kelatorphan 50 µg + ICI10⁻⁵M	98+6

a) Slices were continuously superfused with a medium containing [³H]DA (44), * $p < 0.05$, n=6.

B

SC pretreatment	intracaudate treatment	locomotor a) activity
Saline	Saline	40+6
ICI 2.5 mg/kg	Saline	34+8
TPZ 5 µg/kg	Saline	39+7
Saline	DAGO 2.5 µg	3+3*
Naloxone 0.3 mg/kg	DAGO 2.5 µg	37+8
Saline	DTLET 1 µg	165+51*
ICI 2.5 mg/kg	DTLET 1 µg	56+16
TPZ 5 µg/kg	DTLET 1 µg	48+11
Saline	Kelatorphan 20 µg	180+39*
ICI 2.5 mg/kg	Kelatorphan 20 µg	61+22
TPZ 5 µg/kg	Kelatorphan 20 µg	53+10

a) measured in the open field (62). * $p < 0.05$. TPZ = Thioproperazine.

enkephalins at the level of the respiratory neurons. In order to discard possible differences in the ability of peptides with various µ/δ selectivity to cross the blood brain barrier, these agonists were administered in rodents by an introcerebroventricular route. In the mouse hot plate test (54) as well as in the tail flick test (55) a statistically significant correlation has been found between the antinociceptive effects of these peptides and their affinity for the μ site. Thus, the μ agonist DAGO is about 500 to 1000 times more potent than the δ ligands DSTBULET or DPLPE.

In agreement with these findings, the apparent pA_2 values in a naloxone antagonist trial were in the same range for both selective μ and δ agonists (56). All these results strongly indicate a preferential involvement of μ receptors in supraspinal analgesia. At the spinal level both μ (DAGO) and δ (DPDPE or DSTBULET) ligands are able to control pain stimuli. However the slopes of the dose-response relationship for μ and δ agonists were found to differ, and morphine and DAGO caused 100% analgesia while the δ agonists produced a maximal 60% effect (57).

The nucleus caudatus (NC) and nucleus accumbens (NA), forebrain structures which are involved in emotional as well as in cognitive and motor functions, receive a rich innervation from dopaminergic neurons located in the ventral tegmental area (VTA) and substantia nigra (SN) respectively. A high density of both enkephalins, opioid receptors and "enkephalinase" has been observed in these structures. In relation to this neuroanatomical organization various pharmacological and biochemical studies have shown that morphine and the opioid‑peptides enkephalin are involved in the control of behavior such as arousal, locomotion self-administration, self-stimulation, learning and memory functions, through modulation of the motor (nigrostriatal) and limbic (mesocorticolimbic) DA systems.

Thus injection of enkephalin analogues into the VTA induces an increase in DA metabolism in the NA and a potentiation of the behavioral effect of DA injected in this structure (58,59). These pharmacological responses were shown to be reversed by opioid as well as by dopamine antagonists. Interestingly, the same effects were induced by foot-shock stress or local infusion in the VTA of thiorphan, suggesting a tonic control of the DA mesocorticolimbic pathway by endogenous enkephalins (Fig. 3). Chronic administration into the VTA of DALA, an enkephalin analogue, results in enhanced behavioural responses (hyperactivity) to subsequent local infusion of this peptide. This sensitization process, similar to that induced by amphetamine, occurred also after daily infusion of thiorphan or environmental stress (60). In addition a cross-sensitization between all these hyperactivity-inducing factors has been demonstrated. At this time the receptor type(s) responsible for these DA-dependent effects remains to be definitely characterized.

The nucleus accumbens is one of the major targets of the DA neurons localized in the VTA. Direct injection of modified

Figure 3. Behavioural effects of local administration of thiorphan in the VTA of rats and concomitant increased turnover of DA in the Nucleus Accumbens (NA).

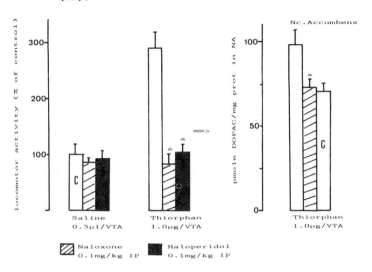

enkephalins such as DAGO, DTLET, DSTBULET, BUBU, DPDPE or kelatorphan into this area led to opposite effects i.e. hypoactivity in the case of the μ agonist vs behavioral activation in the case of δ agonists and endogenous enkephalins protected from degradation (61). These responses were antagonized by naloxone or ICI 174,864 but not by DA antagonists, in accordance with a lack of change in DA release and metabolism after opioid administration (Fig. 4). However the occurrence of indirect interactions between dopaminergic and opioidergic terminals in the NA has been clearly demonstrated by 6-OH-DA-induced lesions of the DA neurons of the VTA or chronic neuroleptic treatment which both potentiate the behavioral effects of opioid infusion into the NA (62).

Injection of DTLET or kelatorphan into the rat striatum also induced an increase in locomotor activity (63) but in contrast to the situation encountered in the NA, this effect was reversed by the DA antagonist thioproperazine and could be related to a specific δ-induced increase in the spontaneous and K⁺-induced release of newly synthesized [³H]DA (64). Finally chronic administration of haloperidol or 6-OH-DA-induced lesions of DA neurons of the substantia nigra was shown to induce a strong

Figure 4. Opposite effects on locomotion of DAGO (μ-agonist) and
DTLET (δ-agonist) or kelatorphan after intra-accumbens
administration in rats.

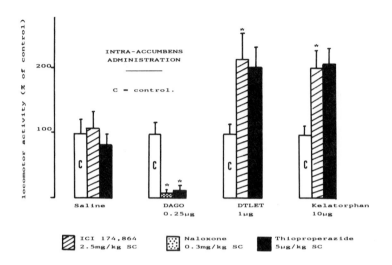

enhancement of both pro-enkephalin mRNA and Met- and
Leu-enkephalin levels in the striatum (65). This suggests the
occurrence of a tonic inhibition of the striatal opioid neurons
by the striatonigral dopaminergic input.

6 OPIOID PEPTIDES : THERAPEUTICAL PERSPECTIVES

The endogenous opioid peptides enkephalins were
characterized in the brain only ten years ago. It is therefore
not surprising that no drug issuing from this discovery has yet
been commercially developed. Nevertheless this is often felt as
a check because there were initial high hopes of rapidly
developing analgesics devoid of the major side effects of
morphine. In this line the first clinically tested modified
enkephalins : FK 33824, DADLE and Metkephamid were unsuccessful,
even though the latter compound was reported to be promising for
intrathecal analgesia in parturition pain (66). It should be
borne in mind however, that these compounds were prepared before
the crucial results reported in this review were obtained. New
therapeutic applications issuing from our knowledge of the

opioidergic systems can now be proposed. As regards "physiological analgesia", it is clear that potent antinociceptive responses can only be obtained by complete inhibition of enkephalin metabolism. Preliminary chronic icv administration of enkephalin-degrading enzyme inhibitors, and especially kelatorphan and derivatives have shown a lack of tolerance and dependence and no significant signs of respiratory depression. This could be explained by a tonic (or phasic) release of enkephalins in the brain areas involved in pain control (S.G., P.A.G., thalamus ...) while this active process could be absent in the locus coeruleus (site of adrenergic neurons) (67) and in the Nucleus Tractus Solitarius. With kelatorphan and derivatives, the strength of analgesia obtained by the icv route is not very different from that obtained with morphine albeit at higher doses and is stronger than that obtained with classical analgesics such as aspirin, paracetamol or derivatives of glafenine. The possible clinical use of blockers of enkephalin metabolism is mainly conditioned by the development of orally active inhibitors.

Furthermore both selective δ agonists and mixed inhibitors could represent a novel approach to analgesia following direct spinal administration. Moreover the additive effects of a μ agonist and a δ agonist or kelatorphan may allow the clinical use of lower doses of μ agonists so avoiding some of the side effects produced by high doses of these substances.

On the other hand the presence of opioid peptides in human brain and the observed euphoriogenic and anxiolytic properties of opiates suggest that a defectively operating endorphinic system may be involved in the pathogenesis of various mental illnesses. This hypothesis is supported by : i) the well-known desinhibitory potency of laudanum tincture in humans, ii) the clinically observed antipsychotic effects of opiates (methadone, buprenorphine, β-endorphin and related peptides...), iii) the involvement of endogenous enkephalins in behavioral reinforcement (reward system). One can speculate that induction of enkephalinergic system activation may play a therapeutic role in the treatment of endogenous depression, very likely by modulation of aminergic systems which are implicated in the etiology and pharmacotherapy of depression.

According to these assumptions, the now well-demonstrated interrelationships between the opioidergic and DAergic systems in the mesocorticolimbic and nigrostriatal pathways provide strong support for a crucial role of endogenous opioids in the control of mood. Amphetamine enhances release of central dopamine, and chronic use of this compound results in psychotic symptoms resembling schizophrenia. As discussed before, several recent studies have shown a cross-sensitization between amphetamine, stressful stimuli and the opioid peptide DALA (68). This suggests that a hypersecretion of endogenous opioid peptides in the mesocorticolimbic pathway could induce an exaggerated behavioral response to stressful environmental stimuli, whose repetition could induce psychotic symptoms (69 and ref. cited therein). Conversely depression might result from a deficiency in the release of enkephalins in the brain, minimizing the rewarding and euphoriogenic effects conveyed by these opioid peptides. Likewise drug abuse could be caused by a deficiency in the internal opioid-controlled rewarding system. Interestingly, chronic treatment with clinically-used neuroleptics induces a behavioral supersensitivity to administration of opioids into the NA, and this effect has been shown to be maximal after 2-3 weeks (62). This delay corresponds to the first appearance of the antipsychotic effects of the neuroleptics, suggesting that alterations in the opioidergic system, very likely via its interrelations with the DAergic pathway, could be taking place in a neuronal system critically involved in the control of mood.

Opioid behavioral activation appears to be preferentially under the control of δ receptors. Thus it appears more and more interesting to design highly potent and selective δ agonists and antagonists which could cross the blood brain barrier. These compounds could be interesting in schizophrenia (antagonists) or in depression (agonists).

Since the pioneer work of De Wied et al., the memory-enhancing effect of peptides are now well-established. In line with this it is important to note that in Alzheimers disease, the brain structures enriched in plaques correspond to areas where the concentration of opioid peptides is highest (70). Likewise, deficiencies in opioid peptides have been reported in some degenerative diseases such as Parkinson, Huntington chorea (71) etc... Therefore it appears now critical to develop new biochemical tests such as quantitative evaluation of mRNA's for opioid peptides and their targets, measurement of axonal transport of receptors and peptidases etc... and neuropeptide-specific pharmacological assays able to give us indications of the role of peptides in the control of "psychological homeostasis".

Finally owing to their peptidase susceptibility, peptides are often discarded for the design of CNS-active drugs. This situation is now changing and linear or cyclic peptides bearing modified peptide bonds are currently being designed. Despite their peptidic nature these peptidase-resistant compounds could be able to cross the blood brain barrier as illustrated by the Sandoz compound FK 33824, much more active than morphine after oral administration (72).

REFERENCES

1. J. Hughes, Ed. in Opioid Peptides, Br. Med. Bull, 1983, 39, 1.
2. T. Hokfelt, V.R. Holets, W. Staines, B. Meister, T. Melander, M. Schalling, M. Schultzberg, J. Freedman, H. Bjorklund, L. Olson, B. Lindh, L.G. Elfvin, J.M. Lundberg, J.A. Lindren, B. Samuelsson, B. Pernow, L. Terenius, C. Post, B. Everitt and M. Goldstein, Progress in Brain Research, T. Hökfelt, K. Fuxe and B. Pernow, Eds, Elsevier Science Publishers, 1986, vol 68, p.33.
3. B.P. Roques and M.C. Fournié-Zaluski, NIDA Research Monograph Series 70, Opioid Peptides : Molecular, Pharmacology, Biosynthesis and Analysis, R.S. Rapaka and R.L. Hawks, Eds, 1986, p.128.
4. J.M. Zajac, G. Gacel, P. Dodey, J.L. Morgat, P. Chaillet, J. Costentin and B.P. Roques, Biochem. Biophys. Res. Commun., 1983, 111, 390.
5. R.R. Bott and D.R. Davies, in Peptides, Structure and Function, Proc. 8th Amer. Pept. Symp., 1983, p. 531.
6. I. Ghosh and V.S.R. Rao, Int. J. Biol. Macromol., 1982, 4, 130.
7. P. Roy, M. Delepierre, M. Wagnon, D. Nisato and B.P. Roques, Int. Pept. Prot. Res., 1987, 30, 44.
8. C.S. Craik, W.J. Rutter and R. Fletterick, Science, 1983, 220, 1125.
9. I.A. Wilson, D.H. Haff, E.D. Getzoff, J.A. Trainer, R.A. Lerwer and S. Brenner, Proc. Natl. Acad. Sci. USA, 1985, 83, 5255.
10. A.B. Schreiber, P.O. Couraud, C. André, B. Vray and A.D. Strosberg, Proc. Natl. Acad. Sci., 1980, 77, 7385.
11. J.Y. Couraud, E. Escher, D. Regoli, V. Imboff, B. Rossignol and P. Pradelles, J. Biol. Chem., 1985, 260, 9461.
12. R.J. Massey, Nature, 1987, 328, 457.
13. A.S.V. Burgen, G.C.K. Roberts and J. Feeney, Nature, 1975, 253, 753.
14. B. Maigret, M.C. Fournié-Zaluski, S. Prémilat and B.P. Roques, Mol. Pharmacol., 1986, 29, 314.
15. B.P. Roques, C. Garbay-Jaureguiberry, R. Oberlin, M. Anteunis and A.K. Lala, Nature, 1976, 262, 778.
16. M.C. Fournié-Zaluski, J. Belleney, B. Lux, C. Durieux, D. Gérard, G. Gacel, B. Maigret and B.P. Roques, Biochemistry, 1986, 25, 3778.
17. S.J. Paterson, L.E. Robson and H.W. Kosterlitz, Opioid Receptors, in The Peptides, Udenfried S. and Meienhofer, J., Eds, Academic Press, London, 1984, Vol. 6, p. 147.
18. T. Ishida, M. Kenmotsu, Y. Mino, M. Inoue, T. Fujiwara, K. Tomita, T. Kimura and S. Sakakibara, Biochem. J., 1984, 218, 677.
19. M.C. Fournié-Zaluski, G. Gacel, B. Maigret, S. Prémilat and

B.P. Roques, Mol. Pharmacol., 1981, 20, 484.

20. B.P. Roques, G. Gacel, M.C. Fournié-Zaluski, B. Senault and J.M. Lecomte, Eur. J. Pharmacol., 1979, 60, 109.

21. P.W. Schiller, in The Peptides : Analysis, Synthesis, Biology, S. Udenfried and J. Meienhofer, Eds, Academic Press, 1984, Vol 6, 219.

22. B.K. Handa, A.C. Lane, J.A.H. Lord, B.A. Morgan, M.T. Rance and C.F.C. Smith, Eur. J. Pharmacol., 1982, 78, 385.

23. B.P. Roques, Annales d'Endocrinologie, 1986, 47, 88.

24. H.T. Mosberg, R. Hurst, V.J. Hruby, K. Gee, M.J. Yamamura, J.J. Gillian and J.F. Burks, Proc. Natl. Acad. Sci., 1983, 80, 5871.

25. P. Delay-Goyet, J.M. Zajac, P. Rigaudy, B. Foucaud and B.P. Roques, FEBS Lett., 1985, 183, 499.

26. P. Delay-Goyet, C. Seguin, V. Daugé, G. Calenco, J.L. Morgat, G. Gacel and B.P. Roques, NIDA Research Monograph, Series 75, Progress in Opioid Research, 1987, p. 197.

27. J.T. Pelton, W. Kazmierski, K. Gulya, H. Yamamura and V.J. Hruby, J. Med. Chem., 1986, 29, 2370.

28. R. Cotton, M.G. Gilles, L. Miller, J.S. Shaw and D. Timms, Eur. J. Pharmacol., 1984, 97, 331-332.

29. J.M. Zajac and B.P. Roques, J. Neurochem., 1985, 44, 1605-1614.

30. R. Quirion, J.M. Zajac, J.L. Morgat and B.P. Roques, Life Sci., 1983, 33, 227-230.

31. P. Delay-Goyet, J.M. Zajac, F. Javoy-Agid, Y. Agid and B.P. Roques, Brain Res., 1987, 414, 8.

32. P. Delay-Goyet, V. Kayser, G. Guilbaud and B.P. Roques, in preparation.

33. J.D. Belluzi, N. Grant, V. Garsky, D. Sarantakis, C.D. Wise and L. Stein, Nature, 1976, 260, 625.

34. B. Malfroy, J.P. Swerts, A. Guyon, B.P. Roques and J.C. Schwartz, Nature, 1978, 276, 523.

35. M.A. Kerr and A.J. Kenny, Biochem. J., 1974, 137, 477.

36. A.F. Monzingo and B.W. Matthews, Biochemistry, 1982, 21, 3390.

37. B.P. Roques, M.C. Fournié-Zaluski, E. Soroca, J.M. Lecomte, B. Malfroy, C. Llorens and J.C. Schwartz, Nature, 1980, 288, 286.

38. B.P. Roques, E. Lucas-Soroca, P. Chaillet, J. Costentin and M.C. Fournié-Zaluski, Proc. Natl. Acad. Sci. USA, 1983, 80, 3178.

39. T. Benchetrit, M.C. Fournié-Zaluski and B.P. Roques, Biochem. Biophys. Res. Commun., 1987, 147, 1034-1040.

40. A. Devault, C. Lazure, C. Nault, H. Le Moual, N.G. Seidah, M. Chretien, P. Kahn, J. Powell, J. Mallet, A. Beaumont, B.P. Roques, P. Crine and G. Boileau, EMBO J., 1987, 6, 1317.

41. T. Benchetrit, J.P. Mornon, G. Boileau, P. Crine and B.P. Roques, Biochemistry, 1987, in press.

42. A. Beaumont and B.P. Roques, Biochem. Biophys. Res. Commun., 1986, 139, 733-739.
43. M.C. Fournié-Zaluski, A. Coulaud, R. Bouboutou, P. Chaillet, J. Devin, G. Waksman, J. Costentin and B.P. Roques, J. Med. Chem., 1985, 28, 1158.
44. G. Waksman, R. Bouboutou, J. Devin, S. Bourgoin, F. Cesselin, M. Hamon, M.C. Fournié-Zaluski and B.P. Roques, Eur. J. Pharmacol., 1985, 117, 233.
45. S. Bourgoin, D. Le Bars, F. Artaud, A.M. Clot, R. Bouboutou, M.C. Fournié-Zaluski, B.P. Roques, M. Hamon and F. Cesslin, J. Pharmacol. Exp. Ther., 1986, 238, 360.
46. G. Waksman, E. Hamel, M.C. Fournié-Zaluski and B.P. Roques, Proc. Natl. Acad. Sci., 1986, 83, 1523.
47. R. Matsas, A.J. Kenny and A.J. Turner, Neuroscience, 1986, 18, 991.
48. G. Waksman, E. Hamel, P. Delay-Goyet and B.P. Roques, EMBO J., 1986, 5, 3163.
49. M.C. Fournié-Zaluski, P. Chaillet, R. Bouboutou, A. Coulaud, P. Chérot, G. Waksman,, J. Costentin and B.P. Roques, Eur. J. Pharmacol., 1984, 102, 525.
50. A.H. Dickenson, A.F. Sullivan, M.C. Fournié-Zaluski and B.P. Roques, Brain Res., 1987, 408, 185.
51. A.H. Dickenson, A. Sullivan, C. Feeney, M.C. Fournié-Zaluski and B.P. Roques, Neurosci. Lett., 1986, 72, 179.
52. V. Kayser, M.C. Fournié-Zaluski, B.P. Roques and G. Guilbaud, Brain Res., in press.
53. M.P. Morin-Suron, G. Gacel, J. Champagnat, M. Denavit-Saubié and B.P. Roques, Eur. J. Pharmacol., 1984, 98, 241.
54. P. Chaillet, A. Coulaud, J.M. Zajac, M.C. Fournié-Zaluski, J. Costentin and B.P. Roques, Eur. J. Pharmacol., 1984, 101, 83.
55. V. Daugé, F. Petit, P. Rossignol and B.P. Roques, Eur. J. Pharmacol., 141, 171-178.
56. F.G. Fang, H.L. Fields and N.M. Lee, J. Pharm. Exp. Ther., 1986, 238, 1039.
57. A.H. Dickenson, A.F. Sullivan, R. Know, J.M. Zajac and B.P. Roques, Brain Res., 1987, 413, 36.
58. L. Stinus, G.F. Koob, N. Ling, F.E. Bloom and M. Le Moal, Proc. Natl. Acad. Sci. USA, 1980, 77, 2323.
59. P.W. Kalivas, E. Winderlon, D. Stanley, G. Breese and A.J. Prange, Jr, J. Pharmac. Exp. Ther., 1983, 227, 229.
60. P.N. Kalivas and R. Richarson-Carlson, Am. J. Physiol., 1986, 251, R243.
61. V. Daugé, P. Rossignol and B.P. Roques, J. Psychopharmacology, in press.
62. L. Stinus, D. Nadaud, J. Jauregui and A.E. Kelley, Biol. Psychiatry, 1986, 21, 34.
63. B.P. Roques, V. Daugé, G. Gacel and M.C. Fournié-Zaluski, in Biological Psychiatry, Developments in Psychiatry, C. Shagass, R.C. Josiassen, W.H. Bridger, K.J. Weiss, D. Stoff

and G.M. Simpon, Eds, Elsevier New-York, 1985, vol 7, p.287.
64. F. Petit, M. Hamon, M.C. Fournié-Zaluski, B.P. Roques and J.
    Glowinski, Eur. J. Pharmacol., 1986, 126, 1.
65. W. Scott Young III, T.I. Bonner and M.R. Brann, Proc. Natl.
    Acad. Sci. USA, 1986, 83, 9827.
66. R.C.A. Frederickson, in Adv. in Pain Res. and Ther., K.M.
    Foley and C.E. Inturrisi, Eds, Raven Press, New-York, 1986,
    vol 8, p.293.
67. J.J. Williams, J.C. MacDonald, R.A. North and B.P. Roques,
    J. Pharm. Exp. Ther., 1987, in press.
68. P.W. Kalivas, R. Richardson-Carlson and G. Van Orden, Biol.
    Psychiatry, 1986, 21, 939.
69. A.J. Mac Lennan and S.F. Mayer, Science, 1983, 219, 1091.
70. A.C. Cuello, Br. Med. Bull., 1983, 39, 11.
71. Y. Agid and F. Javoy-Agid, Trends in Neurosci., 1985, 8(6),
    30.
72. D. Römer, H.H. Büscher, R.C. Hill, J. Pless, W. Bauer, F.
    Cardinaux, A. Closse, D. Hauser and R. Hughenin, Nature,
    1977, 268, 547.

Pharmacological Consequences of Conformational Restriction in Opioid Peptides

P. W. Schiller

LABORATORY OF CHEMICAL BIOLOGY AND PEPTIDE RESEARCH,
CLINICAL RESEARCH INSTITUTE OF MONTREAL, 110 PINE AVENUE
WEST, MONTREAL, QUEBEC, CANADA H2W 1R7

1 INTRODUCTION

Following the demonstration of the existence of opioid receptors in the early seventies, an intensive search for an endogenous ligand of these receptors culminated in the discovery of the enkephalins, two pentapeptides with opioid activity[1] (Figure 1). Subsequent research revealed that most of the opioid peptides known to date are derived from three precursor proteins: proopiomelanocortin, proenkephalin A and proenkephalin B (for a review, see ref. 2). Furthermore, the results of several pharmacological studies performed with both opiates and opioid peptides led to the concept of multiple opioid receptors. At present, the existence of at least three different receptor classes (μ, δ, $\varkappa$), differing from one another in their structural requirements towards opioid ligands, is generally accepted[3]. Unfortunately, none of the endogenous opioid peptides is very selective for a particular receptor class. Thus, metorphamide displays only moderate selectivity for the μ-receptor and the enkephalins show quite limited preference for the δ-receptor; both the dynorphins and the neoendorphins preferentially bind to the $\varkappa$-receptor but also display significant affinity for the μ- and the δ-receptor. Certain opioid peptides derived from casein (β-casomorphins, morphiceptin) or isolated from frog skin (dermorphins) show some selectivity for the μ-receptor. Both opiates and opioid peptides display a

large spectrum of biological activities, including
analgesia, respiratory depression, tolerance and
physical dependence, euphoria, hypothermia, effects on
gut motility, cardiovascular effects, etc. In order to
associate a particular receptor class with a distinct
biological function, it is of great importance to
develop opioid receptor ligands with high selectivity
for a particular receptor type. To this end, numerous
opioid peptide analogs have been synthesized and
characterized since the discovery of the enkephalins.
In most cases the analog design was based on various
amino acid substitutions, additions or deletions.
The use of these classical design principles led to a
few compounds which indeed showed improved receptor
selectivity. For example, H-Tyr-D-Ala-Gly-Phe(NMe)-
Gly$_5$ol (DAGO)[4] and H-Tyr-Pro-Phe(NMe)-D-Pro-NH$_2$ (PLO-
17)[5] show quite high preference for the μ-receptor,
whereas H$_z$Tyr-D-Ser(or Thr)-Gly-Phe-Leu-Thr-OH (DSLET
or DTLET)[6] display considerably improved δ-receptor
selectivity.

CYCLIC OPIOID PEPTIDE ANALOGS

The structural flexibility of the natural opioid
peptides and of many of their linear analogs (see ref.
7) is most likely the reason for their lack of
selectivity, since it permits conformational adaptation
to more than one opioid receptor type. Therefore, it
has recently been attempted to enhance receptor
selectivity through incorporation of various types of
conformational constraints into opioid peptides. The
most drastic restriction of the overall conformational
freedom can be achieved through peptide cyclizations.
In particular, cyclizations of enkephalin via side
chains of appropriately substituted amino acid residues
have been successful. For example, substitution of a
D-α, ω-diamino acid (Daa) in the 2-position of the
[Leu^5]enkephalin sequence and cyclization of the ω-
amino group to the C-terminal carboxyl group resulted
in cyclic analogs, H-Tyr-cyclo[-D-Daa-Gly-Phe-Leu-]
(Figure 1, analogs 2-5), showing high potency at the μ$_8$
receptor and considerable μ-receptor selectivity.
Corresponding open-chain analogs, H-Tyr-D-Xxx-Gly-Phe-
Leu-NH$_2$ (Xxx = Apr, Abu, Orn or Lys) turned out to be
non-selective and it can thus be concluded that the μ-
receptor selectivity of these cyclic analogs is a

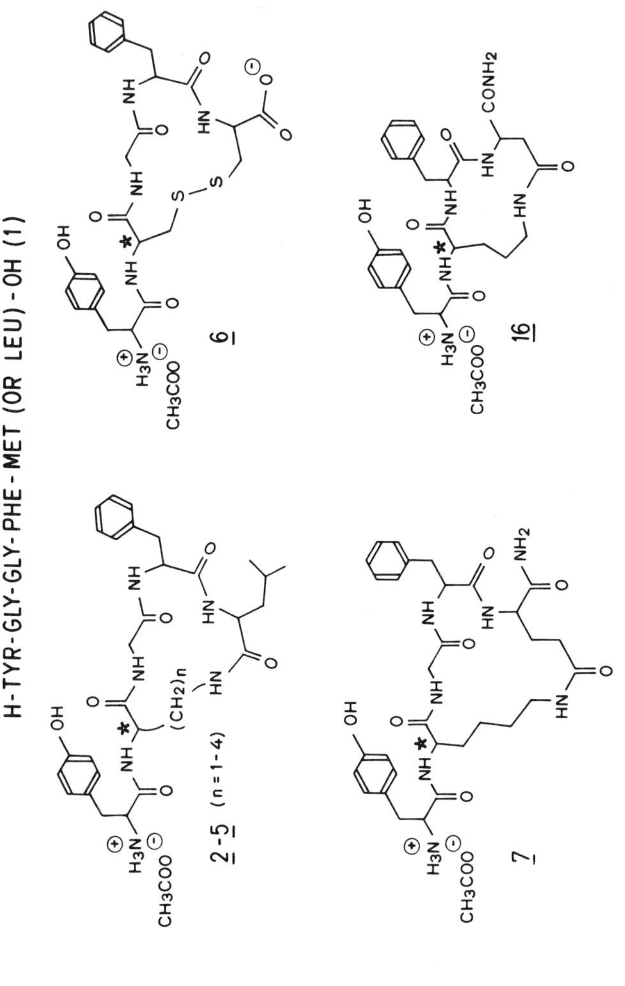

Figure 1. Structural formulas of cyclic opioid peptide analogs

direct consequence of the conformational restriction introduced through ring closure.[9] The cystine-containing cyclic peptides, H-Tyr-D-Cys-Gly-Phe-D(or L)-Cys-X (X = NH_2 or OH) (Figure 1, structure 6) are examples of side chain-to-side chain cyclized enkephalin analogs. The cystine-bridged enkephalin-amide analogs (X = NH_2) were found to be more than two orders of magnitude as potent as [Leu[5]]enkephalin in the guinea pig ileum (GPI) assay,[10] but were non-selective. On the other hand, the corresponding analogs with a free C-terminal carboxyl group showed about the same moderate δ-receptor selectivity as the natural enkephalins.[11] Structure-activity studies with analogs of H-Tyr-cyclo[-D-Lys-Gly-Phe-Leu-] and H-Tyr-D-Cys-Gly-Phe-Cys-NH_2 revealed that these cyclic opioid peptides have the same configurational requirements in positions 1, 2, 4 and 5 as linear enkephalins.[12] In addition, substitution of a nitro group in para position of the aromatic ring of Phe[4] in the cyclic analogs produced the same drastic potency enhancement as in the case of linear enkephalins.[12] These findings strongly suggest that cyclic and linear enkephalin analogs have the same mode of binding to the receptor.

The activity profile of cyclic enkephalin analogs can be further manipulated through the performance of various peptide bond replacements. Thus, reversal of two amide bonds in H-Tyr-cyclo[-D-A_2bu-Gly-Phe-Leu-] (3) (A_2bu = α, γ-diaminobutyric acid) resulted in an analog, H-Tyr-cyclo[-D-Glu-Gly-gPhe-D-Leu-] which in comparison with 3 showed the same affinity for the μ-receptor but three times lower affinity for the δ-receptor.[13] Thus, this so-called retro-inverso analog turned out to be three times more μ-selective than the cyclic parent peptide (3). On the other hand, replacement of the 4-5 position peptide bond in H-Tyr-cyclo[-D-Lys-Gly-Phe-Leu-] (5) with a thiomethylene ether linkage resulted in a complete loss of the μ-receptor selectivity shown by the parent peptide.[14] The lack of selectivity observed with the cyclic pseudopeptide may be due to the greater flexibility of the 16-membered ring structure as a consequence of the peptide bond substitution. Replacement of the two half-cystine residues in H-Tyr-D-Cys-Gly-Phe-D(or L)-Cys-OH (6) with penicillamine residues led to a compound with greatly improved δ-receptor selectivity,[15] most likely because of the enhanced

rigidity of the 14-membered ring structure due to the
presence of the gem dimethyl group in the β-position of
the penicillamine side chains. These examples
demonstrate that additional structural modifications of
cyclic opioid peptide analogs can lead to pronounced
shifts in the receptor selectivity profile.

None of the cyclic analogs had significant
affinity for the ϰ-receptor. Furthermore, these cyclic
analogs were found to be highly resistant to
enzymolysis and were able to induce a long-lasting
analgesic effect in rats (i.c.v. administration).

Cyclic Lactam Analogs of Enkephalin and of Dynorphin A

Side chain-to-side chain cyclized peptide analogs
can also be obtained through appropriate substitution
of a Lys (Orn) and Glu (Asp) residue and subsequent
amide bond formation between the side chain amino and
carboxyl groups of these residues. Cyclic lactam
analogs of this type can be synthesized by the solid-
phase method according to a recently published
protection scheme.[16] Aside from the desired cyclic
monomers, cyclization of the peptide still attached to
the resin also leads to the formation of side chain-
linked antiparallel cyclic dimers. The extent of
cyclodimerization has been shown to depend on
conformational factors rather than on the level of
resin substitution.[17]

The cyclic enkephalin analog H-Tyr-D-Lys-Gly-Phe-
Glu-NH$_2$ (Figure 1, 7) showed high affinity for both
the μ- and the δ-receptor and, therefore, turned out
to be non-selective (Table 1). Interestingly, the two
structurally related open-chain analogs, H-Tyr-D-Nle-
Gly-Phe-Gln-NH$_2$ (7a) and H-Tyr-D-Lys(For)-Gly-Phe-Abu-
NH$_2$ (7b) (Abu = α-aminobutyric acid), showed a
pronounced[18] preference for μ-receptors over δ-
receptors. This finding represents the first
reported example of a peptide cyclization resulting in
a loss of receptor selectivity. Obviously, the
relatively flexible 18-membered ring structure of 7
permits conformational adaptation to the distinct
topographies of the μ- and the δ-receptor, resulting in
efficient binding to both receptor types. The μ-
receptor selectivity of linear correlates 7a and 7b is
a consequence of their relatively poor affinity for the

Table 1. μ- and δ-Receptor Affinities of Cyclic Lactam Analogs of Enkephalin and Dynorphin A

No.	Compound	K_i^μ [nM]	K_i^δ [nM]	K_i^δ/K_i^μ
7	H-Tyr-D-Lys-Gly-Phe-Glu-NH$_2$	1.31	0.690	0.527
7a	H-Tyr-D-Nle-Gly-Phe-Gln-NH$_2$	0.628	23.4	37.3
7b	H-Tyr-D-Lys(For)-Gly-Phe-Abu-NH$_2$	2.76	66.3	24.0
7c	(H-Tyr-D-Lys-Gly-Phe-Glu-NH$_2$)$_2$	5.33	23.2	4.35
8	H-Phe-D-Lys-Gly-Phe-Glu-NH$_2$	4.69	86.4	18.4
9	H-Tyr-D-Orn-Gly-Phe-Asp-NH$_2$	0.762	2.71	3.56
10	H-Tyr-D-Orn-Gly-Phe-Asp-Arg-Arg-Ile-NH$_2$	0.0555	0.893	16.1
11	H-Tyr-Gly-Gly-Phe-Orn-Arg-Arg-Asp-Arg-Pro-Lys-Leu-Lys-NH$_2$	6.84	253	37.0
12	H-Tyr-Gly-Gly-Phe-Orn-Arg-Arg-Ile-Arg-Asp-Lys-Leu-Lys-NH$_2$	38.6	323	8.37
13	H-Tyr-Gly-Gly-Phe-Orn-Arg-Arg-Ile-Arg-Pro-Lys-Leu-Asp-NH$_2$	3.39	163	48.1
14	Dynorphin A-(1-8)	1.63	4.51	2.77
15	Dynorphin A-(1-13)	3.95	3.74	0.947
1	[Leu^5]enkephalin	9.43	2.53	0.268

δ-receptor. This may be due to the fact that these linear peptides assume an average conformation which is not well tolerated by the δ-receptor and the formation of which is prevented in 7 by cyclization. The cyclic dimer, $(H-Tyr-D-Lys-Gly-Phe-Glu-NH_2)_2$ (7c)[19] had eight times lower μ-receptor affinity than the cyclic monomer and about the same δ-receptor affinity as 7a. Therefore, it is moderately μ-selective with a K^δ_i/K^μ_i ratio lying in between those of the cyclic and linear monomers. The conformational constraints in the dimer are clearly different from those in the cyclic monomer and can be assumed to be responsible for the difference in receptor selectivity observed between these two compounds. Alternatively, it could also be argued that the interaction of one of the dimer's individual enkephalin chains with either the μ- or the δ-receptor might be affected by the presence of the second covalently linked chain which could cause steric interference or interact with accessory binding sites.

In linear enkephalins omission of the hydroxyl group of the tyrosine residue always resulted in a drastic potency loss. Interestingly, substitution of Phe for Tyr^1 in cyclic analog 7 led to a compound (8) which retained μ-receptor affinity twice as high as that of [Leu^5]enkephalin. Similarly high μ-receptor affinities had previously been observed with Phe^1 analogs of cyclic opioid peptides 5 and 6[12]. The difference in the effect of Phe^1 substitution on the potency of cyclic versus linear opioid peptides may reflect different binding processes. The conformation of the semi-rigid cyclic analogs in solution may closely resemble the receptor-bound conformation and, therefore, these compounds may interact with the receptor more or less according to Fischer's "lock-and-key" model.[20] On the other hand, the flexible linear analogs are likely to undergo conformational changes in the course of the binding process which perhaps can be best described by the "zipper" model.[21] In a "lock-and-key" type binding process the tyrosyl hydroxyl group may no longer be as crucial as in the case of a "zipper" type interaction where perhaps initial hydrogen bond formation between the hydroxyl group and an acceptor moiety of the receptor may be necessary for the subsequent conformational adaptations to occur.

In an effort to reduce the conformational flexibility of the ϰ-receptor selective opioid peptide dynorphin A (H-Tyr-Gly-Gly-Phe-Leu-Arg-Arg-Ile-Arg-Pro-

Lys-Leu-Lys-Trp-Asp-Asn-Gln-OH), we synthesized several cyclic lactam analogs of dynorphin A-(1-8) and dynorphin A-(1-13). In the μ-receptor representative binding assay ([^{3}H]DAGO displacement) cyclic dynorphin analog 10 (H-Tyr-D-Orn-Gly-Phe-Asp-Arg-Arg-Ile-NH$_2$) was 170 and 29 times more potent than [Leu^5]enkephalin and linear dynorphin A-(1-8), respectively. The K$^\delta_i$/K$^\mu_i$ ratio of 16 obtained for compound 1 is indicative of a moderate preference for μ-receptors over δ-receptors. The corresponding cyclic enkephalin analog 9 displayed about 14 times lower μ-receptor affinity than 10 and was less μ-selective. The extremely high μ-receptor affinity of 10, as compared to 9, may be due to the two additional positive charges on the arginine residues in positions 6 and 7. The high positive charge (+ 3) of cyclic dynorphin analog 10 may promote its accumulation in the anionic fixed charge membrane compartment which$_{22}$ according to Schwyzer's membrane compartment concept, contains primarily μ-binding sites. In addition, the cyclic segment of analog 10 fulfils the conformational requirements at the μ-receptor. Thus, its very high μ-receptor affinity may be due to both its net positive charge and the μ-receptor compatible conformation of its enkephalin segment. Cyclic dynorphin analogs 11, 12 and 13 contain ring structures in the middle and C-terminal region of the peptide chain. In the binding assays these three cyclic analogs showed μ-receptor affinities comparable to those of [Leu^5]enkephalin and dynorphin A-(1-13) but weak affinities for the δ-receptor and, therefore, are also μ-selective. In the GPI assay K$_e$-values for naloxone as antagonist above 10 nM are indicative of a κ-receptor interaction, whereas the μ-receptor interaction is characterized by K$_e$-values below 5 nM. Thus, the κ-receptor selective linear dynorphins 14 and 15 displayed K$_e$-values of 13.4 and 29.3 nM, respectively. On the other hand, K$_e$-values below 5 nM were determined with compounds 10-13, indicating that these analogs no longer interact significantly with κ-receptors but have become μ-agonists instead.$_{22}$ According to the membrane compartment concept^{22} dynorphin A binds to the κ-receptor with its enkephalin segment ("message sequence") assuming an α-helical conformation. Obviously, the ring structure contained in cyclic dynorphin analog 10 would prevent formation of an α-helical structure of its enkephalin segment. The loss of κ-receptor selectivity observed with 10 is therefore in agreement with this model. The lack of κ-receptor binding shown by cyclic dynorphin analogs 11-13 could be due to the fact that the performed cyclizations may have resulted in overall folded conformations which are no longer compatible with the conformational requirements of the κ-receptor. Other conformational constraints will have to be built into dynorphin in order to obtain analogs with retained or improved κ-receptor selectivity which will provide insight into the bioactive conformation of dynorphin at the κ-receptor.

Cyclic Opioid Peptides Containing a Phenylalanine Residue in the 3-Position

Whereas the enkephalins, dynorphins and β-endorphin contain a phenylalanine residue in position 4 of the peptide sequence, the Phe residue in the dermorphins (H-Tyr-D-Ala-Phe-Gly-Tyr-Pro(or Hyp)-Ser-NH$_2$) and in the β-casomorphins (H-Tyr-Pro-Phe-Pro-NH$_2$ (morphiceptin), H-Tyr-Pro-Phe-Pro-Gly-NH$_2$ (β-casomorphin-5), etc.) is located in the 3-position. It was therefore of interest to synthesize and characterize a series of cyclic opioid peptide analogs which also contain a Phe residue in position 3. The cyclic lactam analog H-Tyr-D-Orn-Phe-Asp-NH$_2$ (16) contains a rather rigid 13-membered ring structure. In the receptor binding assays compound 16 showed very high μ-receptor selectivity (K_i^δ / K_i^μ = 213) as a consequence of its very weak affinity for the δ-receptor[17] (Table 2). In comparison with 16 the corresponding open-chain analog H-Try-D-Nva-Phe-Asn-NH$_2$ (16a) (Nva = norvaline) had about the same μ-receptor affinity but five times higher affinity for the δ-receptor and, therefore, is about five times less μ-selective. Based on the results obtained so far it appears that cyclic opioid peptides containing relatively small and rigid ring structures (13- to 14-membered) are more selective than their linear correlates, whereas in the case of cylic analogs with larger and more flexible ring structures (e.g. H-Tyr-D-Lys-Gly-Phe-Glu-NH$_2$ (7)) the opposite may be true. Compared to cyclic monomer 16, the corresponding side chain-linked antiparallel cyclic dimer, (H-Tyr-D-Orn-Phe-Asp-NH$_2$)$_2$ (16b), had about 2.5 times lower affinity for the μ-receptor but 50 times higher affinity for the δ-receptor and, therefore, is non-selective. Expansion of the 13-membered ring structure in 16 to a more flexible 15-membered one was achieved through synthesis of the analog H-Tyr-D-Lys-Phe-Glu-NH$_2$ (17).[23]

Table 2. μ- and δ-Receptor Affinities of Cyclic Opioid Peptide
Analogs Containing a Phenylalanine Residue in the 3-
Position

No.	Compound	K_i^μ [nM]	K_i^δ [nM]	K_i^δ/K_i^μ
16	H-Tyr-D-Orn-Phe-Asp-NH$_2$	10.4	2,220	213
16a	H-Tyr-D-Nva-Phe-Asn-NH$_2$	11.7	441	37.7
16b	(H-Tyr-D-Orn-Phe-Asp-NH$_2$)$_2$	25.6	42.2	1.65
17	H-Tyr-D-Lys-Phe-Glu-NH$_2$	1.43	4.36	3.05
18	H-Tyr-Orn-Phe-Asp-NH$_2$	4,830	> 60,000	> 12.4
19	H-Tyr-D-Orn-D-Phe-Asp-NH$_2$	3,040	>380,000	>125
20	H-Tyr-D-Orn-Phe-D-Asp-NH$_2$	21.7	422	19.4
21	H-Tyr-D-Orn-Phe(pNO$_2$)-Asp-NH$_2$	273	5,060	18.5
22	H-Tyr-D-Orn-Phg-Asp-NH$_2$	445	3,570	8.02
23	H-Tyr-D-Orn-Hfe-Asp-NH$_2$	11.9	849	71.3
24	H-Tyr-D-Orn-Phe(NMe)-Asp-NH$_2$	3,570	28,100	7.87
25	H-Tyr-D-Asp-Phe-Orn-NH$_2$	9.55	1,320	138
26	H-Tyr-D-Asp-Phe-A$_2$bu-NH$_2$	24.8	4,170	168
27	H-Tyr-D-Cys-Phe-Cys-NH$_2$	11.0	373	33.9
1	[Leu^5]enkephalin	9.43	2.53	0.268

The latter analog had seven times higher affinity for the μ-receptor than analog 16 and nearly 500 times higher δ-receptor affinity. These data indicate that cyclic analog 17 is highly potent but non-selective. Comparison of the K_i^δ/K_i^μ ratios of 16 and 17 reveals that ring expansion by two methylene groups produced a 70-fold shift in selectivity. This result indicates that subtle variation in the degree of conformational restriction can drastically affect the selectivity profile.

Configurational inversion at the ornithine residue of cyclic analog 16 resulted in a compound (18) with greatly reduced affinity for both the μ- and the δ-receptor. Whereas a similar potency drop had been observed upon substitution of an L-amino acid residue in the 2-position of linear dermorphin-related peptides,[24] L-configuration is required at the Pro residue of morphiceptin.[25] Substitution of D-phenylalanine in position 3 of compound 16 (analog 19) also reduced the affinity for both μ- and δ-receptors by more than two orders of magnitude. This observation is again in agreement with the drastic potency loss resulting from configurational inversion at the Phe[3] residue in a dermorphin tetrapeptide.[26] Again, the corresponding D-Phe[3] substitution in β-casomorphin had the opposite effect, as indicated by the increased potency of H-Tyr-Pro-D-Phe-Pro-Gly-OH in comparison with H-Tyr-Pro-Phe-Pro-Gly-OH.[27] Inversion of the configuration of Asp[4] in parent peptide 16 led to a compound (20) with only slightly reduced affnity for μ-receptors and with five times higher δ-receptor affinity. This relatively minor change in the activity profile is again in agreement with the observation that substitution of D-leucine in position 4 of the dermorphin - related analog H-Tyr-D-Arg-Phe-Leu-OH did not have a major effect on analgesic potency.[24] On the other hand, the morphiceptin analog [D-Pro[4]]morphiceptin has been reported to be nearly 20 times more potent in the GPI assay than its diastereomer, morphiceptin.[27] Taken together, the receptor binding data obtained with analogs 18, 19 and 20 indicate that cyclic analog 16 has the same configurational requirements in positions 2, 3 and 4 of the peptide sequence as dermorphin-related peptides, whereas β-casomorphins show opposite configurational requirements in all three positions. Substitution of a nitro group in para position of the Phe[3] aromatic ring in cyclic peptide 16 (analog 21) produced a drastic drop in μ-receptor affinity, as had been the case with dermorphin and morphiceptin.[28]

Shortening of the Phe[3] side chain (analog 22) led to a 40-fold drop in μ-receptor affinity without much effect on δ-receptor affinity and, consequently, the μ-receptor selectivity of 22 is considerably lower. These data indicate that the more restricted orientational freedom of the phenylglycine residue may no longer permit an optimal interaction of its aromatic ring with the complementary μ-receptor subsite. On the other hand, lengthening of the side chain in the 3-position of compound 16 resulted in a cyclic analog (23) with unchanged μ-receptor affinity but nearly three times higher δ-receptor affinity. It thus appears that due to chain lengthening the aromatic ring of homophenylalanine has somewhat better access to a hydrophobic subsite of the δ-receptor. The observation that the length of the side chain of the aromatic residue in position 3 of 16 is important for receptor affinity and selectivity suggests that μ- and δ-receptors differ from one another in the relative spatial disposition of the binding sites for the Tyr[1] tyramine moiety and the Phe[3] aromatic ring.

Further conformational restriction of cyclic analog 16 was achieved through N-methylation of the Phe[3] residue, resulting in a compound (24) with very low affinity. The poor binding properties of analog 24 may be due either to the additional conformational constraint introduced by incorporation of the methyl group or simply to the bulkiness of the latter group which may produce steric interference at the receptor. N^α-methylation at the 3-position residue in a dermorphin-related peptide analog had also resulted in a considerable potency loss in the GPI assay.[29] Again, the opposite effect was obtained upon methylation of the Phe[3] residue in morphiceptin, as indicated by the observation that H-Tyr-Pro-Phe(NMe)-Pro-NH$_2$ is about twice as potent as morphiceptin at the μ-receptor.[5] Transposition of the Orn and Asp residues in parent peptide 16 resulted in a compound (25) with almost identical μ- and δ-receptor affinities. Shortening of the side chain in position 4 of 25 through substitution of α, γ-diaminobutyric acid produced an analog (26) containing a more rigid 12-membered ring. This ring contraction resulted in a 2.5 fold decrease in μ-receptor affinity and a slightly larger decrease in δ-receptor affinity. Therefore, the μ-receptor selectivity of 26 is higher than that of 25. The cystine-containing cyclic analog (27) has an even

smaller, 11-membered ring structure. Despite its
smaller ring size analog 27 is less μ-selective than
cyclic peptides 16, 25 and 26. This may be due to the
fact that the side chain-connecting disulfide linkage
in 27 is more flexible than the corresponding amide
linkages in compounds 16, 25 and 26. This enhanced
flexibility may permit better adaptation to the δ-site
despite the smaller ring size.

Taken together these structure-activity data
indicate that analogous structural modifications in
cyclic peptide 16 and in dermorphin-related peptides
produce qualitatively the same effect on biological
activity, whereas corresponding modifications in β-
casomorphin (morphiceptin) in general had the opposite
effect. The fact that the configurational requirement
of compound 16 and of dermorphin-related peptides in
positions 2, 3 and 4 are identical is compatible with
the assumption that the cyclic analog and dermorphin
interact with the same receptor subsites. The opposite
configurational requirements in the corresponding
positions of β-casomorphin in conjunction with the
opposite effects of Phe[3] N-methylation suggest that the
mode of binding of β-casomorphins (morphiceptin) may be
different from that of cyclic peptide 16 and of
dermorphin.

A theoretical conformational analysis recently
performed with H-Tyr-D-Orn-Phe-Asp-NH$_2$ (16) revealed
that the lowest energy conformer is characterized by a
tilted stacking[30] interaction between the Tyr[1] and Phe[3]
aromatic rings (Figure 2). This conformational
analysis was extended to several analogs of 16
described in the present paper. The potent μ-selective
analogs 20 and 25 showed a stacking arrangement of the
two aromatic rings similar to that observed in the
lowest energy conformer of 1. On the other hand, three
analogs with low μ-receptor affinity (18, 22 and 24)
were shown to be unable to adopt this type of stacking
configuration. These results suggest that a specific
tilted stacking interaction of the aromatic rings in
the 1- and 3-position of cyclic analog 16 may represent
an important structural requirement for binding at the
μ-receptor.

Comparison of Cyclic Opioid Peptide Potencies in μ-
Receptor Representative Binding Assay and Bioassays

The cyclic analogs were also tested using in vitro
bioassays based on inhibition of electrically evoked

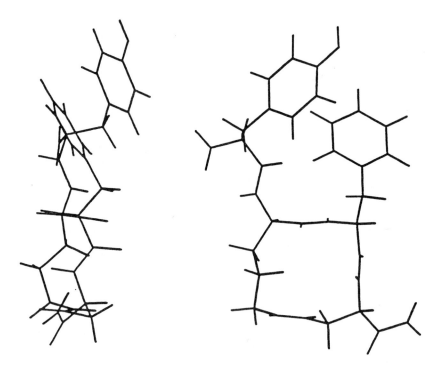

Figure 2.   Molecular graphics display of lowest energy
             conformer of   H-Tyr-D-Orn-Phe-Asp-NH$_2$ ($\underline{16}$)
             (two different views)

contractions of the GPI (μ-receptor representative) and
of   the   mouse   vas   deferens   (MVD,   δ-receptor
representative).   In general the bioassay structure-
activity   data   were   found to be   in   good   qualitative
agreement   with   the   receptor binding   data.   However,
most of the cyclic analogs showed much higher   potency
in   the   GPI assay than was expected on   the   basis   of
their   μ-receptor affinities determined in the   binding
assay   ([^{3}H]DAGO displacement).   A typical example   is
cyclic   analog $\underline{7}$ which in comparison with   its   linear
correlate   ($\underline{7a}$) has about half the μ-receptor   affinity
but   is nine times more potent in the GPI   assay.   The
same   discrepancy has been observed with numerous other
cyclic   opioid   peptides.   These observations   can   be

interpreted to indicate that the cyclic analogs may display an increased "efficacy" ("intrinsic activity") at the μ-receptor in comparison with their linear correlates. On the basis of thermodynamic considerations the cyclic analogs should have higher receptor affinity than their corresponding open-chain analogs because in the case of the cyclic peptides the loss of internal rotational entropy upon binding is smaller than in the case of the linear ones. The observation that in fact the opposite is the case suggests that in comparison with the linear analogs a larger part of the receptor binding energy of the cyclic peptides may be used to lower the energy requirement for the conformational transition of the receptor in the ground state to its excited state. It has been proposed that the "efficacy" of a compound is related to the rate at which excited receptors are formed as well as to the final equilibrium between[31] excited receptors and receptors in the ground state. On the basis of their structural features — D-amino acid in position 2 and C-terminal carboxamide function — both the cyclic and the linear peptides investigated are stable against enzymatic degradation under the conditions used in the bioassay and binding assay. The potency differences observed between the GPI assay and the [^{3}H]DAGO binding assay can therefore not be due to a different extent of peptide degradation in the two tissues. The effects of all these cyclic and linear analogs on the GPI were completely naloxone-reversible and the apparent dissociation constants (K_e) for naloxone as antagonist were all in the range from 0.8 to 2.1 nM. These low values are typical for μ-receptor interactions and rule out the possibility that the enhanced potencies observed with the cyclic analogs in the GPI assay are due to an additional interaction with the ϰ-receptor. Furthermore, the cyclic analogs showed an even more pronounced "efficacy" enhancement in the rat vas deferens assay which represents an alternative bioassay for the measurement of a μ-receptor-mediated opioid effect.[32] It could also be argued that these potency discrepancies are due to the fact that μ-receptors in rat brain membrane preparations differ somewhat in their structural requirements from μ-receptors in peripheral tissues. The observation that the relative potencies observed with a large number of linear enkephalin analogs in the GPI assay correlate very well with the relative affinities determined in the [^{3}H]DAGO binding assay (Figure 3) argues against this possibility. In contrast to the cyclic enkephalin analogs, morphine and some morphine-related opiates

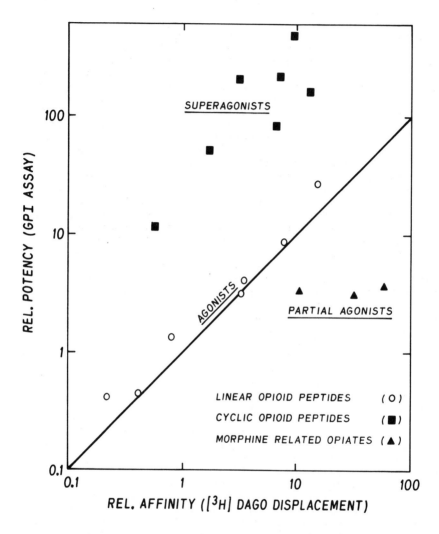

Figure 3.   Correlation between relative potencies
             determined with μ-receptor representative
             bioassay and binding assay

(levorphanol, butorphanol, etc.) were found to have lower potency in the GPI assay than was expected on the basis of their μ-receptor affinities (Figure 3). This finding confirms[32] the observation that morphine is a partial agonist[32], whereas many of the cyclic opioid peptide analogs appear to be superagonists. It should be emphasized, however, that at the present time we consider the proposed "efficacy" enhancement of cyclic enkephalin analogs merely as a strong working hypothesis which needs to be corroborated by further experiments.

CONCLUSIONS

The obtained results indicate that conformational restriction through appropriate cyclizations via side chains permits the manipulation of receptor affinity and receptor selectivity of opioid peptides in a more drastic manner than is in general possible through conventional peptide analog design based on simple amino acid substitutions. Furthermore, these studies also led to the hypothesis that conformational restriction may alter the "efficacy" ("intrinsic activity") of peptides. Whereas many of the cyclic opioid peptides described in this paper might represent examples for an "efficacy" enhancement, introduction of conformational restrictions into peptides may also lead to an "efficacy" loss, as illustrated by the antagonist properties of [Pen[1]]oxytocin.[33] Morevoer, cyclic opioid peptides were shown to be extremely resistant to enzymatic degradation and to produce long lasting effects <u>in vivo</u> (see ref. 12). The detailed conformational characterization of these semi-rigid cyclic analogs can be expected to provide further insight into the conformational requirements and the distinct topographies of the different opioid receptor classes. In conclusion, the results of studies on cyclized peptide analogs undertaken in the past few years suggest that the concept of conformational restriction has various interesting implications for peptide drug design and that it may represent a first promising step towards the goal of developing peptide mimetics.

ACKNOWLEDGMENT

This work was supported by operating grants from the
Medical Research Council of Canada (Grant MT-5655), the
Quebec Heart Foundation and the National Institute on
Drug Abuse (Grant 1 R01 DA-04443-01).

REFERENCES

1.   J. Hughes, T.W. Smith, H.W. Kosterlitz, L.A.
     Fothergill, B.A. Morgan and H.R. Morris,
     Nature (London) 1975, 258, 577.
2.   V. Höllt, Ann. Rev. Pharmacol. Toxicol. 1986, 26,
     59.
3.   S.J. Paterson, L.E. Robson and H.W. Kosterlitz,
     'The Peptides: Analysis, Synthesis, Biology', S.
     Udenfriend and J. Meienhofer, eds., Academic
     Press, Orlando, 1984, Vol. 6, Chapter 5, p. 147.
4.   B.K. Handa, A.C. Lane, J.A.H. Lord, B.A. Morgan,
     M.J. Rance and C.F.C. Smith, Eur. J. Pharmacol.
     1981, 70, 531.
5.   K.-J. Chang, E.T. Wei, A. Killian and J.-K.
     Chang, J. Pharmacol. Exp. Ther. 1983, 227, 403.
6.   J.M. Zajac, G. Gacel, F. Petit, P. Dodey, P.
     Rossignol and B.P. Roques, Biochem. Biophys. Res.
     Commun. 1983, 111, 390.
7.   P.W. Schiller, 'The Peptides: Analysis,
     Synthesis, Biology', S. Udenfriend and J.
     Meienhofer, eds., Academic Press, Orlando, 1984,
     Vol. 6, Chapter 7, p. 219.
8.   J. DiMaio, T.M.-D. Nguyen, C. Lemieux and P.W.
     Schiller, J. Med. Chem. 1982, 25, 1432.
9.   P.W. Schiller and J. DiMaio, Nature (London)
     1982, 297, 74.
10. P.W. Schiller, B. Eggimann, J. DiMaio, C.
     Lemieux and T.M.-D. Nguyen, Biochem. Biophys.
     Res. Commun. 1981, 101, 337.
11. P.W. Schiller, J. DiMaio and T.M.-D. Nguyen,
     'Proc. 16th FEBS Meeting', Y.A. Ovchinnikov, ed.,
     VNU Science Press, Utrecht, The Netherlands,
     1985, p. 457.
12. P.W. Schiller and J. DiMaio, 'Peptides: Structure
     and Function', V.J. Hruby and D.H. Rich, eds.,
     Pierce, Rockford, IL, 1983, p. 269.
13. J.M. Berman, M. Goodman, T.M.-D. Nguyen and P.W.
     Schiller, Biochem. Biophys. Res. Commun. 1983,
     115, 864.

14. J.V. Edwards, A.F. Spatola, C. Lemieux and P.W. Schiller, Biochem. Biophys. Res. Commun. 1986, 136, 730.

15. H.I. Mosberg, R. Hurst, V.J. Hruby, K. Gee, H.I. Yamamura, J.J. Galligan and T.F. Burks, Proc. Natl. Acad. Sci. USA 1983, 80, 5871.

16. P.W. Schiller, T.M.-D. Nguyen and J. Miller, Int. J. Peptide Protein Res. 1985, 25, 171.

17. P.W. Schiller, T.M.-D. Nguyen, C. Lemieux and L.A. Maziak, J. Med. Chem. 1985, 28, 1766.

18. P.W. Schiller, L.A. Maziak, C. Lemieux and T.M.-D. Nguyen, Int. J. Peptide Protein Res. 1986, 20, 493.

19. P.W. Schiller, T.M.-D. Nguyen, C. Lemieux and L.A. Maziak, FEBS Lett. 1985, 191, 231.

20. E. Fischer, Ber. Dtsch. Chem. Ges. 1894, 27, 2985.

21. A.S.V. Burgen, G.C.K. Roberts and J. Feeney, Nature (London) 1975, 253, 753.

22. R. Schwyzer, Biochemistry, 1986, 25, 6335.

23. P.W. Schiller, T.M.-D. Nguyen, L.A. Maziak and C. Lemieux, Biochem. Biophys. Res. Commun. 1985, 127, 558.

24. Y. Sasaki, M. Matsui, H. Fujita, M. Hosono, M. Taguchi, K. Suzuki, S. Sakurada, T. Sato, T. Sakurada and K. Kisara, Chem. Pharm. Bull. 1985, 33, 1528.

25. K.-J. Chang, A. Killian, E. Hazum, P. Cuatrecasas and J.-K. Chang, Science 1981, 212, 75.

26. S. Salvadori, M. Marastoni, G. Balboni and R. Tomatis, Il Farmaco Ed. Sci. 1985, 40, 454.

27. H. Matthies, H. Stark, B. Hartrodt, H.-L. Ruethrich, H.-T. Spieler, A. Barth and K. Neubert, Peptides 1984, 5, 463.

28. P.W. Schiller, T.M.-D. Nguyen, J. DiMaio and C. Lemieux, Life Sci. 1983, 33, 319.

29. F.M. Casiano, W.R. Cumiskey, T.D. Gordon, P.E. Hansen, F.C. McKay, B.A. Morgan, A.K. Pierson, D. Rosi, J. Singh, L. Terminiello, S.J. Ward and D.M. Wescoe, 'Peptides: Structure and Function', V.J. Hruby and D.R. Rich, eds., Pierce, Rockford, IL, 1983, p. 311.

30. B.C. Wilkes and P.W. Schiller, Biopolymers, in press.

31. T.J. Franklin, Biochem. Pharmacol. 1980, 29, 853.

32. C.F.C. Smith and M.J. Rance, Life Sci. 1983, 33, 327.

33. H. Schulz and V. DuVigneaud, J. Med. Chem. 1966, 9, 647.

'Peptoids' from CCK-8

D. C. Horwell

PARKE-DAVIS RESEARCH UNIT, ADDENBROOKE'S HOSPITAL SITE,
HILLS ROAD, CAMBRIDGE CB2 2QB, UK

There are now more than 50 endogenous peptides found in mammalian central and peripheral nervous tissue that are cited as neurotransmitters or neuromodulators. A major challenge to medicinal chemists is to design small molecule equivalents (peptoids) of these peptides. These will provide selective agonists and antagonists in order to aid characterisation of the numerous activities of the peptides, and to produce credible drug candidates.

A peptoid, in its broadest sense, is here defined as a credible drug candidate that mimics or blocks the action(s) of an endogenous peptide. A small molecule mimic of the hormone/neuromodulator/neurotransmitter peptide would offer the potential of a credible drug candidate with oral bioavailability, a pharmaceutically useful half-life and ease of manufacture on a plant scale (Fig. 1).

Some of the more familiar neuropeptides, together with their most common fragments found in mammalian CNS tissues are given in Table 1. Since the discovery

PEPTOID: A credible drug candidate that mimics or blocks the action(s) of an endogenous peptide.

CREDIBLE DRUG CANDIDATE:
1. Demonstrable efficacy on the peptide/hormone/ neuromodulator/neurotransmitter/enzyme/protein system of interest
2. Oral bioavailability
3. Pharmaceutically useful half-life
4. Manufacturable on a plant scale

Figure 1 The design of 'Peptoids'

Table 1 Some mammalian CNS neuropeptides

Putative Neuropeptide	Amino Acid Residues in Most Common Form
ACTH	39
Angiotensin	8
Bombesin	14
Bradykinin	9
Calcitonin	32
Cholecystokinin	8
Dynorphin A	17
β-endorphin	31
Leu-enkephalin	5
Met-enkephalin	5
Glucagon	29
GHRH	44
LHRH (GnRH)	10
α-MSH	13
Neuromedin B	32
Neuropeptide K	36
Neuropeptide Y	36
Substance P	11
TRH	3
VIP	28
Vasopressin	9
Oxytocin	9

of the apparent departure from Dales rule, that
norepinephrine coexists in sympathetic ganglia with
somatostatin, further work has shown that many
neuropeptides cotransmit with "classical"
neurotransmitters (Table 2), and also with other
neuropeptides. The discovery of peptoids should help
to unravel the complexities of these cotransmitter
systems.

Table 2 Some cotransmitters with neuropeptides from the
nervous system

"Classical" Neurotransmitter	Peptide Cotransmitter/ Modulator	Anatomical Location
Acetylcholine	— CCK	Myenteric plexus
	— Neurotensin	Myenteric plexus
	— Somatostatin	Myenteric plexus, cerebral cortex
	— Substance-P	Myenteric plexus, pons
Dopamine	— CCK	Vent. tegmental area
	— Enkephalin	Carotid body
	— Neurotensin	Vent. tegmental area
GABA	— CCK	Myenteric plexus
	— Somatostatin	Myenteric plexus, cerebral cortex
	— VIP	Myenteric plexus, cerebral cortex
	— Substance-P	Myenteric plexus, pons
	— Neurotensin	Myenteric plexus
Norepinephrine	— Enkephalin	Superior cervical ganglia
	— Neuropeptide Y	Medulla oblongata, locus coeruleus
	— Neurotensin	Adrenal medulla
	— Somatostatin	Sympathetic ganglia
5-HT	— Substance-P	Medulla oblongata

At present there is no example where the peptide structure has been systematically modified to give a truly "designed" receptor active peptoid. Is it possible to design such agents? Some examples of "peptoids" and their corresponding peptides are given in Table 3. Perhaps the best known example is morphine, a naturally occurring alkaloid that mimics the action of the endogenous opioid peptides, such as enkephalin. Recently we have discovered the chemically simple 4-benzothiophenyl cyclohexyl acetamide derivative, PD-117302, which is a potent and super-selective kappa opioid compound[1]. Hence PD-117302 may be considered a peptoid of the poorly kappa selective endogenous peptide, dynorphin 1-17.

Table 3 Examples of peptoids

PEPTIDE

1. Asp-Phe-OMe (aspartame)
2. Tyr-Gly-Gly-Phe-Leu (Met)-
   (Enkephalin)

3. <Glu-Trp-Pro-Arg-Pro-Gln-Ile-Pro-
   Pro-(SQ. 20,881)

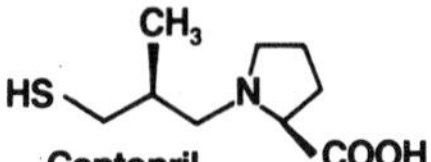

4.

5. Tyr—Gly—Gly—Phe—Leu—
   Arg—Arg—Ile—Arg—Pro—Lys—
   Leu—Lys—Trp—Asp—Asn—Gln—OH
   (Dynorphin—1-17)

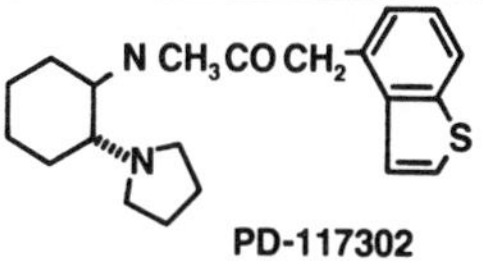

PEPTOID

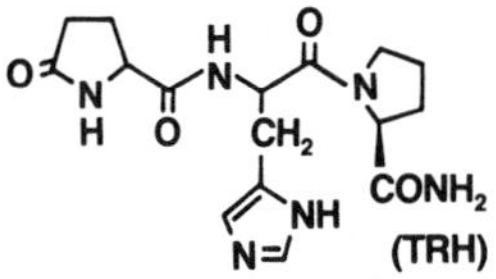

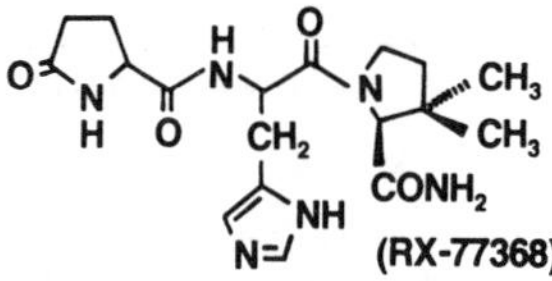

Sugar, saccharin
Morphine
Meperidine (pethidine)

Captopril
(A modified Ala-Pro)

(RX-77368)

PD-117302

Peptoids, therefore, may be derived from several sources, given in Fig. 2. In order to design peptoids, it is vital that a multidisciplinary approach is adopted, involving close liaison between chemists, biochemists and pharmacologists. Some of the multidisciplinary problems associated with the evolution of SAR of peptides are given in Figure 3. A strategy to design peptoids from peptides is outlined in Figure 4. The working hypothesis assumes that i,

1. 'Me-too' drugs

2. Natural products

3. Company or other files

4. Molecular modelling:
    I. Energy minimisation of peptide or peptide fragment
    II. Based on X-ray of peptide/receptor/enzyme complex

5. *De novo* organic synthesis based on structure of peptides

Figure 2 Sources of peptoids

1. Relevance of assay to anticipated pharmacological profile
   e.g. GPI/MVD/ for μ and δ opioid activity, respectively.

2. Different criteria and protocols in laboratories.

3. Indeterminant errors in measurements.
   e.g. metabolising enzymes, tachyphalaxis.

4. Purity of peptides.

5. Wrong conclusions:
   e.g. those based on synchnological rather than rhegnylogical organisation of peptides, i.e. assumption that primary structure is relevant to the information presented to, and read by, the receptor.

Figure 3 Some problems with literature SAR of peptides

1. **WORKING HYPOTHESES**
    A. Robust binding and bioassay available
       —avoids complication of pharmacokinetics
       when evolving SAR
    B. Minimum requirement for competitive
       antagonist: peptide has identifiable binding
       site(s)
    C. Minimum requirement for agonist: peptide has
       identifiable binding *and* functional site(s)
    D. Only a fraction of the protein/peptide hormone
       receptor site is utilized by most therapeutically
       useful small organic molecules
       i.e. Stokes diameter receptors = 12-70 Å
       Diameter by X-ray of known binary complexes
       <12Å for most therapeutically useful
       'small molecule' drugs (M.W.<700)
    E. Only 3 of the numerous ligands in
       protein/peptide hormones need be
       used to contribute to binding energy
       leading to observed receptor
       dissociation constants in the range
       $10^{-9}-10^{-12}$M
       [Gibbs Helmholtz equation relates $\triangle$G
       to K_D]

2. **PROPOSED STRATEGY**
    A. Identify minimum peptide fragment active in binding or bioassay by:
       I. Preparing peptide fragments by solution and solid phase synthesis
          (Merrifield). These should include examples of both continuous and
          non-continuous fragments
       II. Critical appraisal of literature SAR

       $\downarrow$

    B. Use conventional medicinal chemistry techniques to enhance binding
       affinity 10-100 fold, e.g. QSAR

       $\downarrow$

    C. Introduce groups that optimise lipophilicity and stabilise the peptide
       bonds.

Figure 4 A strategy to design peptoids from peptides

an appropriate and robust binding and bioassay is
available to evolve the peptide SAR, thus minimising
the complications of pharmacokinetic parameters; ii, a
minimum requirement for a competitive antagonists is
that the peptide has identifiable binding sites (e.g.
the amino acid side chains); iii, a minimum
requirement for an agonist is that the peptide has
identifiable binding and functional sites; iv, only a
fraction of the receptor protein material is used to
recognise the substrate; v, only 3 of the numerous
ligands need to be used to contribute to the binding
energy leading to observed receptor dissociation
constants in the range $10^{-9}-10^{-12}$M ^{2}.

With these concepts in mind, a suggested strategy
is first to identify the minimum peptide fragment
active in the binding or bioassay of choice. This
should include synthesis of both continuous and
non-continuous fragments of the peptide. The minimum
active fragment would be, say, $\simeq 10^{-6}$M for a receptor
binding affinity and a $pA_2 \simeq 6$ for bioassay of an
antagonist. Once this fragment, preferably a di- or
tri-peptide has been identified, it is within the
experience of conventional medicinal chemistry
techniques (e.g. QSAR) to enhance this activity 10-100
fold by systematic modification of the now limited
functional groups. For example, the benzene ring of a
phenylalanine residue can be systematically
substituted with simple groups, or the methyl group of
an alanine residue can be homologated or branched, and
optimised by techniques such as Hansch analysis. This
will give a novel compound with activity in the range
found with most therapeutically useful receptor active

small molecule drugs e.g. $K_i \simeq 10^{-7}$-10^{-8}M, $pA_2 \simeq$ 7-8.
Finally, further modification to give optimal
lipophilicity and enhance stability to acid and
enzymatic degradation, such as replacement of the
amide bond by isosteres or the [L]-amino acid by a
[D]-amino acid, will give an agent likely to be orally
bioavailable and have a therapeutically useful
half-life.

The application of this strategy to CCK-8 is in
progress[3]. Examination of fragments of CCK-8 by the
mouse cerebral cortex binding assay (Table 4) has
shown that the minimum fragment with nanomolar
affinity is the tetrapeptide Trp-Met-Asp-Phe-NH$_2$. The
Trp- and Phe-residues together are essential
components for the receptor recognition process.
Furthermore, the Asp-Tyr(SO$_3$H)-Met-Gly- fragment can
be replaced by the simple N-blocking groups Boc- or
Amoc with no loss of affinity (Table 5). Whilst
maintaining the N-Boc terminal of the tetrapeptide,
systematic replacement of the Met-31 and Asp-32 side
chains with H clearly shows that the full tetrapeptide CCK31-33 is needed for nanomolar affinity, as
the Gly-modified tetrapeptides have reduced affinity
of at least 1000-fold. The spacer groups between the
Trp- and Phe- residues were then systematically
replaced by methylene groups. Significantly, the
simpler dipeptides BocTrp-Phe-NH$_2$ and Boc-β
-Ala-PheNH$_2$ maintained pentamolar binding affinity (7
x 10^{-5} and 2.5 x 10^{-5}M resp.) comparable to the
modified tetrapeptides. Hence the dipeptide
BocTrp-Phe-NH$_2$ is considered a continuous message
mimic of the non-continuous message imparted by CCK-4

Table 4 Structure-activity relationships of CCK fragments at
        the CNS CCK receptor

FRAGMENTS OF THE C-TERMINAL OCTAPEPTIDE CCK 26-33

Fragment		CCK receptor affinity [K_i (nM)]
CCK 26-33 (Sulphated)	Asp-Tyr (SO$_3$H) Met-Gly-Trp-Met-Asp-Phe-NH$_2$	2.5
CCK 27-33	Tyr (SO$_3$H) Met-Gly-Trp-Met-Asp-Phe-NH$_2$	2.4
CCK 27-32	Tyr (SO$_3$H) Met-Gly-Trp-Met-Asp-NH$_2$	Inactive 10^{-4}M
CCK 30-33	Trp-Met-Asp-Phe-NH$_2$	3.1
CCK 31-33	Met-Asp-Phe-NH$_2$	100,000
CCK 30-32	Trp-Met-Asp-NH$_2$	Inactive 10^{-3}M

> **CONCLUSION:** (i) C-terminal tetrapeptide retains nanomolar affinity
>
> (ii) Trp and Phe residues *together* are essential for high binding affinity to central CCK-receptors

Table 5 Modified CCK fragments screened in mouse cerebral
        cortex receptor binding assay

X-Trp-Y-Phe-NH$_2$

The effects of altering X and Y were investigated.

THE EFFECT OF ALTERING THE N-TRYPTOPHAN PROTECTING GROUP

with Y = Met-Asp as in CCK-4

X	Structure	CCK receptor affinity (K_i)
H	H-Trp-Met-Asp-Phe-NH$_2$	3.1 × 10^{-9}M (CCK-4)
Amoc (t-amyloxycarbonyl)	Amoc-Trp-Met-Asp-Phe-NH$_2$	2.7 × 10^{-10}M
Boc (t-butyloxycarbonyl)	Boc-Trp-Met-Asp-Phe-NH$_2$	3.0 × 10^{-10}M

> **CONCLUSION:** Asp—Tyr (SO$_3$H)—Met—Gly— of CCK-8 can be replaced by simple groups like Boc and retain nanomolar affinity.

Table 5 (continued)

THE EFFECT OF ALTERING THE SPACER GROUPS Y

1. **ALTERING AMINO ACID SPACERS**

Y	Structure	CCK receptor affinity (K_i)
Gly-Gly	Boc-Trp-Gly-Gly–Phe-NH$_2$	5×10^{-3}M
Gly-Asp	Boc-Trp-Gly-Asp-Phe-NH$_2$	2.0×10^{-6}M
Met-Gly	Boc-Trp-Met-Gly-Phe-NH$_2$	2.7×10^{-5}M
Met-Asp	Boc-Trp-Met-Asp-Phe-NH$_2$	3.0×10^{-10}M

2. **ALTERING SPACERS BY METHYLENE GROUPS**

Y	Structure	CCK receptor affinity (K_i)
—	Boc-Trp——Phe-NH$_2$	7×10^{-5}M
—CH$_2$—	Boc-Trp-Gly—Phe-NH$_2$	10^{-4}M
—(CH$_2$)$_2$—	Boc-Trp-Bala–Phe-NH$_2$	2.5×10^{-5}M
—(CH$_2$)$_3$—	Boc-Trp-Gaba-Phe-NH$_2$	5×10^{-3}M
—(CH$_2$)$_4$—	Boc-Trp-Dava-Phe-NH$_2$	10^{-3}M

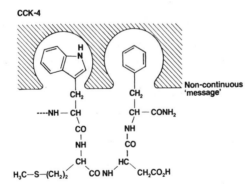

Figure 5

(Fig. 5). Systematic chemical modification of this simple dipeptide as outlined above is now viable to produce potential drug candidates.

Some of the potential therapeutic uses of CCK peptoids are shown in Table 6. The most viable of these clinical indications will become apparent when such CCK peptoids become available.

Table 6 Effects of CCK and potential therapeutic indications of CCK peptoids

Effects of CCK	Therapeutic Indications of CCK-Peptoids
A. PERIPHERAL NERVOUS SYSTEM	
(i) Gall bladder contraction	—
(ii) Stimulation of pancreatic secretion	—
a. amylase; b. insulin	
(iii) Stimulation of gastric acid secretion	Anti ulcer, Hypergastrinaemia (antagonist)
B. CENTRAL NERVOUS SYSTEM	
(i) Induces satiety	Anti obesity
(ii) Analgesia	Pain
(iii) Modulation of dopamine	Schizophrenia Parkinsons disease

References

1. D.C. Horwell, U.S. Patent No. 4656182, (1987).
2. P.S. Farmer, Drug Design Vol. X, Chapter 3 (Ed. E. Ariens), Academic Press (1980).
3. D.C. Horwell, A. Beeby, C.R. Clark, J. Hughes, J. Med. Chem., 1987, 30, 729.

Enzyme Inhibition: Approaches to Drug Design

R. B. Silverman

DEPARTMENT OF CHEMISTRY AND DEPARTMENT OF
BIOCHEMISTRY, MOLECULAR BIOLOGY, AND CELL BIOLOGY,
NORTHWESTERN UNIVERSITY, EVANSTON, ILLINOIS 60208, USA

1 THE IMPORTANCE OF ENZYME INHIBITION TO DRUG DESIGN

Many diseases or, at least, symptoms of diseases arise
from a deficiency or excess of a specific metabolite in
the body, an infestation of a foreign organism, or
aberrant cell growth. If the metabolite deficiency or
excess can be normalized, and if the foreign organisms
and aberrant cells can be nullified, disease states will
be remedied. All of these situations can be effected by
specific enzyme inhibition. The concentration of a substrate for an enzyme will increase and the product concentration will decrease upon specific enzyme inhibition
(provided alternative metabolic pathways are not available). If an enzyme involved in a vital metabolic process of a foreign organism or aberrant cell is inhibited,
this can prevent their growth or replication. For example, inhibition of monoamine oxidase leads to the increase in the concentration of various biogenic amine
substrates, and this has an antidepressant effect.
Xanthine oxidase inhibition blocks the formation of the
product uric acid, and results in a uricosuric effect.
Inhibition of bacterial transpeptidase prevents cross-
linking of the cell wall and, thereby, has an antibacterial effect. Tumor cell thymidylate synthetase inhibition shuts down de novo biosynthesis of deoxythymidylate,
a DNA precursor, thus destroying the tumor.

In order to minimize side effects, enzyme inhibition
should be totally specific, i.e., only one enzyme affected. Highly selective inhibition, however, is a more
realistic objective. In some cases there are known metabolic differences between foreign organisms and their

hosts. In other instances there are substrate specificity differences between enzymes from the two sources.
These differences can be taken advantage of in the design
of specific enzyme inhibitors as potential drugs. Unfortunately, when dealing with various organisms and especially when dealing with tumor cells, the enzymes that
are essential for foreign organism or tumor cell growth
are also vital to human health. Inhibition of these
enzymes can destroy human cells as well. Nonetheless,
this approach is utilized in various types of chemotherapy. The reason that this approach is effective is that
foreign organisms and tumor cells replicate at a rate
much faster than do most normal human cells (those in the
gut, the bone marrow, and the mucosa, however, are exceptions). Consequently, rapidly proliferating cells have
an elevated requirement for DNA precursors. Therefore,
their uptake systems for these molecules are more efficient than those for normal cells, as are the enzymes
responsible for metabolism of the precursors to DNA. The
selective toxicity, in this case, then, derives from a
kinetic difference rather than a qualitative difference
in the metabolism. An ideal situation with regard to a
foreign organism or aberrant cell would be to inhibit an
enzyme that is essential for the growth of the foreign
organism or aberrant cell but which is either not essential for human health, or, better yet, not even present
in humans. This form of selective toxicity would not
require the careful administration of drugs that would be
necessary if the inhibited enzyme were important to human
metabolism as well.

Target enzymes selected for drug design are ones
whose inhibition would lead to the desired therapeutic
effect. In most cases a potential drug is designed for
an enzyme whose inhibition is known to produce a specific
physiological effect. A more daring approach is to design inhibitors of enzymes whose inhibition has not yet
been established as leading to a desired therapeutic effect. The latter approach, however, could be more pro-
fitable since, presumably, none of the competitors in the
field would be taking this approach. In order to embark
on a program aimed at inhibition of an enzyme in the latter category, it first would be necessary to study the
physiological effect of inhibition of that enzyme.

In comparison with receptors and carrier proteins,
enzymes are more straightforward targets for rational
drug design. Enzyme purification is a much simpler task
than receptor purification, and a homogeneous preparation
can be obtained which may be used to elucidate the active

site structure upon which drug design would be based.
Furthermore, enzyme catalysis can be utilized in the design of transition state analogues, some slow, tight-
binding inhibitors, and mechanism-based enzyme inactivator analogues (vide infra). Whereas effective receptor
antagonists often bear no structural similarity to the
agonists, enzyme inhibitors are generally very similar in
molecular structure to substrates or products of the target enzyme. Thus, lead compounds are more readily
obtainable when enzymes are targets.

2   TYPES OF ENZYME INHIBITORS

Any compound that slows down or prevents enzyme catalysis
from occurring is an enzyme inhibitor. Inhibitors can be
grouped into two general classifications, namely, reversible and irreversible inhibitors. As the name implies,
inhibition of enzyme activity by a reversible inhibitor
is reversible, suggesting non-covalent interactions between the enzyme and the inhibitor. This is not strictly
the case, however; a reversible, covalent bond may be
formed. An irreversible inhibitor (also called an inactivator) is one that prevents the return of enzyme activity for an extended period of time, suggesting covalent
bond formation. Likewise, it also is possible for non-
covalent interactions to be so effective that the enzyme-
inhibitor complex is, for all intents and purposes, virtually irreversible.

Reversible Enzyme Inhibitors

The most common enzyme inhibitor drugs are the reversible type, particularly ones that compete with the
substrate for active-site binding (competitive reversible
inhibitors). Because they are reversible, the equilibri-
um, $E + I \rightleftharpoons E \cdot I$, which is set up rapidly, depends upon
the concentrations of the enzyme, the substrate, and the
inhibitor. Since the enzyme concentration is usually low
and fixed, the equilibrium constant, and, therefore, the
$E \cdot I$ concentration, will depend principally upon the inhibitor concentration. When the inhibitor concentration
diminishes, so will the $E \cdot I$ complex concentration dimin-
ish, and the effect of the inhibitor can be overcome by
substrate. If the enzyme inhibitor is a drug, the maximal pharmacological effect will occur when the drug
concentration is maintained at a saturating level at the
site of the enzyme. As the drug is metabolized, the
concentration of the $E \cdot I$ complex will drop. Repeated
administration of the drug over an extended period of

time is required to maintain the inhibitor concentration. In order to increase the potency of reversible inhibitors, and, thereby permit a lower dose of the drug to be administered, they should be designed so that binding interactions with the active site are optimized (i.e., low K_i values). Three types of reversible inhibitors, simple competitive inhibitors, transition state analogues, and slow, tight-binding inhibitors, are discussed in more detail below.

    <u>Simple</u> <u>Competitive</u> <u>Reversible</u> <u>Inhibitors</u>. These compounds have a strong molecular similarity to substrates for the target enzyme, which allows for tighter binding to the enzyme. An example of this type of inhibitor is the angiotensin-converting enzyme (ACE) inhibitor and antihypertensive drug, captopril (1).

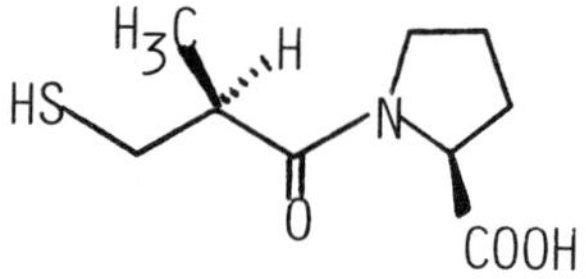

(1)

Angiotensinogen is an α-globulinprotein, produced by the liver, which is hydrolyzed by the enzyme renin to the decapeptide angiotensin I (Scheme 1). This biologically-

angiotensinogen $\xrightarrow{\text{renin}}$ Asp·Arg·Val·Tyr·Ile·His·Pro·
                                           Phe·His·Leu
                                           (angiotensin I)

                                           ACE $\downarrow$

Asp·Arg·Val·Tyr·Ile·His·Pro·Phe   His·Leu

        (angiotensin II)

                    Scheme 1

inactive peptide is further hydrolyzed at the penultimate peptide bond by ACE in the lungs and blood vessels to give the dipeptide His·Leu and the octapeptide angiotensin II. This octapeptide acts as a hormone with very potent vasoconstriction properties, and also is the physiological stimulus for the release from the adrenal gland of aldosterone, which induces sodium ion and water reten-

tion. Both vasoconstriction and sodium ion/water retention lead to an increase in blood pressure. To make
matters worse, in addition to cleaving angiotensin I to
angiotensin II, ACE also hydrolyzes the C-terminal dipeptide from the potent vasodilator bradykinin (Arg·Pro·-
Pro·Gly·Phe·Ser·Pro·Phe·Arg). Consequently, ACE both
generates a potential hypertensive agent, angiotensin II,
and destroys a potent hypotensive agent, bradykinin.
This leads to an increase in blood pressure. Therefore,
inhibition of ACE has a dramatic antihypertensive effect.

Since ACE is a zinc-containing peptidase with properties similar to that of carboxypeptidase A, the known
active site structure of that enzyme was used as a model
for the active site of ACE. The three most important
features of the active site are a positively-charged
group to bind the C-terminal carboxyl group, a group with
affinity for the C-terminal peptide bond, and a Zn(II)
ion to coordinate the carbonyl of the scissile peptide
bond; other interactions, however, also are important
(Figure 1). According to this model a succinyl amino

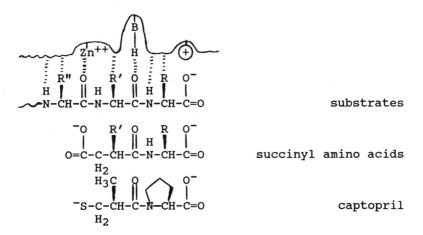

Figure 1 Proposed binding interactions of substrates,
succinyl amino acids, and captopril to ACE

acid should interact with all three groups. This was
found to be the case; however, when the Zn(II)-coordin-
ating carboxyl group was replaced by a sulfhydryl group,
a dramatic increase in binding affinity was observed.
The K_i value for captopril was determined to be 0.0017
μM.[1] The model for this binding action is depicted in
Figure 1. An important feature of an effective inhibitor

if it is to become useful as a drug, is specificity for the target enzyme. The K_i values of captopril for two related enzymes, carboxypeptidase A and B, are 620 μm and 250 μm, respectively, therefore, indicating that captopril is highly selective.

Transition State Analogues. Captopril exhibits two unwanted side effects; namely, it can produce rashes and cause loss of taste, both of which, however, are reversed upon withdrawal of the drug. These side effects were traced to the thiol group in the molecule. Related studies for inhibitors of ACE at Merck[2] led to the finding that enalaprilat (2, R = H), which has a K_i for ACE of

(2)

0.0002 μM, is devoid of side effects.* Because of the exceedingly low K_i value for enalaprilat, it has been suggested that it is not just a simple competitive reversible inhibitor, but, rather, is a transition state analogue.[3-5]

An enzyme increases the rate of a reaction by stabilizing the transition state, which it does by changing its conformation in such a way that the strongest interactions occur between the substrate and enzyme active site at the transition state of the reaction. Therefore, a stable compound whose structure resembles that of the substrate at a postulated transition state (or transient intermediate) of the reaction will bind much more tightly than will the substrate in the ground state. This is a transition-state analogue. A compound whose structure mimicks the transition state structure of a multi-substrate reaction is a multi-substrate transition state analogue. If a hypothetical ACE-catalyzed hydrolysis of a peptide is considered (Scheme 2), the structural similarities of the transition state of this reaction to enalaprilat can be envisioned. An enzyme conformational change during the transition state may increase the binding interactions.

---

*Enalaprilat is poorly absorbed when administered orally. Consequently, it is given in a prodrug form as the ethyl ester (2, R = Et), known as enalapril, which is hydrolyzed *in vivo* to enalaprilat.

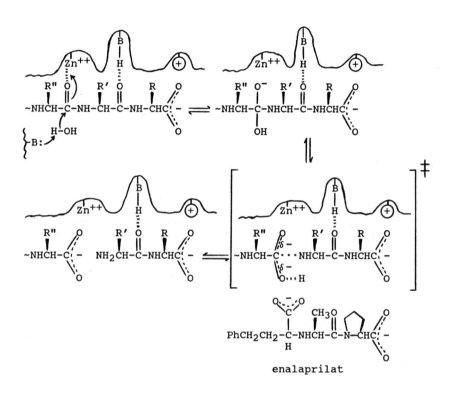

Scheme 2

A related approach was taken recently at Upjohn[6] for the design of transition state analogues of renin, the enzyme shown in Scheme 1 that catalyzes the hydrolysis of angiotensinogen to angiotensin I. In this case a glycol (3, R = OH) and an alcohol (3, R = H) were proposed as mimicks of the tetrahedral intermediate formed during hydrolysis of angiotensinogen (Scheme 3). Both compounds were potent renin inhibitors.

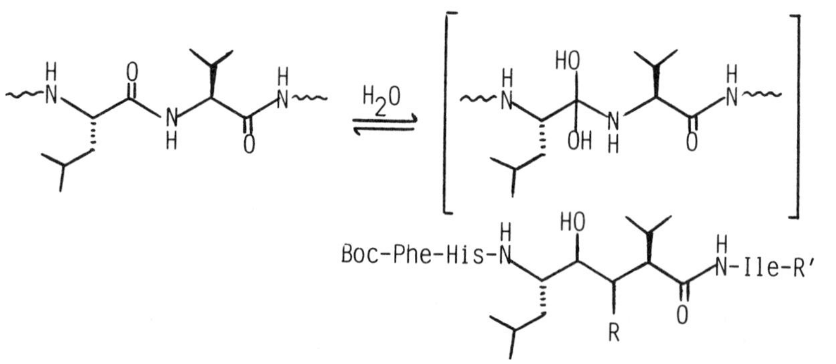

(3)

Scheme 3

Slow, Tight-Binding Inhibitors. These inhibitors
bind slowly and either non-covalently[7,8] or covalently[9,10] and interact very strongly with the active site.
There may be several reasons for the slowness and tight-
ness of binding. A conformational change, possibly as a
result of a change in the protonation state of the enzyme, may be involved.[11] Displacement of an essential
water molecule by the inhibitor also may be important.[8]
When a covalent bond is involved, a slowly reversible
adduct may be responsible.

A recent example of this type of inhibition in drug
design was reported by the group at Stuart Pharmaceuticals.[10] Trifluoromethyl ketones were synthesized as
inhibitors of human leukocyte elastase for the treatment
of emphysema. A hypothesis for the cause of emphysema is
that there is an imbalance in certain proteases and protease inhibitors in the lungs. Leukocyte elastase and
cathepsin G, both serine proteases, are believed to be
released by neutrophils in the lungs to digest dead lung
tissue and destroy foreign bacteria. Inhibitors of these
enzymes also are released to prevent these enzymes from
destroying elastin and lung connective tissue. When, for
various reasons, the protease inhibitors are deficient,
uncontrolled proteolysis of lung connective tissue can
occur, resulting in emphysema. Inhibitors of elastase
and cathepsin G would substitute for the natural inhibitors. A trifluoromethyl group adjacent to a ketone
stabilizes the hydrate form. Consequently, it was reasoned[9,10] that attack of the active site serine would
produce a stable, but reversible, enzyme-inhibitor
hemiketal adduct (Scheme 4).

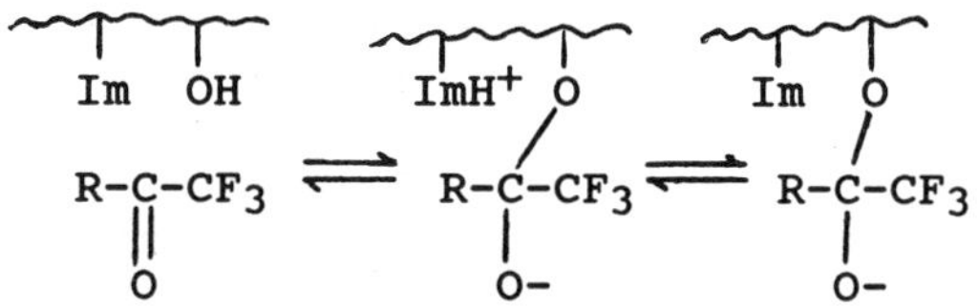

$$\text{Im} \quad \text{OH} \qquad \text{ImH}^+ \quad \text{O} \qquad \text{Im} \quad \text{O}$$

$$R-C-CF_3 \;\rightleftharpoons\; R-C-CF_3 \;\rightleftharpoons\; R-C-CF_3$$

$$\underset{O}{\|} \qquad\qquad \underset{O-}{|} \qquad\qquad \underset{O-}{|}$$

Scheme 4

Irreversible Enzyme Inhibitors

In theory, an improved approach to drug design would be the use of a specific irreversible enzyme inhibitor drug. In this case, once the enzyme reacts with the inactivator, a process that, ideally, could require only one inactivator molecule per enzyme active site, it would not be necessary to maintain a high drug concentration in the body. Of course, the gene encoding the inactivated enzyme will produce new enzyme, so additional drug will be required, but protein synthesis could take days. Under these circumstances it would not be necessary to maintain a steady concentration of inhibitor, because once the enzyme reacted with the inactivator, the excess inactivator could be removed and the enzyme would remain inactive. This could translate into administration of fewer and smaller doses of a drug. As discussed below, however, many irreversible inactivators are reactive compounds (called affinity labeling agents), and they can be quite toxic. One may wonder what the effect on meta-bolism would be if an enzyme were inhibited for an ex-tended period of time. Consider the case of aspirin, an irreversible inhibitor of prostaglandin synthetase (cy-clooxygenase). If the quantity of aspirin consumed in the United States were averaged over the entire popu-lation, then every man, woman, and child would be taking approximately 200 mg of aspirin a day, enough to shut down all of human prostaglandin biosynthesis for the entire country permanently!

The term irreversible is a loose one. As long as the enzyme remains non-functional long enough to produce the desired pharmacological effect, it is considered irreversibly inhibited. This, then, could include transition state analogues and slow, tight-binding inhibitors. The two principal types of irreversible inhibitors are reactive compounds called affinity labeling agents and unreactive compounds termed mechanism-based enzyme inactivators. As described above, there also are reversibly covalent inhibitors such as certain slow, tight-binding inhibitors.

Affinity Labeling Agents. These are covalent
inactivators that have a structural similarity to a
substrate for a target enzyme, but which contain a
reactive functional group, e.g., an α-haloketone or a
reactive ester. Subsequent to E·I complex formation,
they react with active-site nucleophiles, generally via
S_N2 alkylation or acylation mechanisms. Because of the
reactivity of this class of inactivators, not only can
they react with the target enzyme, but they also can
react with other enzymes and biomolecules, making them
potentially quite toxic (many cancer chemotherapy drugs,
for example, are affinity labeling agents). Consequently, they are not generally as useful in drug design
as other types of enzyme inhibitors.

There are several principal reasons why these
reactive molecules are, nonetheless, effective in drug
design. First, once the inactivator forms an E·I complex,
a unimolecular reaction ensues which can be many orders
of magnitude more rapid than non-specific bimolecular
reactions with other proteins. Furthermore, the inactivator may form an E·I complex with other enzymes, but if
there is no nucleophile near the reactive functional
group, no reaction will take place. Third, in the case
of antitumor agents, mimicks of DNA precursors are
rapidly transported to the appropriate site and, therefore, they are concentrated at the desired target.

The key to the design of affinity labeling agents as
drugs is specificity of binding. If the molecule has a
very low K_i for the target enzyme, then E·I complex formation will be favored, and the selective reactivity will
be enhanced. This class of inactivators also can be more
selective if the reactive functional group is modulated
and is situated at a position that is involved in the
normal catalytic mechanism.

A variation on affinity labeling in which enzyme
modification occurs only in the presence of a substrate
which converts the enzyme into an active form for inactivation has been termed syncatalytic enzyme modification.[12]

An example of affinity labeling drugs that utilizes
a moderately reactive functional group located at the
normal catalytic site is the penicillins. The early
penicillins were natural products isolated from
Penicillium strains. These are ideal drug candidates
because they react with bacterial transpeptidase, the

enzyme responsible for catalyzing the cross linking of
the cell wall peptidoglycan. Since mammalian cells do
not have cell walls, this enzyme is not present in
humans. It is believed that the penicillins are
structurally related to the terminal D-alanyl-D-alanine
residues of the N-acetylmuramic acid side chain of the
peptidoglycan. This terminal dipeptide is believed to
bind to the active site of transpeptidase which initially
acts as a serine protease, clipping the terminal peptide
bond and making a serine ester (Scheme 5). Cross linking

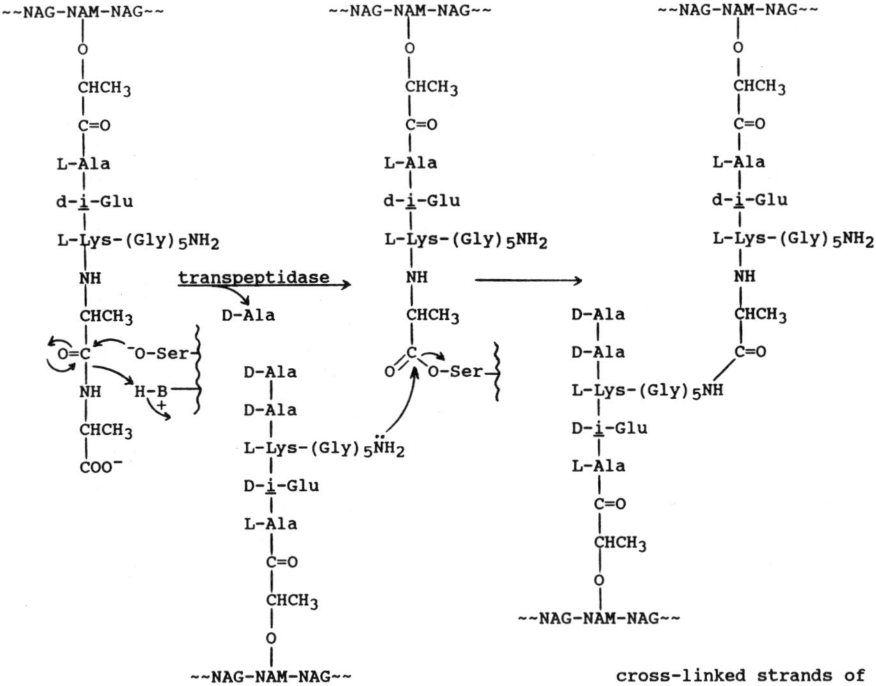

Scheme 5

of this ester with another peptidoglycan strand builds the
cell wall. The penicillins bind at the D-alanyl-D-alanine
binding site such that the reactive β-lactam carbonyl is
attacked by the active site serine (Scheme 6). The bulk
of the penicillin molecule precludes hydrolysis or
amidation. The beauty of the penicillins is that they
are not exceedingly reactive, so few if any non-specific
acylation reactions occur, and they are quite specific.
Penicillins are unusual affinity labeling agents because

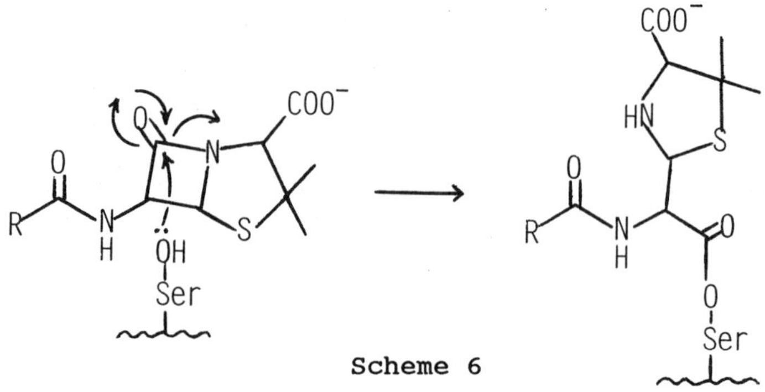

Scheme 6

of their low reactivity and non-toxicity. In fact, if it
were not for allergic responses and problems associated
with drug resistance, penicillins, which contain C, H, N,
O, and S atoms, would be nutritious foods, comprised of
various carboxylic acid derivatives (the RCO-side
chains), cysteine, and the essential amino acid valine.

<u>Mechanism-Based Enzyme Inactivators.</u> Unlike
virtually all affinity labeling agents (penicillins
excepted), a mechanism-based enzyme inactivator is an
unreactive compound that bears a structural similarity to
a substrate or product for the target enzyme. Once at
the active site, the target enzyme converts these
compounds, via their normal catalytic mechanisms, into
products that generally form covalent bonds to the
enzyme. There also are examples where the product forms
a very tight non-covalent complex with the enzyme. The
key feature of mechanism-based enzyme inactivators that
makes them so amenable to drug design is that they are
unreactive compounds. Consequently, non-specific
alkylations of other proteins should not be a problem.
Ideally, only the target enzyme will be capable of
catalyzing the appropriate conversion of the inactivator
to the activated species, and inactivation will result
with every turnover. This latter situation can be quite
important for potential drug design. If the activated
species, which usually is a highly reactive compound, is
released from the active site, it may react with other
proteins. The released product itself may be toxic or it
may be metabolized to a toxic substance. Therefore, the
ideal partition ratio, i.e., the ratio of the number of
inactivator molecules converted to product and released
per inactivation event, is zero. Under these ideal conditions, the inactivator would be a strong drug candidate
because it should be highly specific and low in toxicity.
In fact, α-difluoromethylornithine (eflornithine), a spe-

cific mechanism-based inactivator of ornithine decarboxylase used in the treatment of protozoal infections, has been administered to patients in amounts of 30 g a day for several weeks with only minor side effects.[13] In the case described above where the activated species is released from the enzyme and then inactivates the same enzyme or a different enzyme, it is not a mechanism-based enzyme inactivator; rather, this is a metabolically-activated inactivator.[14]

A variation of mechanism-based inactivation that combines metabolically-activated and mechanism-based inactivation, also has been demonstrated. This approach is a type of prodrug design in which a compound is metabolically converted by one or more enzymes into a mechanism-based inactivator for a different enzyme. These compounds were termed dual enzyme-activated inhibitors[15] when two enzymes were involved in the inactivation. A more general terminology would be multi enzyme-activated inactivators, since there are examples where more than two enzymes are involved.[16-19]

Despite the great potential of mechanism-based enzyme inactivators as drugs, there are no drugs on the American drug market today that were rationally designed as mechanism-based inactivators of specific enzymes. That is not to say that no drugs are mechanism-based enzyme inactivators; only that the ones that are in current medical use were determined ex post facto to be mechanism-based inactivators. A list of current drugs that are mechanism-based enzyme inactivators includes the antidepressant agents, tranylcypromine and phenelzine, the antihypertensive agents, hydralazine and pargyline, and the antiparkinsonian drug, deprenyl (all of which inactivate monoamine oxidase); clavulanic acid, a compound used to protect penicillins and cephalosporins against bacterial degradation (inactivates β-lactamases); the antitumor drug, 5-fluoro-2'-deoxyuridylate and the antiviral agent, 5-trifluoromethyl-2'-deoxyuridylate (both of which inactivate thymidylate synthetase); the uricosuric agent, allopurinol (inactivates xanthine oxidase); the antithyroid drugs, methimazole, methylthiouracil, and propylthiouracil (thyroid peroxidase); and the antibiotic, chloramphenicol, the antifertility drug, norethindrone, the anesthetics, halothane and fluoroxene, the sedative, ethchlorvynol, the diuretic and antihypertensive agent, spironolactone, the pituitary suppressant, danazol, and the hypnotic, novonal (all of which inactivate cytochrome P-450). The drugs that inactivate cytochrome P-450, however, do not derive their

medicinal effect as a result of that inactivation. The
first two rationally-designed mechanism-based enzyme
inactivator drugs to be given U. S. drug approval may be
4-amino-5-hexenoic acid (γ-vinyl GABA; vigabatrin; inactivates γ-aminobutyric acid aminotransferase) and α-di-
fluoromethylornithine (eflornithine; inactivates ornithine decarboxylase), which are in latter stages of
clinical trials for the treatment of seizures and protozoal infections, respectively. Enzymes that already have
been targeted for mechanism-based enzyme inactivation and
the therapeutic goals of inactivation are listed in Table
1. A review of mechanism-based enzyme inactivators that
have been designed for these enzymes and a brief descrip-
tion of why inactivation of each of these enzymes leads
to the desired therapeutic effect was published recently.[20] The mechanisms of inactivation of various enzymes
by mechanism-based enzyme inactivators in current clinical use have been reviewed.[21] A more comprehensive in-
depth review and discussion of mechanism-based enzyme
inactivation in general, its chemistry and enzymology, is
forthcoming.[22] There are numerous excellent examples
that could be chosen to be representative of mechanism-
based enzyme inactivators. However, poetic license
permits me to choose examples from my own laboratory.

The two enzymes on which I will focus are γ-amino-
butyric acid (GABA) aminotransferase and monoamine oxidase (MAO). GABA aminotransferase is the PLP-dependent
enzyme that catalyzes the degradation of the inhibitory
neurotransmitter GABA to succinic semialdehyde. It has
been shown that convulsions can arise from an imbalance
in the brain concentrations of glutamate, an excitatory
neurotransmitter, and GABA, the concentrations of which
are regulated by L-glutamic acid decarboxylase and GABA
aminotransferase. When GABA is injected into the brain
of a convulsing animal, the convulsions cease. However,
peripheral administration of GABA has no effect because
it does not cross the blood-brain barrier. An approach
to the design of new anticonvulsant agents has been to
prepare compounds capable of crossing the blood-brain
barrier that then inactivate GABA aminotransferase specifically. This raises the GABA concentration in the
brain and can lead to an anticonvulsant effect. γ-Vinyl
GABA (vigabatrin) is an example of an effective drug
candidate whose mechanism of action appears to be as just
described. The design of mechanism-based inactivators
depends upon the catalytic mechanism of the target enzyme. The mechanism of GABA aminotransferase is shown in
Scheme 7 (Pyr is the pyridine nucleus of PLP or PMP).
Our initial effort in this area was a series of 4-ami-

TABLE 1

Enzymes with Potential Use in Medicine Already Targeted for
Mechanism-Based Inactivation[a]

Enzyme	Therapeutic Goal
S-adenosylhomocysteine hydrolase	antiviral agent
alanine racemase	antibacterial agent
D-amino acid aminotransferase	antibacterial agent
γ-aminobutyric acid aminotransferase	anticonvulsant agent
arginine decarboxylase	antibacterial agent
aromatase	anticancer agent
L-aromatic amino acid decarboxylase	synergistic with antiparkinson drug
dihydrofolate reductase	anticancer agent; antibacterial agent; antiprotozoal agent
DNA polymerase I	antiviral agent
dopamine β-hydroxylase	antihypertensive agent; pheochromocytoma agent
histidine decarboxylase	antihistamine; anti-ulcer agent
β-lactamase	synergistic with antibiotics
monoamine oxidase	antidepressant agent; antihypertensive agent; antiparkinsonian agent
ornithine decarboxylase	anticancer agent; antiprotozoal agent
serine proteases	treatment of emphysema, inflammation, arthritis, adult respiratory distress syndrome, anticoagulant agent, pancreatitis, certain degenerative skin disorders, and digestive disorders
testosterone 5α-reductase	anticancer agent
thymidylate synthetase	anticancer agent
thyroid peroxidase	antithyroid agent
xanthine oxidase	uricosuric agent

[a]R. B. Silverman, *J. Enz. Inhib.*, in press.

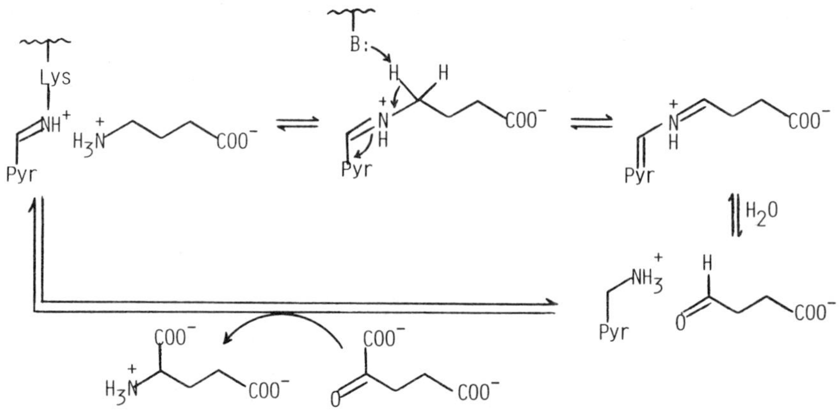

Scheme 7

no-5-halopentanoic acids;[23] the fluoro analogue (4,
Scheme 8) was the most potent entry. On the basis of the
mechanism of GABA aminotransferase, the inactivation of
the enzyme by (4) was proposed[24] to be that shown in
Scheme 8. Experiments that support this mechanism are

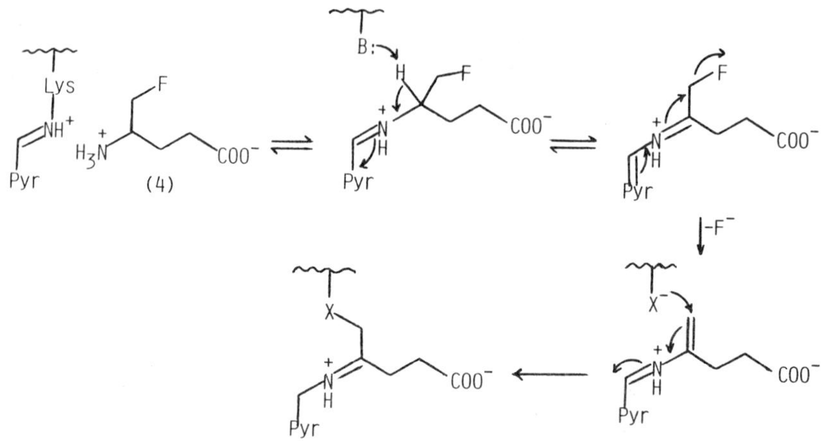

Scheme 8

the following: 1) inactivation only occurs with the PLP
haloenzyme; 2) $[4-^2H]-(4)$ inactivates the enzyme with a
kinetic isotope effect of 5.5; 3) one fluoride ion is
released per active site; 4) $[U-^{14}C]-4$-amino-5-chloropen-
tanoic acid inactivates GABA aminotransferase with the
irreversible incorporation of 1 mole of ^{14}C per active

site of enzyme; 5) the absorption spectrum of the enzyme shows a conversion of PLP to something with the spectrum of PMP during inactivation. The inactivation mechanism was revised,[25] however, to that shown in Scheme 9 based

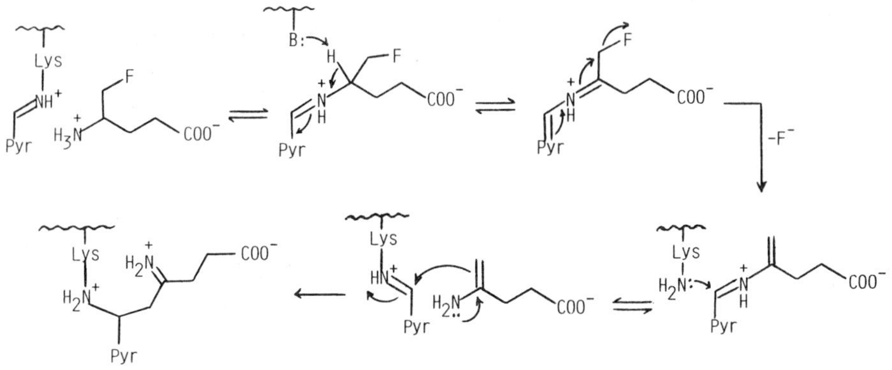

Scheme 9

on the work of Metzler and coworkers.[26,27] Note that all of the data previously collected are consistent with the mechanism shown in Scheme 9 as well as that shown in Scheme 8. In order to differentiate the two mechanisms apo-GABA aminotransferase was reconstituted with [3H] PLP and inactivated with (4); then the pH was raised to 12 and the enzyme was denatured. All of the radioactivity was released from the protein as (5); the mechanism for the generation of (5) is shown in Scheme 10. As mentioned above, an important feature for a mechanism-based enzyme inactivator in drug design is that the partition ratio be low, ideally zero. Compound (4) has a partition ratio of zero. The experiments carried out[24] to confirm the partition ratio were as follows: 1) inactivation of the enzyme by [14C]4-amino-5-chloropentanoic acid produces no [14C] non-amines; 2) 100% inactivation occurs

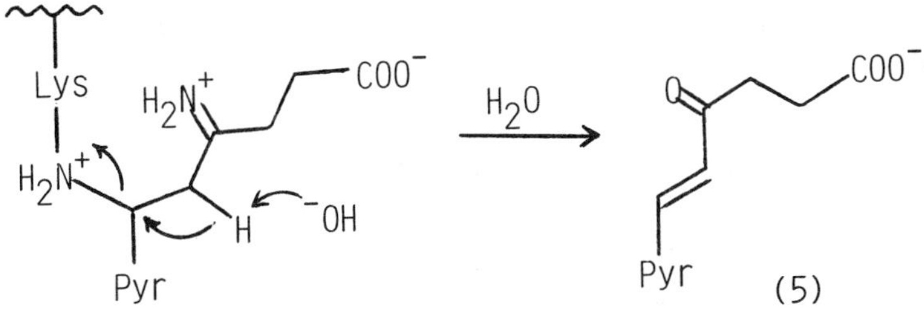

Scheme 10

even in the absence of α-ketoglutarate; 3) no [^{14}C]Glu is produced during inactivation in the presence of [^{14}C]α-ketoglutarate; 4) one fluoride ion is released from (4) per active site inactivated. A similar mechanism for inactivation of GABA aminotransferase by (E)-4-amino-5-fluoropent-2-enoic acid also was suggested; however, this compound has a partition ratio of five.[28,29] Other classes of potential inactivators, namely, 4-amino-3-halobutanoic acids[30] (6) and 4-amino-2-halomethyl-2-butenoic acids[31] (7) are substrates for the enzyme, but do not cause inactivation.

(6)   (7)

We were interested in labeling GABA aminotransferase with a mechanism-based enzyme inactivator that would become attached directly and irreversibly to an active site amino acid residue in order to begin mapping the active site. Allan et al.[32] briefly mentioned in an abstract that 4-amino-2-fluorobut-2-enoic acid (8) was a

(8)

time-dependent inactivator of GABA aminotransferase, but no inactivation mechanism was suggested. A mechanism for inactivation is proposed in Scheme 11. The analogous mechanism, based on the work of Metzler and coworkers,[26,27] is shown in Scheme 12. This mechanism is less attractive than that shown in Scheme 11 for several reasons. The normal catalytic pathway for the enzyme is azallylic isomerization as shown in the deprotonation steps in Schemes 7 and 11. Since direct fluoride ion elimination is not possible, it is not clear what would be the driving force for the isomerization shown in Scheme 12. Furthermore, if enamine (9) were generated, it should undergo facile elimination of fluoride ion (pathway a), thereby deactivating the enamine. Thirdly, enamine (9) is the fluoro analogue of the enamine that would be generated from (6), which is not an inactivator of GABA aminotransferase.

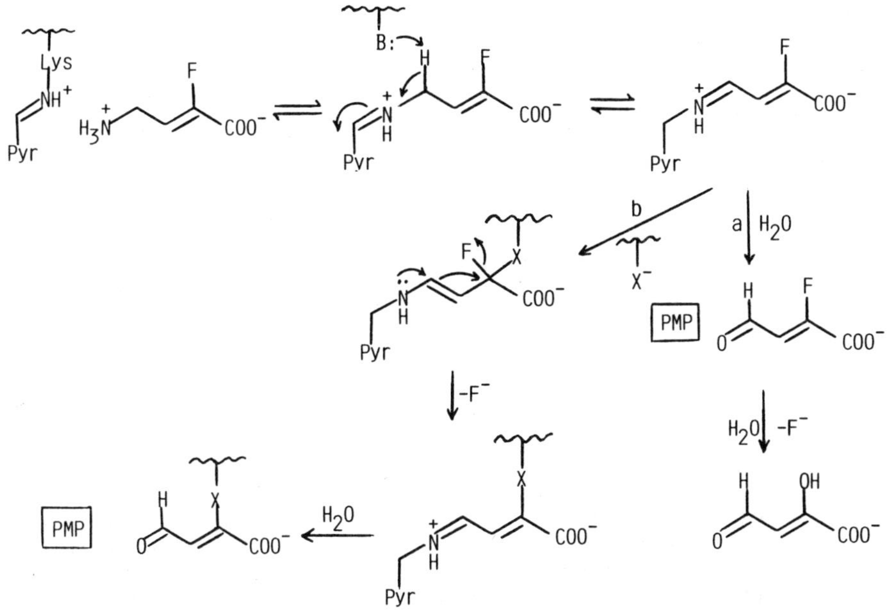

Scheme 11

Incubation of [³H]PLP-reconstituted GABA aminotransferase with (8), followed by treatment as described above for (4), resulted in release of all of the radioactivity as PMP, as expected for the mechanism shown in Scheme 11. The partition ratio, however, is not zero. Titration of the enzyme showed that about 750 molecules of inactivator are required for complete inactivation. Concomitant with inactivation is release of about 750 fluoride ions and approximately 750 transamination events. These results support the bifurcated pathway in Scheme 11. Hydrolysis by pathway a results in both conversion of the PLP to PMP and in loss of fluoride ion. This corresponds to the total requirement of about 750 inactivator molecules for inactivation. Radioactive labeling experiments will determine if attachment to the protein really occurs.

Another enzyme targeted for mechanism-based inactivation in our laboratory is monoamine oxidase (MAO), one of the enzymes responsible for the degradation of biogenic amines. The brain concentration of various biogenic amines was found to be depleted in chronically depressed individuals. Since compounds that inhibit MAO increase the biogenic amine pool, they exhibit an anti-

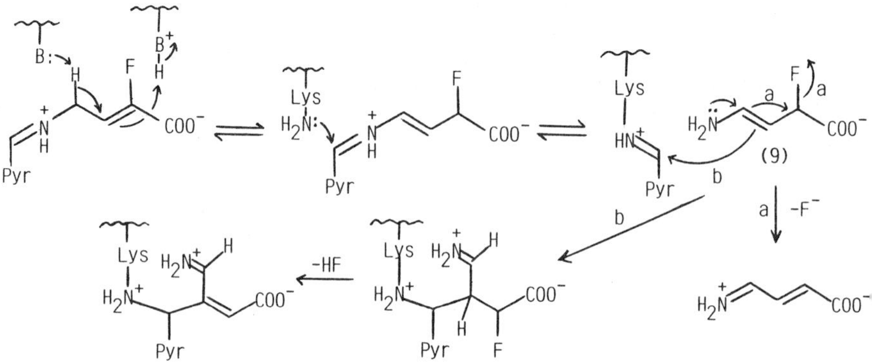

Scheme 12

depressant effect. Unfortunately, a cardiovascular
effect is associated with the use of MAO inhibitors when
foodstuffs containing a high tyramine content are ingested. This is because tyramine triggers the release of
norepinephrine, a potent vasoconstrictor. Since MAO, an
enzyme that degrades norepinephrine, is inhibited, the
blood pressure continues to rise. However, a solution to
this cardiovascular problem may be at hand. MAO exists
in humans in two isozymic forms called MAO A and MAO B.
Although no structural information about the two forms is
known, the principal functional difference appears to be
in their activities for the various biogenic amines.
Selective inhibition of one isozyme should result in an
increase in the biogenic amine concentration; the
remaining isozyme could then function to degrade the
ingested amines. It has now been shown that selective
MAO B inhibition does not exhibit any cardiovascular
effects.[33]

Several years ago we proposed a catalytic mechanism
for MAO that invoked radical intermediates[34] (Scheme 13).

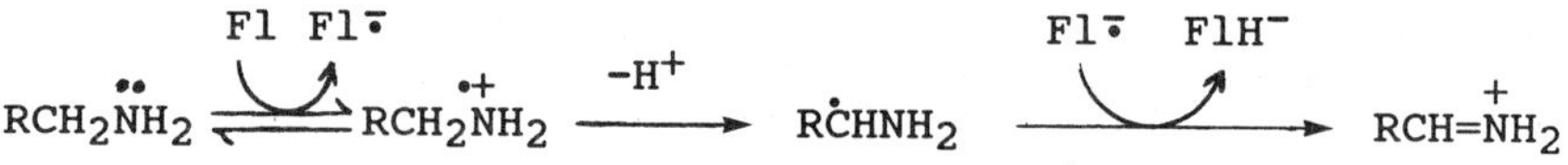

Scheme 13

Although radicals had not been observed by ESR spectroscopy, we believed that they would be too short-lived for
this means of detection and, therefore, sought a chemical
approach to their detection. The first promising results
in support of the radical mechanism were obtained with

the known antidepressant drug, tranylcypromine (10).[35]
Acid treatment of the enzyme inactivated with [7-^{14}C]-
tranylcypromine produced cinnamyldehyde, the expected
product of acid-catalyzed elimination of H-XEnz from (12)
in Scheme 14. One-electron transfer from the amine to

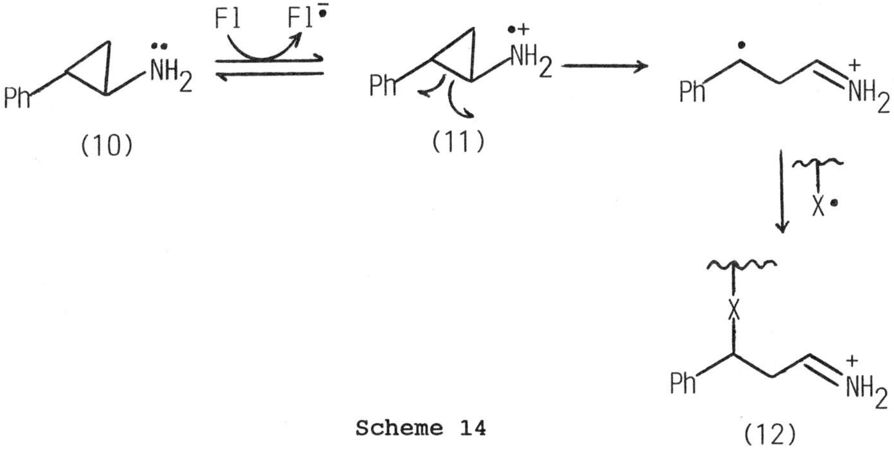

Scheme 14

the flavin would give the cyclopropylaminium radical
cation (11). Maeda and Ingold[36] showed that cyclo-
proylaminyl radicals are exceedingly short lived and
degrade by homolytic cyclopropane ring cleavage. In the
above case C1-C2 bond cleavage produces a benzyl radical
which would be favored over C2-C3 cleavage. Several
other cyclopropylamines were shown to inactivate MAO as
well.[37-45] Further support for a radical mechanism was
obtained by designing an inactivator that would generate
a built-in radical trap if it proceeded by a radical
mechanism.[46] 1-Phenylcyclobutylamine (13) did, indeed,
inactivate MAO one out of 325 turnovers. As depicted in
Scheme 15, the first metabolite generated was 2-phenyl-
1-pyrroline (14), the product of intramolecular radical
cyclization and second-electron oxidation. After a lag
period 3-benzoylpropanal (15) and 3-benzoylpropionic acid
(16) were observed. Compound (15) is the product of MAO-
catalyzed oxidation of 2-phenyl-1-pyrroline (presumably,
of its hydrolysis product, γ-aminobutyrophenone). The
carboxylic acid (16) is, most likely, derived from non-
enzymatic oxidation of (15) by nascent H_2O_2.

3-{4[(3-Chlorophenyl)methoxy]phenyl}-5-[(methyl-
amino)methyl]-2-oxazolidinone methanesulfonate (17,
referred to as MD 780236) was prepared by the group at
Delalande (Rueil-Malmaison) and shown to be a selective

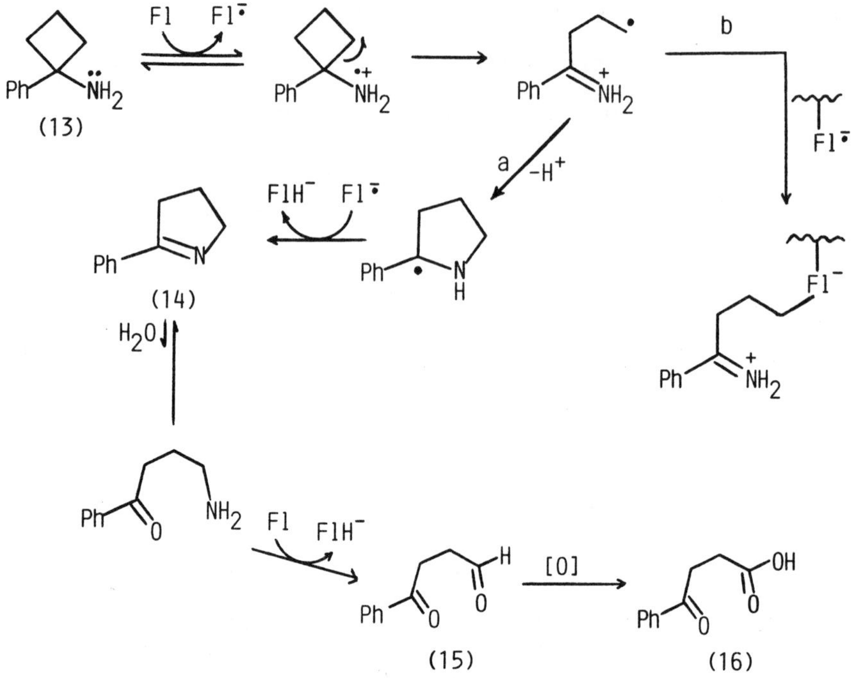

Scheme 15

MAO B inhibitor <u>in</u> <u>vitro</u>, <u>ex</u> <u>vivo</u>, and <u>in</u> <u>vivo</u> in the
rat.[47] After separation of the enantiomers, it was found
that the <u>R</u>-isomer is fully reversible, but the <u>S</u>-isomer
is an irreversible inhibitor.[47,48] A mechanism for
inactivation of MAO by (17) was suggested[48] to involve
oxidation of the amine to the imine followed by selective
active-site nucleophilic attack on the oxidized <u>S</u>-isomer.
This adduct does not appear to be unusually stable so an
alternative inactivation mechanism was proposed[49] (Scheme
16). According to this mechanism the normal radical
intermediate (18) would decompose with loss of CO_2 to
give a new radical (19), which could be trapped by an

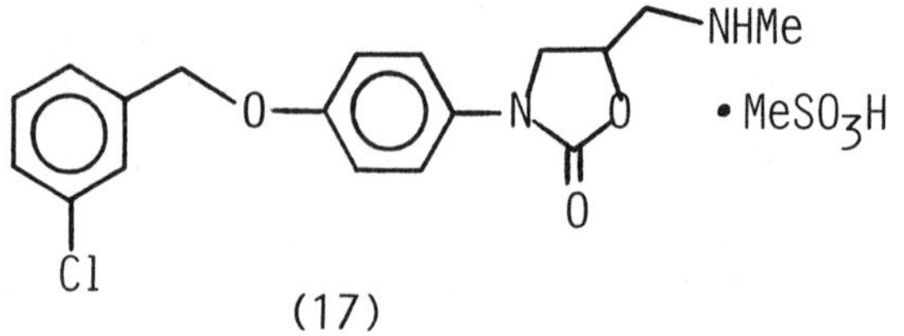

(17)

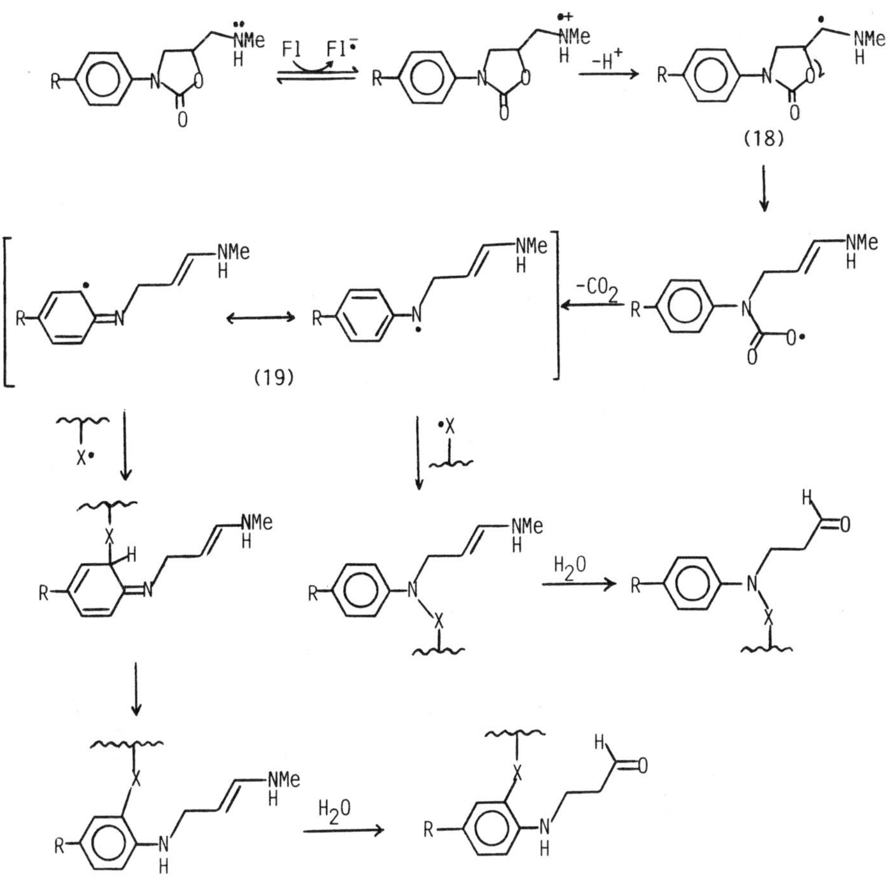

Scheme 16

active site radical at various sites (two examples are
shown in the scheme). A chemical model study was carried
out to test this hypothesis.[49] A radical related to (18)
was generated by heating (20) with tri-n-butylstannane
and a catalytic amount of AIBN. Two products were
obtained, (22) and (23), in a 94:6 ratio (Scheme 17).
The predominance of (22) is not surprising, since it is
well known that high concentrations of Bu₃SnH favor
intermolecular hydrogen atom abstraction over intramolecular processes.[50,51] Control reactions showed that
neither heating (20) without Bu₃SnH and AIBN nor heating
(22) with Bu₃SnH and AIBN resulted in generation of (23).
Loss of CO_2 was detected by precipitation of lead carbonate from a solution of lead acetate. Cleavage of the C-O

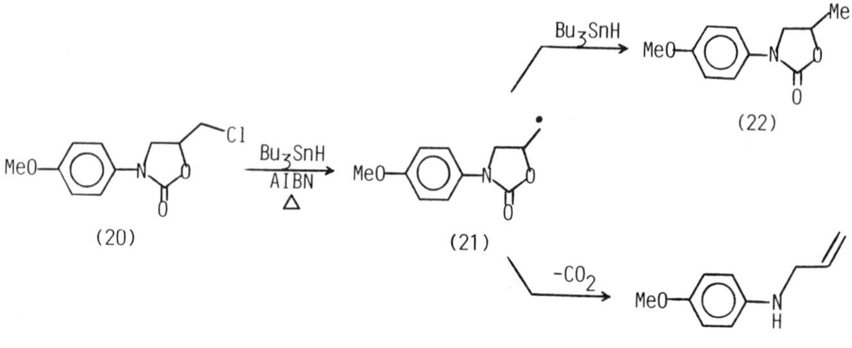

Scheme 17

bond requires interaction of the radical-containing p-orbital with the σ C-O bond. Since the p-orbital also must interact with the amine lone pair orbital, conformational mobility is severely restricted. This may explain the difference in the observation that the R-isomer is a reversible inhibitor of MAO, but the S-isomer is an irreversible inhibitor.

3   CONCLUSION

I have tried to present the essence of enzyme inhibition in drug design; obviously, only the principles have been introduced. The key to the future success of this approach is to continue to uncover important enzymes in humans and microorganisms or tumor cells whose inhibition will lead to new therapeutic benefits, and to find differences between essential enzymes in these systems so that inhibitor selectivities can be targeted.

REFERENCES

1.   D. W. Cushman, H. S. Cheung, E. F. Sabo, and M. A. Ondetti, *Biochemistry*, 1977, *16*, 5484.

2. A. A. Patchett, E. Harris, E. W. Tristram, M. J. Wyvratt, M. T. Wu, D. Taub, E. R. Peterson, T. J. Ikeler, J. ten Broeke, L. G. Payne, D. L. Ondeyka, E. D. Thorsett, W. J. Greenlee, N. S. Lohr, R. D. Hoffsommer, H. Joshua, W. V. Ruyle, J. W. Rothrock, S. D. Aster, A. L. Maycock, F. M. Robinson, R. Hirschmann, C. S. Sweet, E. H. Ulm, D. M. Gross, T. C. Vassil, and C. A. Stone, Nature, 1980, 288, 280.

3. G. E. Lienhard, Science, 1973, 180, 149.

4. R. Wolfenden, Ann. Rev. Biophys. Bioeng., 1976, 5, 271.

5. R. N. Lindquist in "Drug Design", Vol. 5, E. J. Ariëns, ed., Academic Press, New York, 1975, p. 24.

6. S. Thaisrivongs, D. T. Pals, L. T. Kroll, S. R. Turner, and F.-S. Han, J. Med. Chem., 1987, 30, 976.

7. M. J. Sculley and J. F. Morrison, Biochim. Biophys. Acta, 1986, 874, 44.

8. D. H. Rich, J. Med. Chem., 1985, 28, 263.

9. B. Imperiali and R. H. Abeles, Biochemistry, 1986, 25, 3760.

10. R. L. Stein, A. M. Strimpler, P. D. Edwards, J. J. Lewis, R. C. Mauger, J. A. Schwartz, M. M. Stein, D. A. Trainor, R. A. Wildonger, and M. A. Zottola, Biochemistry, 1987, 26, 2682.

11. P. A. Bartlett and C. K. Marlowe, Science (Washington, D.C.), 1987, 235, 569.

12. W. Birchmeier and P. Christen, Meth. Enzymol., 1977, 46, 41.

13. A. Sjoerdsma, J. A. Golden, P. J. Schechter, J. L. R. Barlow, and D. V. Santi, Trans. Assoc. Am. Phys., 1984, 97, 70.

14. S. D. Nelson, J. Med. Chem., 1982, 25, 753.

15. I. A. McDonald, J. M. Lacoste, P. Bey, J. Wagner, M. Zreika, and M. G. Palfreyman, J. Am. Chem. Soc., 1984, 106, 3354.

16. C. Danzin, P. Bey, D. Schirlin, and N. Claverie,

Biochem. Pharmacol., 1982, 31, 3871.

17. C. Danzin, P. Casara, N. Claverie, and J. Grove,
     Biochem. Pharmacol., 1983, 32, 941.

18. A. Wenz, C. Thorpe, and S. Ghisla, J. Biol. Chem.,
     1981, 256, 9809.

19. W. L. Washtien and D. V. Santi, Cancer Res., 1979,
     39, 3397.

20. R. B. Silverman, J. Enz. Inhib., in press.

21. R. B. Silverman in "Protein Tailoring for Food and
     Medical Uses," R. E. Feeney and J. R. Whitaker,
     eds., Marcel Dekker, New York, 1986, Chapter 8.

22. R. B. Silverman, "Mechanism-Based Enzyme
     Inactivation: Chemistry and Enzymology," CRC Press,
     Boca Raton, 1988, in press.

23. R. B. Silverman and M. A. Levy, J. Org. Chem., 1980,
     45, 815.

24. R. B. Silverman and M. A. Levy, Biochemistry, 1981,
     20, 1197.

25. R. B. Silverman and B. J. Invergo, Biochemistry,
     1986, 25, 6817.

26. J. J. Likos, H. Ueno, R. W. Feldhaus, and D. E.
     Metzler, Biochemistry, 1982, 21, 4377.

27. H. Ueno, J. J. Likos, and D. E. Metzler,
     Biochemistry, 1982, 21, 4387.

28. R. B. Silverman, B. J. Invergo, and J. Mathew, J.
     Med. Chem., 1986, 29, 1840.

29. R. B. Silverman and C. George, unpublished results.

30. R. B. Silverman and M. A. Levy, J. Biol. Chem.,
     1981, 256, 11565.

31. R. B. Silverman, S. C. Durkee, and B. J. Invergo, J.
     Med. Chem., 1986, 29, 764.

32. R. D. Allan, G. A. R. Johnston, and B. Twichin,
     Clin. Exp. Pharm. Physiol., 1979, 6, 687.

33. M. B. H. Youdim and J. P. M. Finberg, <u>Modern</u> <u>Probl</u>. <u>Pharmacopsychiatry</u>, 1983, <u>19</u>, 63.

34. R. B. Silverman, S. J. Hoffman, and W. B. Catus III, <u>J</u>. <u>Am</u>. <u>Chem</u>. <u>Soc</u>., 1980, <u>102</u>, 7126.

35. R. B. Silverman, <u>J</u>. <u>Biol</u>. <u>Chem</u>., 1983, <u>258</u>, 14766.

36. Y. Maeda and K. U. Ingold, <u>J</u>. <u>Am</u>. <u>Chem</u>. <u>Soc</u>., 1980, <u>102</u>, 328.

37. R. B. Silverman and S. J. Hoffman, <u>J</u>. <u>Am</u>. <u>Chem</u>. <u>Soc</u>., 1980, <u>102</u>, 884.

38. R. B. Silverman and S. J. Hoffman, <u>Biochem</u>. <u>Biophys</u>. <u>Res</u>. <u>Chem</u>., 1981, <u>101</u>, 1396.

39. R. B. Silverman and R. B. Yamasaki, <u>Biochemistry</u>, 1984, <u>23</u>, 1322.

40. R. B. Silverman, <u>Biochemistry</u>, 1984, <u>23</u>, 5206.

41. R. B. Silverman and P. A. Zieske, <u>Biochemistry</u>, 1985, <u>24</u>, 2128.

42. R. B. Silverman and P. A. Zieske, <u>J</u>. <u>Med</u>. <u>Chem</u>., 1985, <u>28</u>, 1953.

43. M. L. Vazquez and R. B. Silverman, <u>Biochemistry</u>, 1985, <u>24</u>, 6538.

44. R. B. Yamasaki and R. B. Silverman, <u>Biochemistry</u>, 1985, <u>24</u>, 6543.

45. R. B. Silverman and P. A. Zieske, <u>Biochem</u>. <u>Biophys</u>. <u>Res</u>. <u>Chem</u>., 1986, <u>135</u>, 154.

46. R. B. Silverman and P. A. Zieske, <u>Biochemistry</u>, 1986, <u>25</u>, 341.

47. K. F. Tipton, M. Strolin Benedetti, J. McCrodden, T. Boucher, and C. J. Fowler in "Monoamine Oxidase and Disease: Prospects for Therapy with Reversible Inhibitors," K. F. Tipton, P. Dostert, and M. Strolin Benedetti, eds., Academic Press, London, 1984, pp. 155-163.

48. P. Dostert, M. Strolin Benedetti, and C. Guffroy, <u>J</u>. <u>Pharm</u>. <u>Pharmacol</u>., 1983, <u>35</u>, 161.

49.   K. S. Gates and R. B. Silverman, manuscript
      submitted.

50.   A. L. J. Beckwith and G. Moad, <u>J</u>. <u>Chem</u>. <u>Soc</u>. <u>Chem</u>.
      <u>Commun</u>., 1974, 472.

51.   Y. Ueno, K. Chino, and M. Okawara, <u>Tetrahedron</u>
      <u>Lett</u>., 1982, <u>23</u>, 2575.

Studies on Renin Inhibitors

G. Breipohl, R. Geiger, S. Henke, H.-W. Kleeman, J. Knolle,
D. Ruppert, B. A. Schölkens, H. Urbach, A. Wagner,* and
H. Wegmann

HOECHST AG, 6230 FRANKFURT-AM-MAIN 80, POSTFACH 80 03 20,
FEDERAL REPUBLIC OF GERMANY

The renin-angiotensin system (RAS)[1] (Fig. 1) is a complex, mixed enzymatic-hormonal system controlling elec-
trolyte balance, blood volume, and arterial blood pressure. Due to the pioneering research of Page et al.[2] in
USA and Braun-Mendendez et al.[3] in Argentina we are
able to understand the molecular basis of this regula-
ting mechanism. It consists of two main enzymes, renin
and converting enzyme.

Angiotensinogen, a circulating a-globulin, which is
produced by the liver, is cleaved by renin to form the
inactive decapeptide angiotensin I. The cleavage site
is the peptidic bond between Leu and Val.

Removal of the C-terminal histidylleucine by the angio-
tensin-converting enzyme present in a large number of
tissues yields the octapeptide ANG II. This peptide is a
very potent vasoconstrictor, and the physiological stimulus of aldosterone biosynthesis and secretion. Aldosterone, in turn, induces sodium and water retention.
Furthermore sympathetic stimulation and vasopressin release then ensue. All these effects finally lead to
blood pressure elevation.

It has been demonstrated that pharmacological interference of the RAS via ACE-inhibitors can lower blood
pressure in a great number of hypertensive patients[4,5].
Thus inhibition of renin, the more specific of the two
enzymes within this cascade, might be a promising
alternative tool in the treatment of elevated blood
pressure[6].

Renin which catalyses the reaction from angiotensinogen
to ANG I, the rate limiting step in this cascade, is an
aspartic protease mainly produced in the juxtaglomerular cells of the kidney. Stored into intracellular
granules, renin and prorenin are released by a process

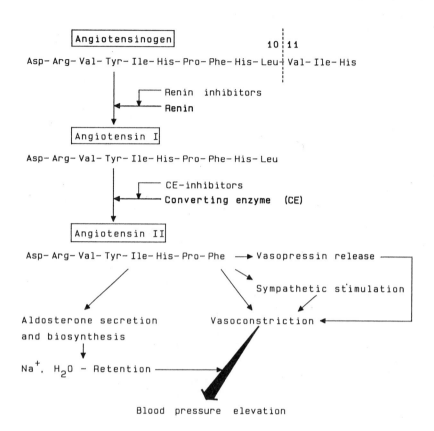

Fig. 1: Renin-Angiotensin System

of degranulation. Fig. 2 shows the simplified biosynthetic pathway of renin in juxtaglomerular cells[7,8].
The nucleotide sequence of the gene has recently been
established consisting of ten exons interrupted by nine
introns. Transcription yields the corresponding
RNA[9,10], and subsequent translation on ribosomes provides the primary product pre-prorenin. This bears a
signal peptide which is rapidly removed from the N-terminal end by the endoplasmic reticulum. Additional
glycosylation leads to inactive prorenin which either
is released or is converted to active renin by N-terminal cleavage of a 46 amino acid unit and conformational
change. Inactive prorenin represents about 90 % of cir-

culating renin[7,8]. Regarding the conversion of inactive prorenin to active renin, C-terminal cleavage of the two dibasic residues Lys-Arg has been discussed[11], as well. Whereas the kidney is probably the only source of active renin[12], because it disappears from blood after

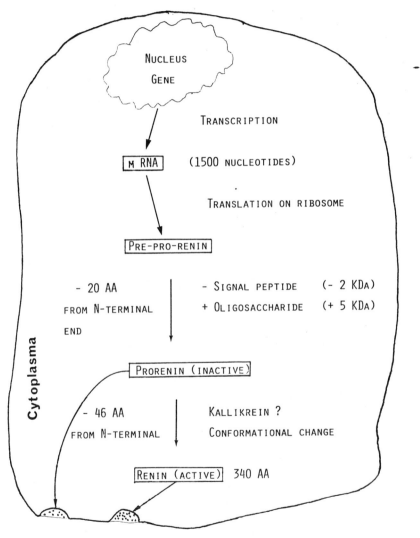

Fig. 2
BIOSYNTHETIC PATHWAY OF ACTIVE RENIN IN JUXTAGLOMERULAR CELLS

bilateral nephrectomy, a number of extrarenal tissues
obviously are capable of synthesizing renins[1] including
the brain[13], the adrenal cortex, the uterus, the wall
of large blood vessels.
Before I start discussing our own research, I briefly
will go through the development of the area of renin
inhibitors. Initially renin inhibitors were based upon
the fully active C-14 segment of angiotensinogen 1
(Fig. 3). The scissile bond is indicated. In 1968 L.T.
Skeggs[14] et al. determined the minimum substrate sequence for human 2 and dog renin 3 with weak inhibitory
activity. Burton et al.[15] started off from the minimum
sequence replacing the scissile bond by Phe-Phe. Final
extension to a decapeptide by N-terminal Pro and C-terminal Lys yielded the much more active inhibitor 4.
RIP stands for renin inhibitory peptide. Szelke et

IC_{50} [μM]

Substrate and reduced substrate analogues   Human renin

H-Asp- Arg- Val- Tyr- Ile- His- Pro- Phe- His- Leu↓ Val- Ile- His- Asn-R   –

  1   Angiotensinogen (human)

       H-His- Pro- Phe- His- Leu↓ Val- Ile-His   313

  2   minimum substrate (human)   L.T.Skeggs et al

       H-His- Pro- Phe- His- Leu- Leu- Val- Tyr   200

  3   minimum substrate (dog)   L.T.Skeggs et al

     H-Pro- His- Pro- Phe- His- Phe- Phe- Val- Tyr- Lys   5.9

  4   RIP   Burton et al

                                    R
       H-His- Pro- Phe- His- Leu—— Val- Ile-His   0.19

  5   H 113   Szelke et al

                                    R
     H-Pro- His- Pro- Phe- His- Leu—— Val- Ile- His- Lys   0.01

  6   H 142   Szelke et al

Fig.3: Substrate and reduced substrate analogues 1 – 6

al.[16] substituted the scissile bond by a non-cleavable reduced peptide bond leading to compounds **5,6** with increased potency.

The development of so-called transition-state analogues (Fig. 4) triggered by the isolation of the natural occuring pepstatin by Umezawa et al.[17], and the pioneering research in Szelke's[18], and Boger's group[19] had great impact on renin research of the recent years[20-26]. Szelke et al. replaced the scissile bond of an octapeptide by a "hydroxy ethylene" moiety producing highly potent inhibitor **7**. The left part of the central unit resembles Leu, the right one Val. The natural occuring pepstatin **8** with the unusual amino acid statine **9** (3S, 4S-3-hydroxy-4-amino-6-methylheptanoic acid) inhibits a whole range of aspartic proteases. Led by computer modelling Boger et al. incorporated statine in place of the dipeptide unit around the scissile bond (Leu-Val) into a substrate analogue furnishing the potent inhibitor **10** (SCRIP = statine-containing renin inhibitory peptide). The inhibitory activity could even be improved by 100 times (peptide **11**) introducing the more hydrophobic ACHPA **12**. S,S-configuration at C-3 and C-4 turned out to be important (ACHPA = 3S,4S-4-amino-5-cyclohexyl-3-hydroxypentanoic acid). The high potency of these peptides can be rationalized considering the general mechanism of peptide bond cleavage. X-ray analyses of renin inhibitor-aspartic proteases complexes[27] support this idea.

Fig. 5 schematically shows the active site cleft of renin with the two aspartates delivering a water molecule necessary for the cleavage of the Leu-Val bond. As intermediate or transition-state[28] a water-addition product is formed giving rise to the cleaved substrate. Statine **9** and the other transition-state analogues resemble this transition-state. As the corresponding bond is not cleaved, these peptides have high inhibitory activities.

Most of the known large peptides exhibit relatively short duration of _in vivo_ action probably caused by protease lability and excretion via bile. Insufficient resorption after oral administration is an additional problem. Our working hypothesis to overcome these problems was to replace some of the peptide bonds by proteolytically stable bonds. Decrease of the molecular weight was another goal.

Following this hypothesis we developed our structure-activity studies.

We began our project with a very potent statine-containing tetrapeptide **14** as lead structure, presented by

	Transition-state analogues	IC_{50} [nM] Human renin

7   Boc- His- Pro- Phe- His- N—⟨...⟩ Ile-His   0.7

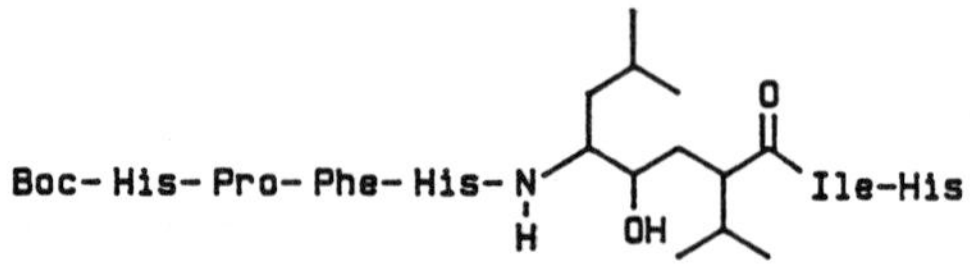

H 261   Szelke et al

8   Iva-Val- Val- Sta- Ala- Sta   6000

Pepstatin

9   Sta = Statine :

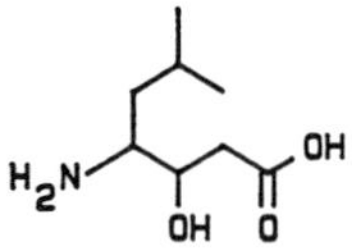

10   Iva-His- Pro- Phe- His- Sta- Leu- Phe- NH$_2$ 13

SCRIP   Boger et al

11   Iva-His- Pro- Phe- His- ACHPA- Leu- Phe- NH$_2$ 0.17

ACRIP   Boger et al

12   ACHPA :   H$_2$N— ⟨...⟩

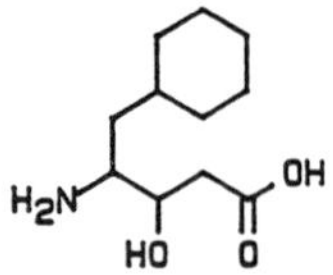

Fig. 4: Transition-state analogues

M.G. Bock et al.[29] at the 9 th American Peptide Symposium in Toronto. We replaced Leu in the original compound by Ile, the amino acid at the corresponding position in human angiotensinogen (Fig. 6). Inhibitory constants were determined versus hog[30] and human plasma renin[31].

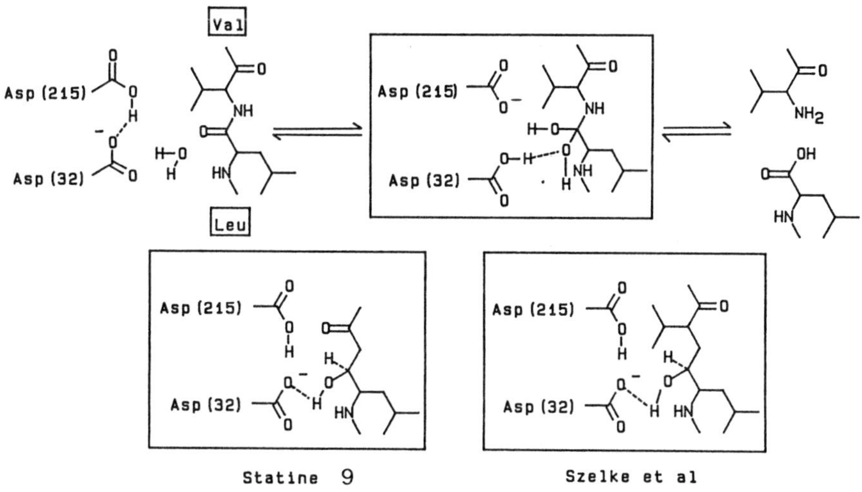

Fig. 5: Active site cleft of renin (schematically)

The N-terminal Phe was then systematically exchanged
(15 - 21). Reduction of the Phe-His amide bond resulted
in a dramatic loss in activity versus human plasma renin. Notice the relatively slight decrease versus hog
renin. A similar result was obtained methylating the
carbamate nitrogen of the Boc-Phe unit. Replacement of
the Boc-residue by a 3-phenyl propyl group led to peptide 17 about 60 times less active then the parent compound. Substituting Phe by an "oxo" or carba-analogue
gave rise to compounds 18,19 with diminished, but at
least in the case of the carba-derivative 18 still acceptable potency. The activity could be retained or
even improved using a bicyclic unusual amino acid-Phe
unit or thienylalanine instead of Phe (20, 21). Obviously the bicyclic amino acid resembles proline of the substrate resulting in maintained activity.
Fig. 7 shows tetrapeptides varied at the position of
His. O-methyl-tyrosine yielded a decreased, but still
acceptable potency, while additional reduction of the
Tyr-Sta-amide bond was not tolerated (22,23). Replacement of His by Phe and 4-Cl-Phe produced peptides 24,25
with a slight loss in activity. Again Thi is an excellent substitute for His. Remarkable is the huge difference in potency using Thi with S- or R-configuration
demonstrating the necessity of chiral purity for opti-
mum inhibition 26. A reduced peptide bond between Phe

		IC_{50} [nM] Hog renin	Human plasma renin	
14	Boc- Phe- His- Sta- Ile- AMP	700	6	
15	Boc-Phe $\overset{R}{\text{——}}$	1600	8500	
16	Boc $\overset{Me}{\text{——}}$ Phe-	>30000	3000	
17	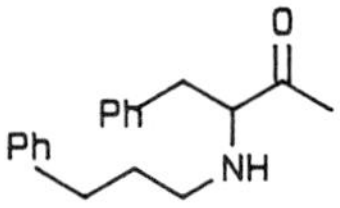	5500	380	
18	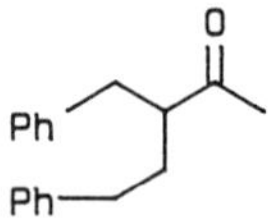	3000	85	
19	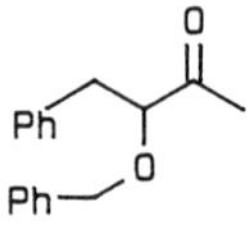	11000	500	
20	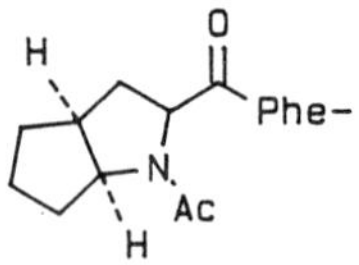	1000	8	(S)
		1400	2	(R)
21	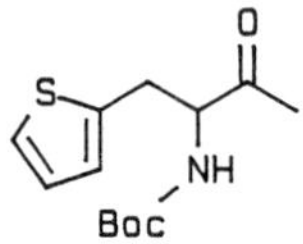	900	3	

AMP : 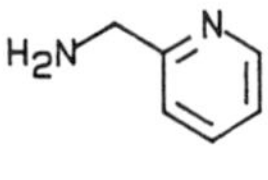

Sta : 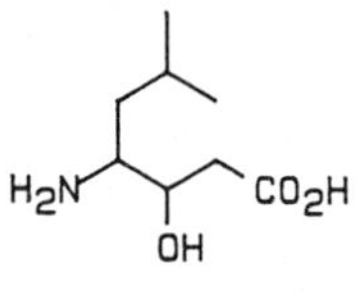

Fig. 6: Substitution of Phe in tetrapeptide 14

		Hog renin	Human plasma renin
		IC_{50} [nM]	
14 Boc-Phe ——— His ——— Sta-Ile-AMP		700	6
22 — Tyr (Me) —		300	72
23 — Tyr (Me) $\overset{R}{————}$		26000	1600
24 —4-Cl-Phe—		210	19
25 —— Phe ——		300	9.5
26 —— Thi ——		140	2.7 (S)
			3400 (R)
27 Boc-Phe $\overset{R}{———}$ Thi ——		4200	1000

Fig. 7: Peptides 22 – 27 modified at His of 14

and Thi resulted in a dramatic fall in biological activity (27). Note that His can be replaced by other aromatic amino acids without dramatic loss in biological
activity. Both the Phe-His and His-Sta amide bond, in
turn, seems to be essential in our series.
We intensively worked on the investigation of transition-state mimics (Fig. 8). Incorporation of the corresponding cyclohexyl or phenyl derivative in place of
Sta[29] gave inhibitors 29,30 with improved potency. The
synthesis of ACHPA 12 ($3\underline{S},4\underline{S}$-4-amino-5-cyclohexyl-3-
hydroxypentanoic acid) and AHPPA, the corresponding
phenylderivative, were reported by Boger et al.[19b] in
1985. Branching at C-2 diminished their activity (31,
32). We assume that in accordance with statine-containing peptides[19a] the more potent inhibitors have $3\underline{S}$-
configuration.
We next explored a retro-inverse statine regarding the
C-terminal amide bond (Scheme 1). For the synthesis of
the appropriate central unit 37, protected Leu 34 was

converted to the corresponding aldehyde **35** according to Castro et al.[32]. Treatment with nitromethane and catalytical amounts of tetramethyl guanidine provided the adduct **36** which subsequently was reduced with raney-nickel and hydrogen. The amine **37** was incorporated into

		IC_{50} [nM]	
		Hog renin	Human plasma renin
28	Boc–Phe–His———Sta———Leu–AMP	92	50
29	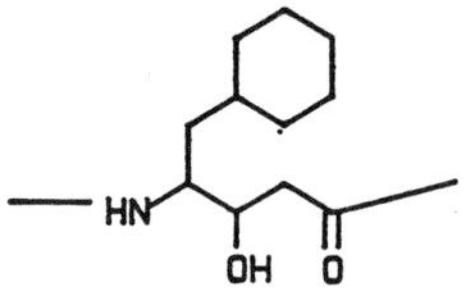 ACHPA	3.3	
30	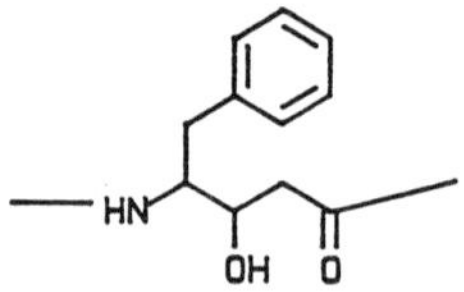 AHPPA	10	
31	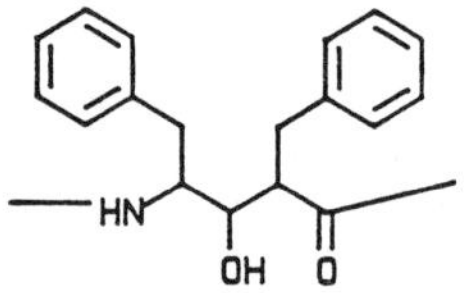	600 (3S) 6500 (3R)	
32	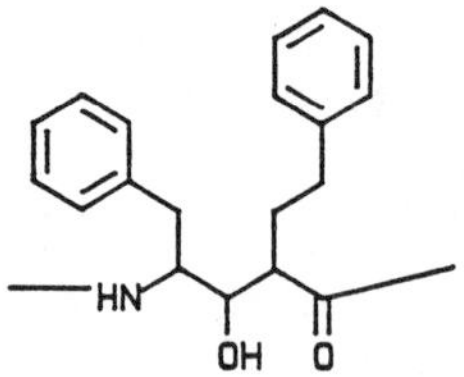	800 (3S) 10000 (3R)	

Fig. 8: Modification of tetrapeptide 28 at the position of sta

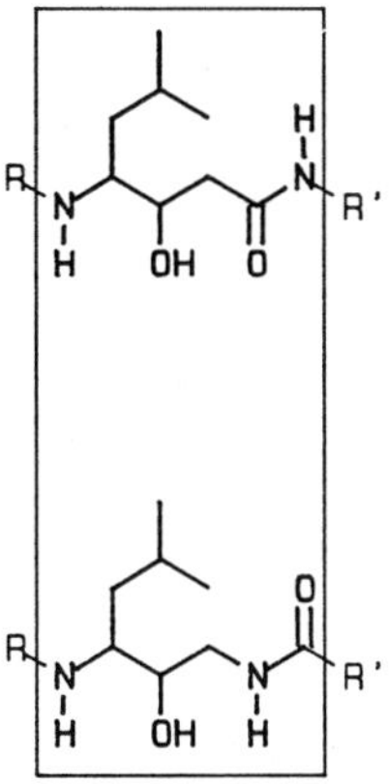

Statine 9

Retro-inverse statine 33

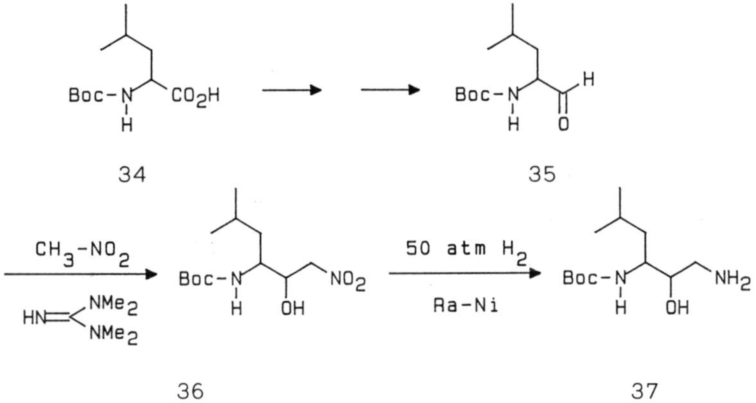

Scheme 1: Synthesis of retro-inverse statine 37

a peptide (Scheme 2). The branched dimethylmalonate 38 was saponified with potassium hydroxyde and the mono-acid was coupled to aminomethyl pyridine using DCC/HOBt[33]. Subsequent treatment of 39 with base yielded the corresponding acid 40 which on reaction with the amine 37 described in the previous Scheme furnished the C-terminal part 41 of the peptide. Deprotection with TFA (trifluoro acetic acid) followed by reaction with

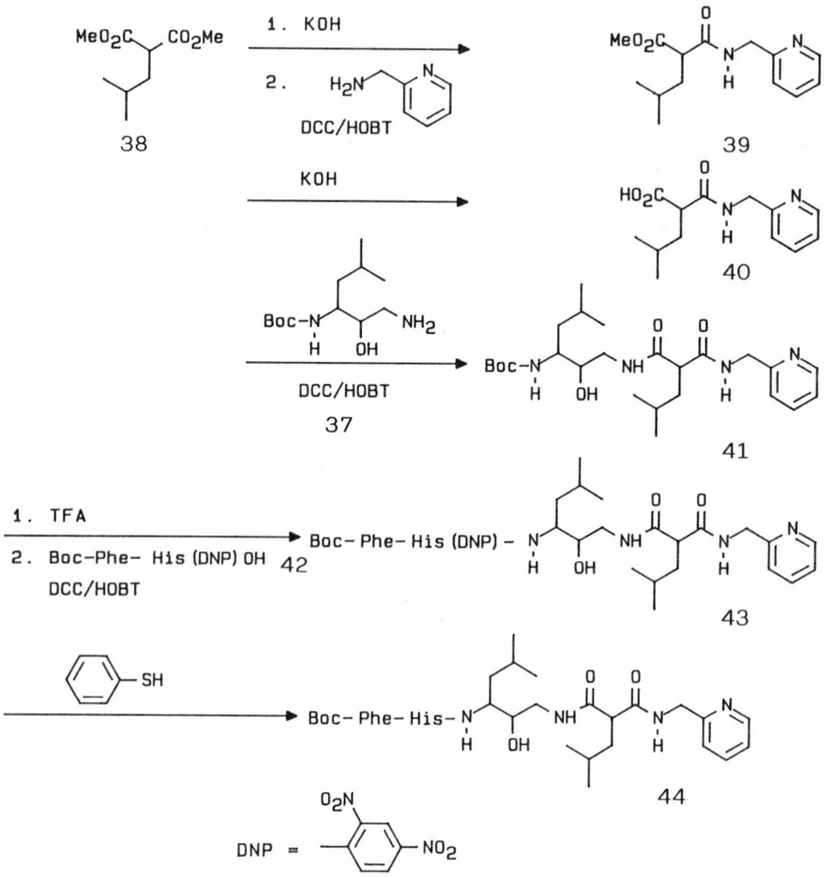

Scheme 2: Synthesis of the retro-inverse statine 37 con-
          taining peptide 44

the dipeptide unit Boc-Phe-His(DNP)-OH **42** gave the protected peptide **43**. The desired inhibitor **44** was obtained after cleavage of the DNP-protecting group with
thiophenol. The weak biological activity[34] (Fig. 9) of
the retro-inverse peptide **44** is somewhat surprising,
because of its similarity to the lead compound **28**.
Discouraged by this result we continued our research
using ACHPA as central unit. Variation of the Leu-moiety

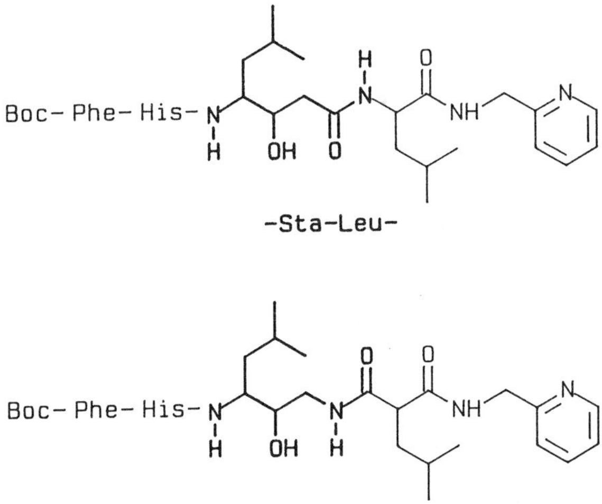

IC$_{50}$ [nM]

Human plasma

renin

28 Boc–Phe–His–Sta–Leu– structure   50

-Sta-Leu-

44 Boc–Phe–His– structure   >1000

Fig. 9: Potency of peptide 44

was carried out in order to replace the C-terminal
amide bond (Fig. 10).
Use of methylphenylpropylamine derived from Ala led to
peptide 46 with decreased potency indicating that a
basic C-terminal unit might improve the interaction
with the enzyme. The increased activity of the pyridine
derivative 47 confirmed this idea. The hydroxyamine 49,
however, is not in agreement. An additional hydroxy
group (48) enhanced the biological activity. Remarkable
is the relatively weak activity of the aminohydroxy derivative 49, when compared with 50 and 51. The latter
ones are derived from Leu and Ile, respectively, and
exhibit high inhibitory activity. Since the angiotensinogen 1, the substrate for renin, bears Ile at the
corresponding position, the higher activity of the latter one can easily be understood. As previously demonstrated, Thi is a well tolerated substitute for Phe.
Therefore it was incorporated into two of the just men-

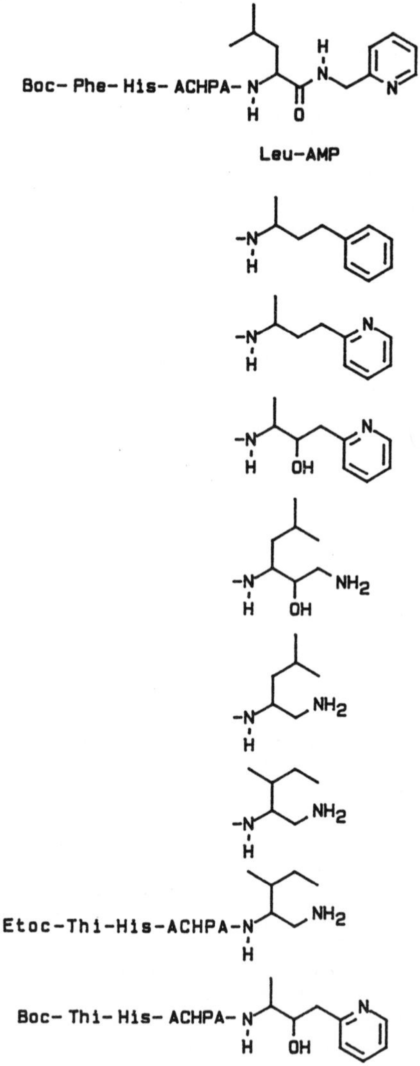

Fig. 10: C-terminal variation of peptide 45

tioned peptides yielding potent inhibitors **52,53**. The loss in activity in case of the Etoc derivative **52** compared to the corresponding Boc-Phe-compound **51** is due to exchange of Boc by Etoc, which is usually associated with a drop in potency.

For the synthesis of peptide **53** (Scheme 3) protected alaninal **55** was treated with the anion of methyl pyridine to give the adduct **56** as diasteriomeric mixture. Deprotection using hydrochloric acid in dimethoxyethane provided the amine **57** which was coupled to Boc-protected ACHPA **58** applying the DCC/HOBt method. Subsequent reaction with hydrochloric acid furnished the corre-

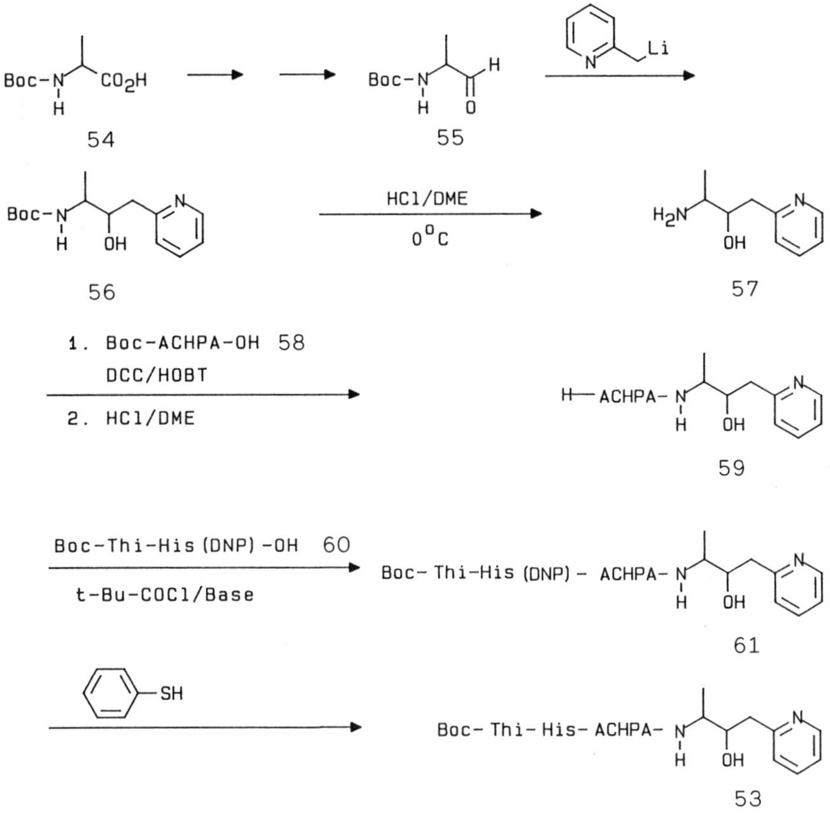

Scheme 3: Synthesis of Thi containing inhibitor 53

sponding amine **59**. This was condensed with the activated dipeptide unit Boc-Thi-His(DNP)-OH **60** using pivaloylchloride[35] to give the tripeptide **61** which after cleavage of the DNP-protecting group yielded the desired product **53**.

We next attempted to substitute Phe by appropriate derivatives in order to reduce the number of proteolytically unstable bonds (Fig. 11). Both incorporation of

IC_{50} [nM]

Human plasma
renin

47	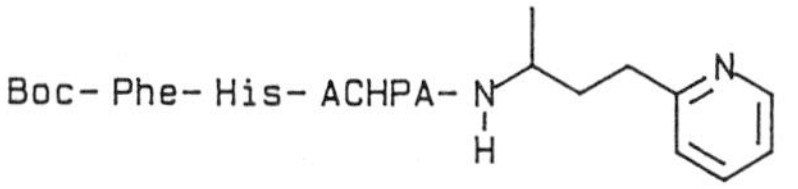 Boc–Phe–His–ACHPA–N	5.2
62	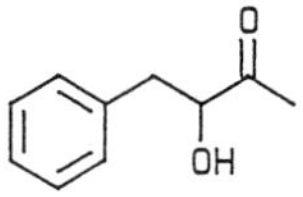	750
63	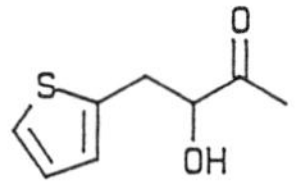	310 >1000
64	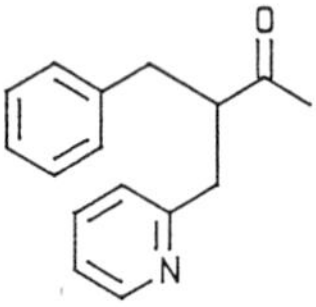	500 >1000
65	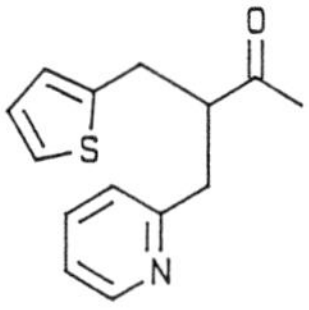	390 >1000
66	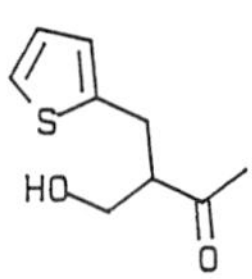	>1000

Fig. 11: N-terminal substitution of the Phe in
peptide 47

hydroxy carboxylates and bis(aryl)carboxylates led to
an unacceptable loss in potency (62 - 66). The two values 63 - 65 correspond to the to diastereomers regarding C-2.
The logical consequence was to concentrate our synthetic efforts on the C-terminal part of the peptides. The
idea was to replace the C-terminal amide bond by an
ethyl moiety. The appropriate intermediates 69 and 71
of the general structure 67 were synthesized. For that
purpose an acidic heterocyclic compound was deprotona-
ted and added to cyclohexylalaninal 68 applying the
same methodology as previously described to provide the
protected amine 69 with n = 1. In a similar manner the
compound 71 with n = 2 was obtained from the epoxide
70. Het stands for heterocycle (**Scheme 4**). The synthesis of the third compound 78 (n = 3) was accomplished
in a four step procedure (**Scheme 5**).
The S,S-ACHPA-ester 72 was sequentially treated with
HCl and ethylbenzimidate hydrochloride to furnish the
corresponding oxazoline 74. As previously mentioned the
S,S-configuration of the transition-state mimic is important for high potency. Since Boger's ACHPA synthesis[19b] yields beside the S,S-72 the S,R-isomer 72a to
the same extent, it was desirable to develop a conversion of the S,R-isomer 72a to an appropriate S,S-deriv-
ative. For that purpose the S,R-aminoalcohol was treated with benzoyl chloride chloride followed by thionyl
providing the desired oxazoline 74. Reduction with
LiAlH$_4$ gave the alcohol 75. Conversion to the tosylate
76 was achieved with tosylchloride. Treatment of the
alcohol with pyridinium bromide under Mitsunobu[36] conditions gave the bromide 77. Reaction with a appropriate anionic species yielded the desired product 78. An
approach to an alternative bromide was accomplished
(**Scheme 6**) converting the S,S-ACHPA 72 to the acetonide
79 with dimethoxypropane. Reduction of the ester moiety
with LiAlH$_4$ gave the alcohol 80. The bromide 81 and the
final adduct 82 was obtained according to the just described method.
Peptides containing the type of compounds and the derivatives with one or two CH$_2$-groups between the oxygen
bearing carbon and heterocycle were synthesized (**Fig.
12**). Besides a simple amino group several heterocyclic
residues were used as C-terminal end. Peptides with one
methylene group between the hydroxy group bearing carbon and heteroatom or heterocycle exhibited only very
weak activity (83 - 86). Marginal improvement of the
potency could be achieved introducing a second CH$_2$-
group (87). Finally a third CH$_2$-group (88) has provided

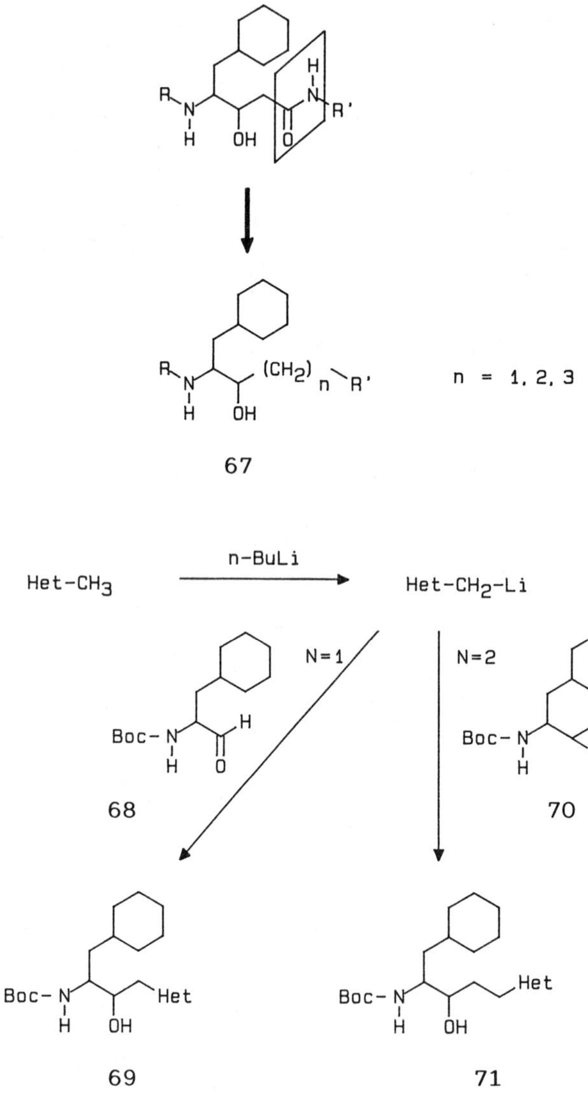

Scheme 4: Synthesis of C-terminal units 69, 71

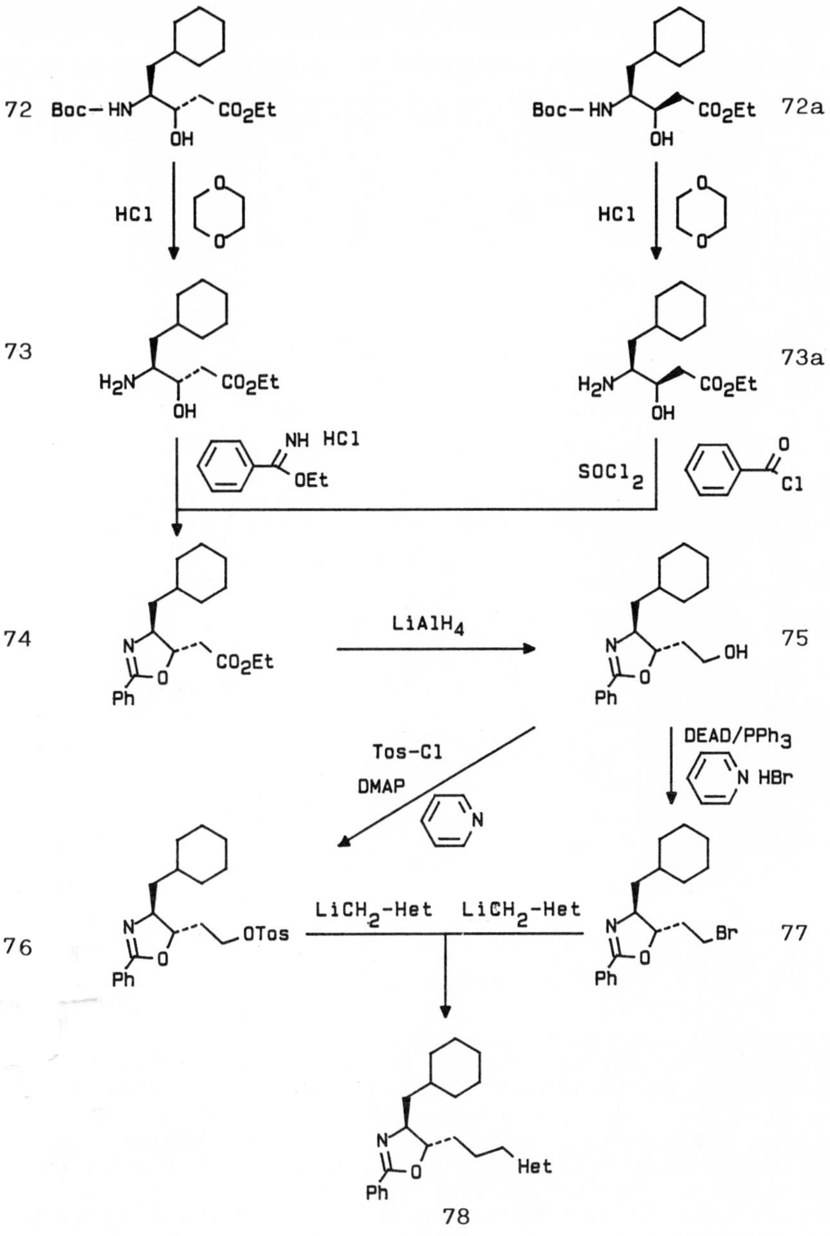

Scheme 5: Synthesis of C-terminal unit 78

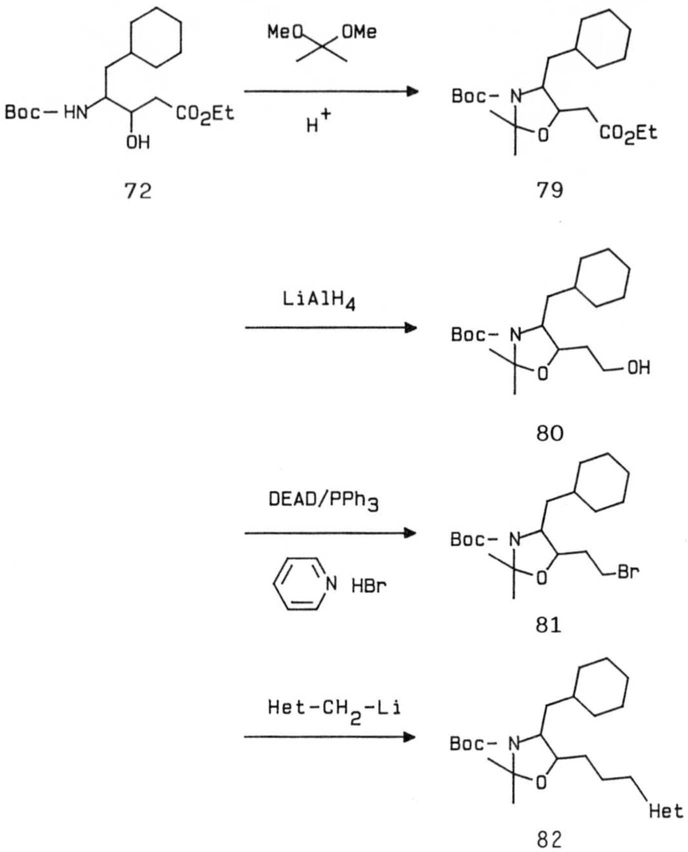

Scheme 6: Synthesis of C-terminal unit 82 via

Boc-ACHPA-OEt 72

a 45-fold boost in activity and acceptable inhibitory
constants were reached. The position of the nitrogen
within the heterocycle seems to play a certain role
(89), furthermore substitution decreased potency (90,
91). A dramatic loss in potency occurred oxidizing the
pyridine nitrogen (92).
Tripeptide 52 and dipeptide 88 were chosen for <u>in vivo</u>

		IC_{50} [nM] Human plasma renin
83	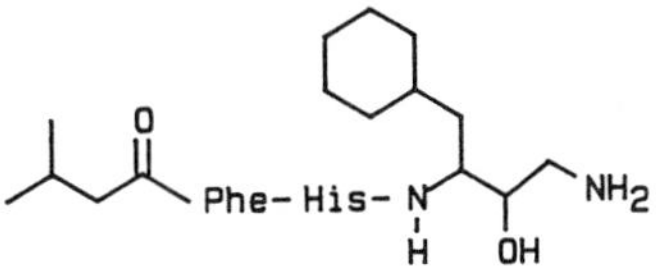	4500 Diastereomeric mixture
84	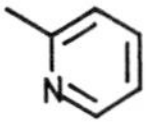	>10000 Diastereomeric mixture
85	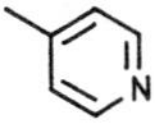	>10000 1800
86	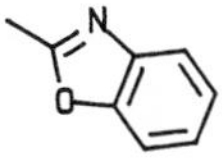	>10000 Diastereomeric mixture
87	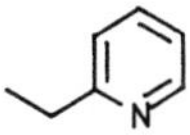	1100 Diastereomeric mixture
88	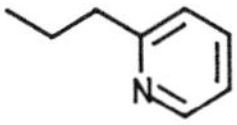	24
89	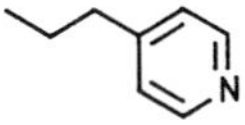	70
90	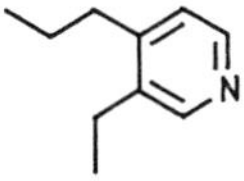	130
91	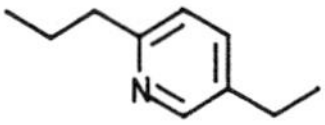	100
92	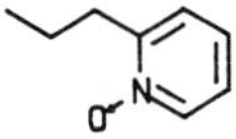	>1000

Fig. 12: Inhibitory activities of dipeptides 83-92

evaluation and were compared with Boc-Phe-His-Sta-Ile-Arg-NH$_2$ 93. In vivo activity was studied in sodium depleted, anesthesized rhesus monkeys. For anesthesia

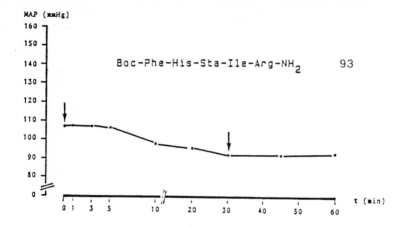

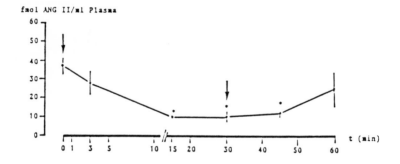

Fig. 13: Influence of peptide 93 on MAP and
         ANG II - concentration after i. v.
         administration to sodium depleted,
         anesthesized rhesus monkeys
         (n = 3; 1 mg/kg/h for 30 min.; ↓
         start and end of infusion)

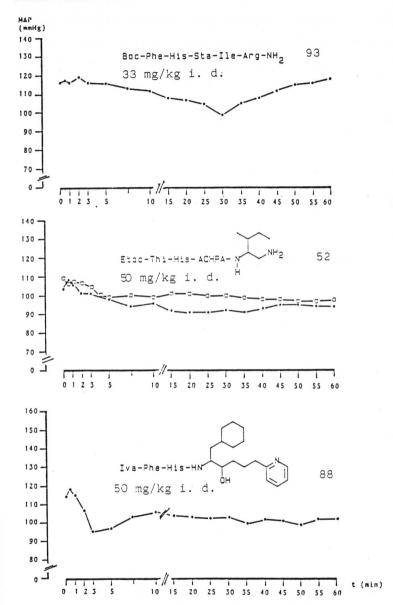

Fig. 14: Influence of peptides 93, 52 and 88 on MAP after
intraduodenal administration to sodium depleted,
anesthesized rhesus monkeys

Nembutal (40 mg/kg i.p.) was used. Sodium depletion was achieved with Furosemide (10 mg/kg/day p.o. for 6 consecutive days, and 30 minutes before administration of the inhibitor 10 mg/kg i.v.). Systemic blood pressure was measured with an electromechanic pressure transducer (Statham P 23 ID). Blood samples for the determination of angiotensin II were withdrawn via a Braunule which was placed into the saphenous vein.

Intravenous adminstration of 93 (1 mg/kg/h for 30 min) led to a drop in both mean arterial blood pressure (MAP) of approximately 10 mm Hg, and in angiotensin II concentration (Fig. 13) which lasted for about 2 hours. In similar manner peptides 93, 52, and 88 reduced MAP after administration into the duodenum via a gastrointestinal fiberscope (Olympus XP 10). Pentapeptide 93 caused a moderate fall in MAP which returned to baseline values within 60 minutes (Fig. 14). A prolonged effect was observed with peptide 52, and 88. Both significantly decreased MAP 60 minutes after administration, indicating an improved bioavailability.

In conclusion, the results described above demonstrate that potent dipeptides can be obtained applying modified transition-state mimics. While alteration of the Phe-His unit turned out to be rather difficult, variation of the post-scissile portion was not as crucial. The three inhibitors selected for <u>in vivo</u> studies lowered blood pressure, both after intravenous and intraduodenal administration. Studies are currently underway to improve duration of action and bioavailability.

Acknowledgement. ANG II-plasma concentrations were kindly determined by Prof. D. Ganten, German Institute for High Blood Pressure Research and Department of Pharmacology, University of Heidelberg, FRG.

1) M.B. Valloton, <u>Trends Pharmacol. Sci.</u>, 1987, <u>8</u>, 69

2) M.C. Khosla, J.H. Page, M.F. Bumpus, <u>Biochem. Pharmacol.</u>, 1979, <u>28</u>, 2867

3) E. Braun-Menendez, J.C. Fasciolo, L.F. Leloir, J.M. Munoz, and A.C. Taquini in "Renal Hypertension" Charles C. Thomas, Springfield, IL, 1946, p. 113

4) B.J. Materson, E.P. Freis, Arch. Intern. Med., 1984, 144, 1947

5) R.O. Davies, J.P. Irvin, D.K. Kramsch, J.F. Walker, and F. Moncloa, Am. J. Med., 1984, 77, 23

6a) S.G. Smith III, A.A. Seymour, E.K. Mazack, J. Boger, and E.H. Blaine, Hypertension, 1987, 9, 150

6b) For a review on renin inhibition see J. Boger, "Annual Report in Medicinal Chemistry", D.M. Bailey (Ed.), Academic Press, Inc., 1985, 20, 257

7) B.J. Morris, D.F. Cantanzaro, J. Hardman, N. Mesterovic, J. Teallam, T. Hart, and J. Shine, J. Hypertension, 1984, 2 (Suppl.), 231

8) F.X. Galen, C. Devaux, A.M. Houot, J. Menard, P. Corvol, M.T. Corvol, M.C. Gubler, F. Mounier, and J.P. Camilleri, J. Clin. Invest., 1984, 73, 1144

9) J.A. Hardman, Y.J. Hort, D.F. Cantanzaro, J.T. Tellam, J.D. Baxter, B.J. Morris, and J. Shine, DNA, 1984, 3, 457

10) P.M. Hobart, M. Fogliano, B.A. O'Connor, I.A. Schäfer, and J.M. Chirgwin, Proc. Natl. Acad. Sci. USA, 1984 81, 5026

11) T. Shinagawa, Y.-S. Do, H. Tam, and W.A. Hsueh, Biochem. Biophys. Res. Commun., 1986, 139, 446

12) N. Glorioso, S.A. Atlas, J.H. Laragh, R. Jewelewicz, J.E. Sealey, Science, 1986, 233, 1422

13) W.F. Ganong, Ann. Rev. Physiol., 1984, 46, 17

14) L.T. Skeggs, K.E. Lentz, J.R. Kahn, H.J. Hochstrasser, J. Exp. Med., 1968, 128, 13

15) J. Burton, R. Cody, J.A. Herd, E. Haber, Proc. Natl. Acad. Sci. USA, 1980, 77, 5476

16) M. Szelke, B. Leckie, A. Hallett, D.M. Jones, J. Sueiras, B. Atrash, A.F. Lever, Nature, 1982, 299, 555

17a) H. Umezawa, Origin Ann. Rev. Microbiol., 1982, 36, 75

17b) T. Aoyagi, S. Kunimoto, H. Morishima, T. Takeuchi, and H. Umezawa, J. Antibiotics, 1971, 24, 687

18) M. Szelke, M. Tree, B.J. Leckie et al., J. Hypertension, 1985, 3, 13

19a) J. Boger, N.S. Lohr, E.H. Ulm, M. Poe, E.H. Blaine, G.M. Fanelli, T.Y. Lin, L.S. Payne, T.W. Schorn, B.I. LaMont, T.C. Vassil, I.I. Stabilito, D.F. Veber, D. Rich, and A.S. Bopari, Nature 1983, 303, 81

19b) J. Boger, L.S. Payne, D.S. Perlow, N.S. Lohr, M. Poe, E.H. Blaine, E.H. Ulm, T.W. Schorn, B.I. LaMont, T.Y. Lin, M. Kawai, D.H. Rich, and D.F. Veber, J. Med. Chem., 1985, 28, 1779

20) S. Thaisrivongs, D.T. Pals, D.W. Harris, M. Kati, and S.R. Turner, J. Med. Chem., 1986, 29, 2088

21a) R. Matsueda, Y. Yabe, H. Kogen, S. Higashida, H. Koike, Y. Iijima, T. Kokubu, E. Murakami, and Y. Imamura, Chem. Lett., 1985, 1041

21b) T. Kokubu, K. Hiwada, A. Nagae, E. Murakami, Y. Morisawa, Y. Yabe, H. Koike, and Y. Iijima, Hypertension 1986, 8 (Suppl. II), II-1

22) R.J. Arrowsmith, K. Carter, J.G. Dann, D.E. Davies, C.J. Harris, J.A. Morton, P. Lister, J.A. Robinson, and D.J. Williams, J. Chem. Soc., Chem. Commun., 1986, 755

23) N. Toda, M. Miyazaki, Y. Etho, T. Kubota, and K. Iizuka, Europ. J. Physiol., 1986, 129, 393

24) H.L. Sham, C.A. Rempel, H. Stein, and J. Cohen, J. Chem. Soc., Chem. Commun, 1987, 683

25) J.R. Luly, J.L. Plattner, H. Stein, N. Yi, J. Soderquist, P.A. Marcotte, H.D. Kleinert, and T.J. Perun, Biochem. Biophys. Res. Commun., 1987, 143, 44

26) C. Cazaubon, C. Carlet, R. Guégan, J.-P. Gagnol, and D. Nisato, Journal of Hypertension, 1986, 4 (Suppl. 6), 459

27) S.I. Foundling, J. Cooper, F.E. Watson, A. Cleasby, L.H. Pearl, B.L. Sibanda, A. Hemmings, S.P. Wood, T.L. Blundell, M.J. Valler, C.G. Norey, J. Kay, J. Boger, B.M. Dunn, B.J. Leckie, D.M. Jones, B. Atrash, A. Hallet, and M. Szelke, Nature, 1987, 327, 349

28) D.H. Rich, J. Med. Chem., 1985, 28, 263

29) M.G. Bock, R.M. Di Pardo, R.M. Evans, B.E. Freidinger, W.L. Whitter, L.S. Payne, J. Boger, E.H. Ulm, E.H. Blaine, D.F. Veber in C.M. Deber, V.J. Hruby and K.D. Kopple (Eds.) Peptides, Structure and Function. Proceedings of the Ninth American Peptide Symposium Pierce Chemical Co., Rockford, IL, 1986, p. 751

30) Inhibition of human plasma renin. Radioimmunoassay for angiotensin I was carried out at pH = 7.4 with a commercial kit (renin MAIA kit; Serono Diagnostics S.A., Coinsins, Switzerland)

31) Inhibition of hog renin was measured using a modified procedure reported by Klickstein, and Wintroub with tetradecapeptide as substrate. L.B. Klickstein, and B.U. Wintroub, Anal. Biochem., 1982, 120, 146

32) J.A. Fehrentz and B. Castro, Synthesis, 1983, 676

33) W. König, and R. Geiger, Chem. Ber., 1970, 103, 788

34) In a very recent paper potent retro-inverted peptides have been described by S.H. Rosenberg et al. using the more hydrophobic cyclohexyl side chain. S.H. Rosenberg, J.J. Plattner, K.W. Woods, H.H. Stein, P.A. Marcotte, J. Cohen, and T.J. Perun, J. Med. Chem., 1987, 30, 1224

35) M. Zaoral, Coll. Czech. Chem. Commun., 1962, 27, 1273

36) M. Alpegiani, A. Bedeschi, and E. Perrone, Gazz. Chim. Ital., 1985, 115, 393

Rational Design of Enzyme Inhibitors Containing Small Rings

C. J. Suckling

DEPARTMENT OF PURE AND APPLIED CHEMISTRY, UNIVERSITY OF STRATHCLYDE, THOMAS GRAHAM BUILDING, 295 CATHEDRAL STREET, GLASGOW G1 1XL, UK

1 INTRODUCTION

Selective and specific enzyme inhibition is more than a fashion in medicinal chemistry although the ways in which such precision of activity is achieved may be affected by changes in enthusiasm according to fashion or experience. Without the commercial pressure to discover drugs, it is possible for the academic scientist interested in medicinal chemistry to indulge in the relative luxury of developing a concept in enzyme inhibition at least to the point at which it is clear that the concept is a useful basis for the design of potential drugs. We have followed this path over the past six years making use of the chemistry of the cyclopropyl group.

The cyclopropane group offers several opportunities for the design of enzyme inhibitors[1]. It is alkene-like in many reactions, but less reactive. This reactivity is latent in most cyclopropane-containing compounds and in the action of specific enzyme inhibitors, the target enzyme interacts with the inhibitor to reveal its reactivity. In cyclopropane chemistry, the presence of a single electron-withdrawing substituent activates the small ring to nucleophilic attack and ring opening but only under extreme conditions. Facile ring opening by nucleophiles, which are the major source of reactivity at enzyme active sites, only takes place in the presence of acid catalysts, either Lewis[2] or Bronsted[3], or if the ring is doubly activated[4]. Such acid catalysis is available at the active sites of many enzymes.

Radical reactions are of particular importance in

the enzyme chemistry of cyclopropanes. Their place in enzyme inhibition centres on the reactions of enzymes capable of promoting single electron transfer leading to a cyclopropylalkyl radical or a heteroatom analogue, for example an aminium radical cation; such a radical then undergoes rapid ring opening to a further radical that interacts with the enzyme or coenzyme[5]. Cyclopropanones have figured in enzyme inhibitors through their ready addition of nucleophiles[6]. Our experiments into latent enzyme inhibition have chiefly concerned the electrophilic properties of cyclopropane derivatives as will be seen but some examples of radical chemistry will also be mentioned.

Several functional groups have been appropriated to form the basis for latent inhibitors of enzymes. Notable are the transformation of alkynes to allenes in fatty acid metabolism[7-9] and the elimination of halides to afford α,β-unsaturated carbonyl analogues[10,11]. An important characteristic of these functionalities is that they can be included in substrate analogues of many enzymes to set up latent inhibitors. Our principal strategy that we hoped would be similarly transposable to many enzymes is illustrated in Figure 1. A substrate analogue containing a cyclopropane ring bearing an electron withdrawing substituent undergoes reaction with the enzyme such that the polarisation is increased sufficiently for nucleophilic attack to occur by a group at the enzyme's active site. As will be described below, this strategy has now been successfully applied to nicotinamide-dependent dehydrogenases and to several peptidases. In contrast, in examples relating to radical chemistry, we have not observed time-dependent or irreversible inhibition of cytochromes P-450 by alkyl or aryl cyclopropanes.

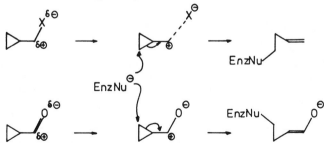

<u>Figure 1</u> The principal strategy used for enzyme inhibition by activation of cyclopropane rings to nucleophilic attack.

2 TEST CASES: NAD-DEPENDENT DEHYDROGENASES

In order to establish the principle of inhibition outlined in figure 1, we chose to investigate an enzyme with a broad substrate specificity, namely horse liver alcohol dehydrogenase (HLADH)[12,13]. A series of alkyl and cycloalkylcyclopropylmethanols (Table 1) was prepared and all were found to inhibit HLADH to some extent. The most effective were bicycloheptan-[4,1,0]-7-methanol (1) and the analogous bicyclohexane derivative which caused inhibition once every 100-200 turnovers.

	K_m $M \times 10^5$	k_{ox} s^{-1}	k_{in} $s^{-1} \times 10^3$	$\frac{k_{ox}}{k_{in}}$
1	1.92	1.70	13.5	126
2	6.38	4.72	5.13	920
3	6.89	4.24	4.33	980
4	6.3	5.64	1.33	4240
5	1.55	19.2	0.4	48000

Table 1 Inhibition of HLADH by some substituted cyclopropylmethanols[12,13].

Unlike inhibition of this enzyme by allylic alcohols[14], inhibition was irreversible. In general, the alcohols were much more effective inhibitors than the corresponding aldehydes and ketones. This observation together with the extremely low inhibitory power of the tetramethylcyclopropylmethanol, (5), suggested that inhibition takes place by nucleophilic attack on the apex of the cyclopropane away from the oxygen-bearing substituent at a stage in the oxidation pathway at which hydride has been substantially removed by NAD but before a carbonyl group has been fully formed (Figure 2). Nucleophilic attack and hydride removal would not be expected to be concerted[15]. Molecular modelling studies[15] suggested that the most likely nucleophile to be alkylated in the inhibition reaction is the hydroxyl group of serine-48 although this remains to be established.

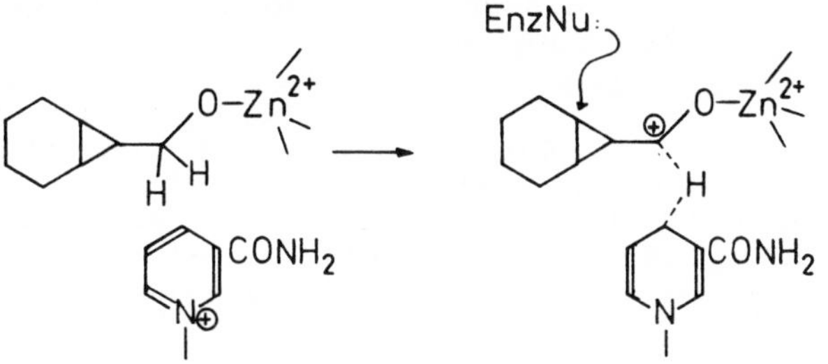

Figure 2 A possible mechanism for the inhibion of HLADH by cyclopropylmethanols.

The role of the active site zinc is also of interest in this reaction, especially in the context of metallopeptidase inhibitors discussed later. The charge on the zinc ion obviously contributes to the polarisation of the cyclopropane ring. An indication of the significance of the zinc ion was obtained in a reaction of the inhibitor (1) with HLADH in the presence of the reducing cofactor NADH; no oxidation could occur under these conditions. Although inhibition was not observed, we isolated trans-2-vinylcyclo-hexanol from a preparative reaction. The most likely explanation for the provenance of this compound is Lewis acid catalysed hydrolysis of (1) at the enzyme's active site (Figure 3); the catalysis is clearly strong enough to promote carbon-oxygen bond cleavage.

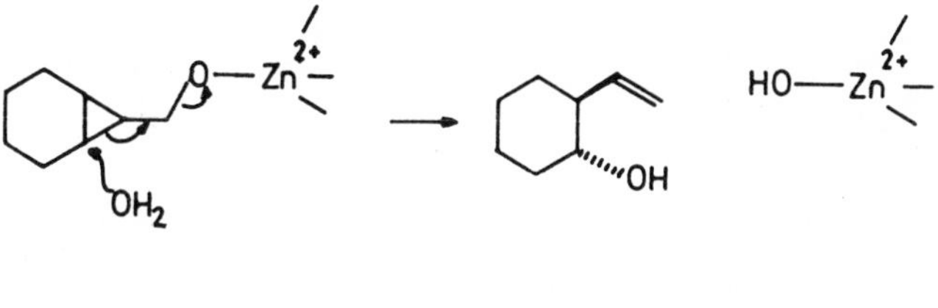

Figure 3 HLADH acting as a hydrolase and a lyase.

If the effect of a Lewis acid were dominant in activating a cyclopropane ring at an enzyme's active site, then perhaps few enzymes would be susceptible to inhibition by weakly electrophilic cyclopropanes. We quickly found, however, that Lewis acid catalysis was not essential for inhibition. Lactate dehydrogenase (LDH) will accept a small number of lactic and pyruvic acids as substrates and accordingly we investigated the obvious substrate analogues (**6**) and (**7**)[12] (Figure 4). LDH employs a histidine residue at the active site to deprotonate the alcohol to be oxidised and no covalent interaction is additionally available to enhance the polarisation of the cyclopropane ring. Nevertheless, irreversible inhibition was observed with the lactic acid analogue (**6**) but not with its oxidised analogue (**7**). The hydroxyacid (**6**) behaved as a substrate (K_m 0.72 mM, k_{ox} 3.83 s^{-1}) and, less rapidly, as an inhibitor (k_i 0.014 s^{-1}). On the other hand, the ketoacid (**7**) was a respectable substrate (K_m 3.73 mM, k_i 61.7 s^{-1}) but time-dependent inhibition was not detectable. The similarity of behaviour with the HLADH inhibitors with regard to the reactivity of alcohols and ketones suggests that a similar inhibition mechanism operates. In this case, the most likely candidate nucleophile is the hydroxyl group of threonine-146[15].

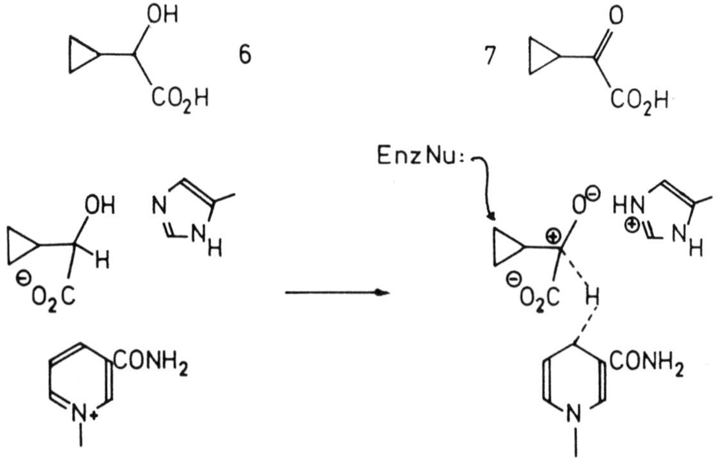

Figure 4 Inhibition of lactate dehydrogenase by cyclpropyl analogues of lactic and pyruvic acids.

3 CYTOCHROMES P-450

The studies of inhibition of nicotinamide dependent dehydrogenases above were carried out in parallel with mechanistic investigations[1,16] and we were led by a similar duality of purpose to examine the properties of aryl cyclopropanes in their reactions with cyctochromes P-450. The extensive work of Ortiz de Montellano in particular has shown that alkynes and alkenes serve as the precursors of radicals that inhibit cytochromes P-450[8,17]. It was interesting to see whether the reactions of cyclopropanes paralleled those of alkenes in this situation. We studied firstly the reactions of aryl cyclopropanes with cytochromes P-450 from rabbit and rat liver. Surprisingly cyclopropylbenzene and 1,2-diphenylcyclopropane proved to be acceptable substrates for the enzyme; cyclopropyl benzene was kinetically competitive with respect to a typical assay substrate, ethoxycoumarin, (K_i 57 nM) and 1,2-diphenylcyclopropane was kinetically non-competitive ($K_i$app 11 μM)[18,19]. No time-dependent inhibition properties were observed with either substrate. Further, we found that benzoic acid was a major product of both reactions and with 1,2-diphenylcyclopropane as substrate, phenylacet-aldehyde also was detected (Figure 5). The enzyme thus brings about an oxidative cleavage of the cyclopropane ring with these substrates. The course of this unusual reaction is not obvious but can be understood if the

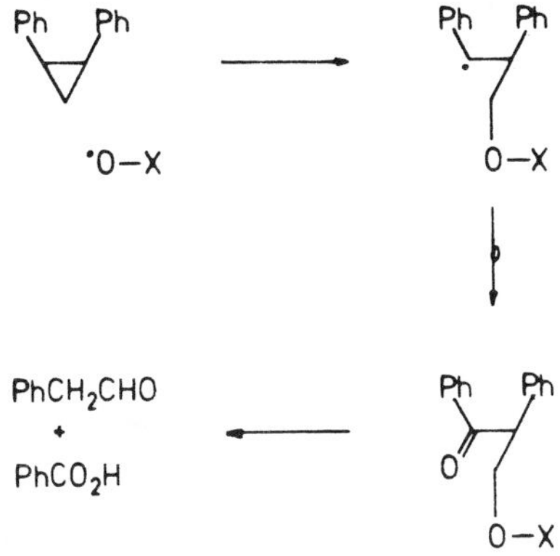

Figure 5 Oxidative cleavage of 1,2-diphenylcyclopropane by cytochromes P-450.

substrate remains bound to cyctochromes P-450 after
initial oxidation through a suitable group at the
active site (represented by X in Figure 5). Since a
substrate remains at the active site, further oxidation
to a carbonyl-contaning intermediate might be expected
and susbsequent oxidative cleavage of the intermediate
(**8**) leads to the observed products.

In contrast to the ring opening observed with
arylcyclopropanes and non-specific cytochromes P-450,
the hydroxylation of alkylcyclopropanes was found to
occur with retention of the cyclopropane ring[20]. The
specific enzyme, cholesterol 7-α-hydroxylase accepted
the methanocholestane (**8a**) as a substrate hydroxylating
it at the 7-position to give (**8b**) without evidence for
substantial inhibition. In agreement with our results,
Ortiz de Montellano reported that methylcyclopropane
was hydroxylated by cytochrome P-450 from rat liver
without inhibition; indeed the only reported case of
this enzyme´s being inhibited by an alkyl or aryl
cyclopropane is with bicyclo[2,1,0]pentane which is
highly strained[21]. This lack of inhibition has been
explained by the evident rapidity of the recombination
of the haem bound oxygen atom and an intermediate alkyl
radical, the so-called ´oxygen rebound´, in the
hydroxylation reaction[20,21]. Nevertheless, alkyl
cyclopropanes have been found to inhibit dopamine
β-hydroxylase[22] but in general, they do not seem to be
suitable as latent inhibitors of cytochrome P-450. In
contrast, nitrogen and oxygen substituted cyclopropanes
are well-known inhibitors of oxidising enzymes[5,23].

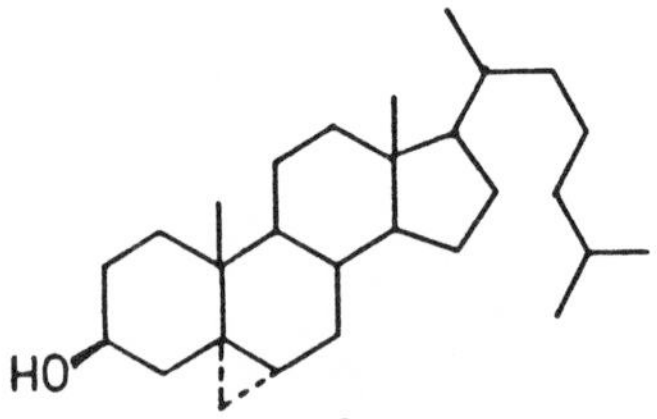

8a

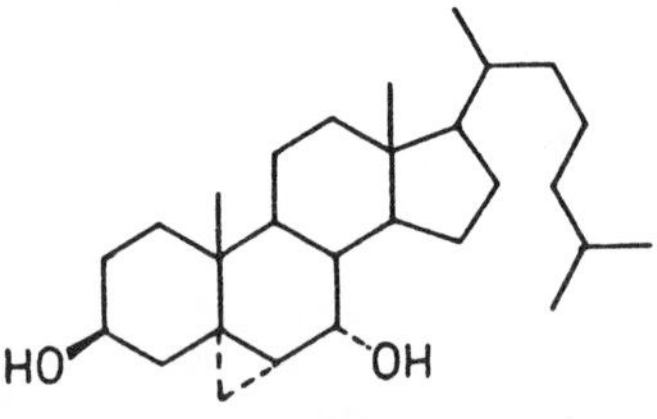

8b

4 DIHYDROOROTATE DEHYDROGENASE

In the latter part of this account, we turn attention to several enzymes of chemotherapeutic significance. Dihydroorotate dehydrogenase (DHODase, Figure 6a)[24] is a key enzyme in the <u>de novo</u> biosynthesis of pyrimidines and is thus a potential target for the action of antiparasitic and anticancer drugs especially. At the start of our work, it was known that barbituric acid was an inhibitor of the enzyme and the obvious cyclopropane-containing analogue was the spirocyclopropyl derivative (**9**, R = H). We have found that this compound is an irreversible inactivator of the enzyme from <u>Clostridium oroticum</u>[25] but so far we have no information about the mechanism of the reaction. One possibility is that (**9**), being a doubly activated cyclopropane, is capable of a direct reaction at the enzyme's active site.

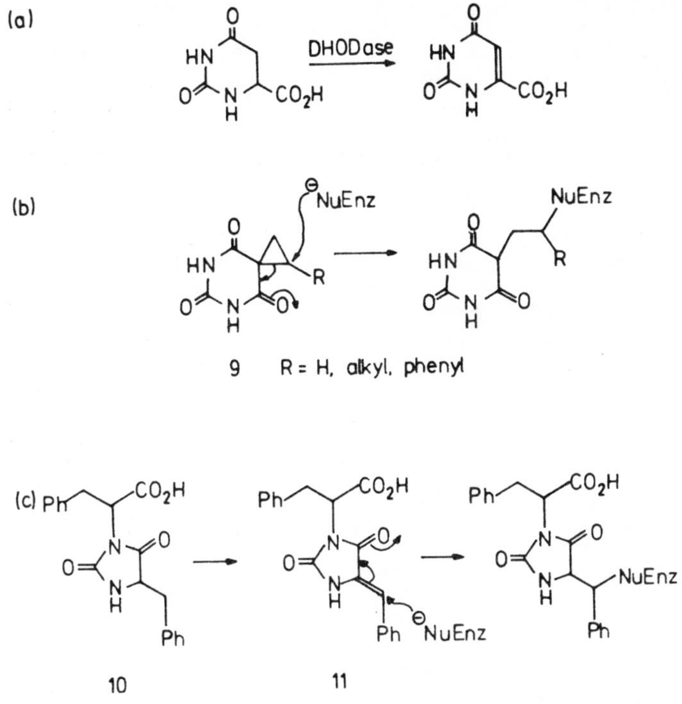

(a)

(b)

9   R = H, alkyl, phenyl

(c)

10   11

Figure 6 Inhibition of DHODase by spirocyclopropyl barbiturates (**9**), R = H, n-Pr, i-Pr, i-Bu, Ph and by hydantoin (**10**).

Alternatively, an electron transfer mechanism in which the flavin prosthetic group takes part could be considered; in this case, cyclopropylalkyl-type radicals would form and inhibition could occur through the ring opened radical.

We have also found that hydantoins derived from phenylalanine (10) are inhibitors of this enzyme[26]. The half life of DHODase (0.7 μM) in the presence of (10, 13 mM) was 40 min. Our working hypothesis is that a dehydrogenation takes place leading to the α, β-unsaturated compound (11) which alkylates the enzyme. It seems clear that the benzene rings have an important role in this reaction. In an effort to correlate the hydantoin series of inhibitors with the cyclopropane-containing compounds, we have prepared a further series of alkyl and aryl substituted spirocyclopropyl barbituric acids (9). All of the substituted cyclopropylbarbiturates were found to inhibit the enzyme[25] with similar potency to the hydantoin (10). For example for (9, R = phenyl, 6.7 mM) the half life of DHODase (0.5 μM) was 41 min. The n-propyl, i-propyl, and unsubstituted derivatives were somewhat poorer inhibitors and further detailed kinetic data is being obtained to clarify the structure-activity relationship.

A non-enzymic observation that may be relevant to the inhibition of DHODase is that the barbiturate (9, R = phenyl) underwent nucleophilic ring opening on recrystallisation from methanol, attack by methoxy taking place at the benzylic carbon (cf. Figure 6, R = phenyl, ´NuEnz´ = MeOH). In contrast, the isopropyl analogue was stable to recrystallisation from methanol and was recovered analytically pure intact. Nevertheless this reaction illustrates the increased reactivity of doubly activated cyclopropanes. Such increased reactivity may, of course, lead to reduced selectivity in inhibition. Although we have found that the hydantoin (10) does not inhibit alcohol or lactate dehydrogenases, we have no information so far concerning the selectivity of the spirocyclopropyl barbiturates (9).

It could be argued that the C. oroticum enzyme is not a good model for chemotherapeutically significant DHODases since it relies upon NAD as a terminal oxidant and does not use an electron transport chain. However we have found that the hydantoin (10) that is active in our enzyme screen is the only one of this series of

compounds to be active against an intact malarial
parasite[27]; this gives us good grounds for continuing
to use the readily available model from C. oroticum.

5   DIHYDROFOLATE REDUCTASE

The position of dihydrofolate reductase (DHFR) as
a target for many classes of enzyme-inhibiting drugs is
well documented[28]. However, the vast majority of DHFR
inhibitors   are   2,4-diaminopyrimidines   that   are
competitive   with   the   natural   substrate.   From   a
combination of X-ray crystallographic[29], nmr[30], and
kinetic studies[31] it has become clear that diamino-
pyrimidine   inhibitors   are   protonated   at   N-1   by   an
aspartic   acid   residue   at   the   active   site   of   DHFR
(Figure 7). In the reaction of dihyrofolate itself,
protonation must occur at N-5 to activate the imine to
reduction   by   NADPH.   Site - specific   mutagenesis
experiments   have   shown   that   the   aspartic   acid   is
essential for effective catalysis[32]. Combining all of
these   strands   of   evidence   led   to   the   design   of   a
cyclopropane-containing irreversible inhibitor of DHFR
in which the cyclopropane is activated by protonation
at N-5 of the pteridine ring as in (12) (Figure 7b).
This   is   a   further   embodiment   of   our   strategy   of
increasing the polarisation of a substituent adjacent
to a cyclopropane ring to increase the electrophilicity
of the cyclopropane. We further reasoned that it was
possible   in   principle   to   extend   the   concept   to
2,4-diaminopyrimidines   such   as   the   pyrimethamine
analogue (13) through protonation at N-1 as suggested
by mechanistic studies[29-31] (Figure 7c).

The pteridine (12) and pyrimidine (13) were duly
synthesised, although not without some difficulty[33],
and tested as inhibitors of DHFR from E. coli. A stark
contrast between the behaviour of the two was evident.
Whereas   the   pyrimidine   (13)   was   a   competitive
inhibitor (K_i 10^{-7}M), the pteridine (12) showed
time-dependent   inhibition   with   good   first   order
kinetics and a rate constant for inhibition at 25°C of
1.4 x 10^{-4} s^{-1}. The contrast was fascinating and we
attempted to evaluate the results with the aid of
molecular graphics. A model of the active site of DHFR
was built from available crystal structures of the
methotrexate complexes[34] and the pyrimidine inhibitor
inserted into the active site in a manner directly
analogous to that of methotrexate so that N-1 is
protonated by the aspartic acid. In this orientation
(Figure 8), the 4-chlorophenyl and cyclopropyl groups

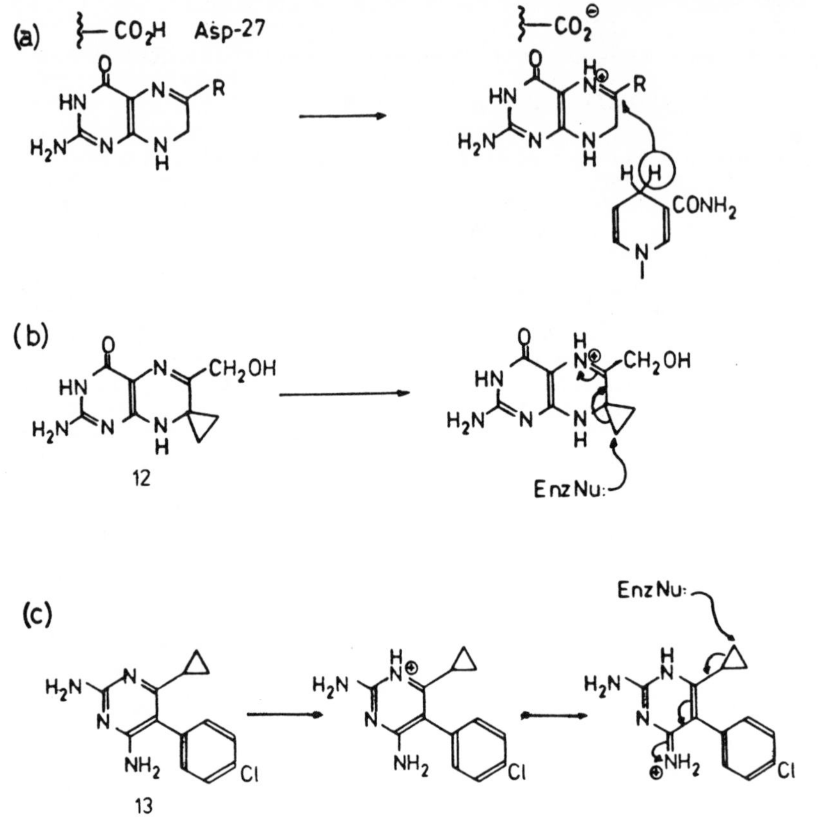

Figure 7 Mechanism based design of DHFR inhibitors.

of (13) fit into a well recognised non-polar pocket
that can accommodate substituents as large as
adamantyl[35]. There are, however, no nucleophiles
readily available within range of the cyclopropane
ring; the pyrimidine (13) binds in a nucleophile-free
zone. On the other hand, the pteridine (12) must find
a suitably reactive nucleophile to cause the observed
time-dependent inhibition. The way in which this
pteridine binds to DHFR is ambiguous because it
contains neither a hydrophobic group nor a
4-aminobenzoylglutamate side chain that would dictate
the likely binding conformation. One can only
speculate, therefore, that the cyclopropane ring could
be attacked by nucleophiles such as threonine-113 or
tyrosine-100 (E.coli numbering, Figure 8).

To investigate these points further, we are synthesising additional cyclopropane-containing pteridines. Whatever the detailed mechanism turns out to be, the pteridine (**12**) is to our knowledge, the first mechanism-based inhibitor of DHFR and the first inhibitor with the natural amino-oxo substitution pattern in the pyrimidine ring; as such it opens the way to the design of a new generation of DHFR inhibitors.

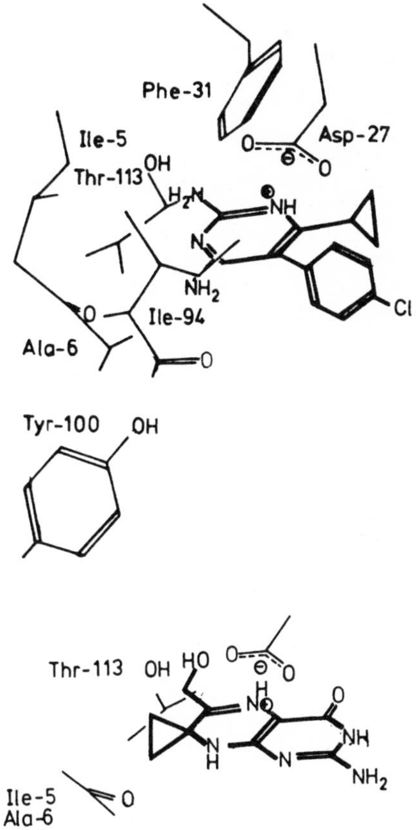

<u>Figure 8</u> Schematic representation of the binding of inhibitors (**13**) and (**12**) to DHFR in orientations modelled on the binding of methotrexate to the <u>E.coli</u> enzyme.

6   INHIBITORS OF PEPTIDASES

The significance of peptidases as targets for drugs was perhaps most forcibly demonstrated by the discovery of the angiotensin converting enzyme inhibitor, Captopril[35]. It was natural, therefore, that we should attempt to apply our concept of selective enzyme inhibition to one of the best characterised metallopeptidases, carboxypeptidase A. Subsequently we have examined representatives of the other main classes of peptidase, serine, cysteine and acid peptidases, and the results of our survey are outlined in the following paragraphs. Unlike DHFR in which we have an inhibitor without a nucleophile, there is a profusion of nucleophiles available especially for those enzymes that catalyse hydrolysis via an acyl enzyme intermediate. If cyclopropane ring opening occurs, it could take place before or after acylation. It is also possible that a long-lived tetrahedral intermediate could form in which the intact cyclopropane ring interferes with attack of water on the carbonyl group in the deacylation step.

Inhibitors of Carboxypeptidase A.

The design of peptidase inhibitors by our strategy follows naturally from the Lewis acid and proton activation believed to be important in HLADH and DHFR respectively. Thus coordination of the carbonyl group of the scissile peptide bond to the active site zinc of carboxypeptidase A (Figure 9) will activate it to nucleophilic attack in a manner resembling the ring opening of methylcyclopropylketone catalysed by zinc dibromide[2]. Alternatively activation by protonation is possible and together, the two strategies should cover all typical peptidases.

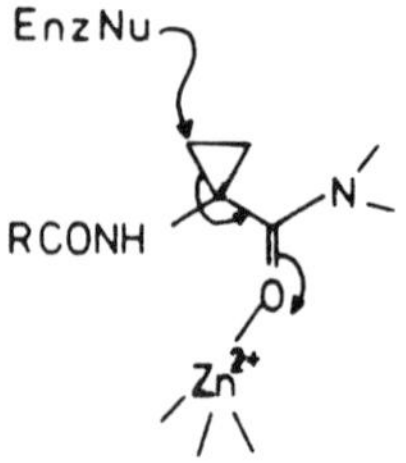

Figure 9 Design of peptidase inhibitors.

For carboxypeptidase A, the obvious initial target is the dipeptide (**14**) built from phenylalanine and 1-aminocyclopropane carboxylic acid. A series of such compounds was prepared with Gly, Phe, and Pro as the C-terminal residues and a range of N-acyl groups including benzoyl, carbobenzoxy, and phenylacetyl. We also investigated simple cyclopropyl carboxamides of these amino acids[36]. The Phe derivative (**14**) was found to be an irreversible time-dependent inhibitor of carboxypeptidase A ($t_{1/2}$ of carboxypeptidase [9.8×10^{-8}M] 4.5 min in the presence of hippurylphenylalanine [0.7 mM] and (**14**) [0.7 mM] at 25°C). The Phe peptide was also a substrate[37], the inhibition reaction occurring about 2.3 times faster than hydrolysis under the conditions used. Proline derivatives are not usually good substrates for carboxypeptidase A and we were surprised to find that the proline derivative (**15**) was a good time-dependent inhibitor ($t_{1/2}$ 3 min under the above conditions) however with no substrate properties as expected. Neither the glycine analogue (**16**) nor the simple cyclopropane carboxamide (**17**) was a time-dependent inhibitor of carboxypeptidase A although they did bind to the enzyme (K_i 0.32 and 0.83 mM respectively). If the binding determinants of Captopril had been sufficient for this type of inhibition, the simple amide (**17**) might have been expected to be effective. We have found, however, that of the compounds so far investigated, only the benzamides have time-dependent and irreversible properties. The substrate used in the enzyme assays was hippurylphenylalanine and these results prompted us to consider how the interaction of the active inhibitors might compare with that of the substrate at the enzyme's active site.

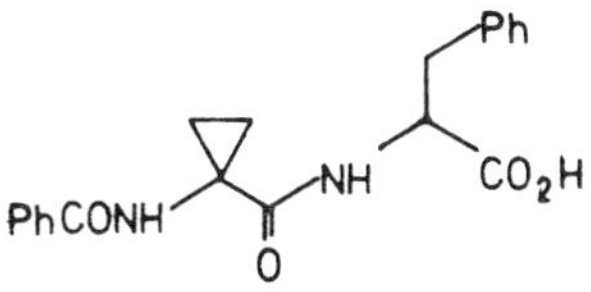

14

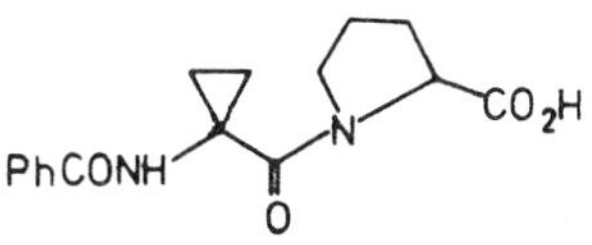

15

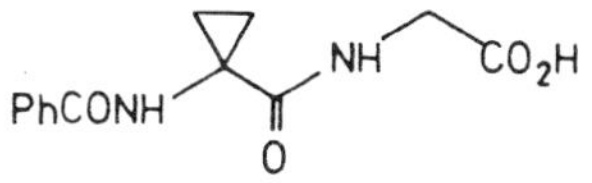

16

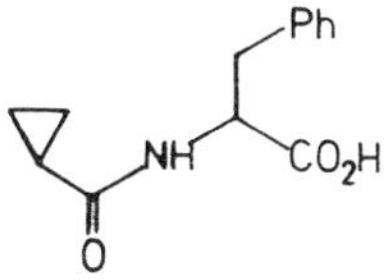

17

Taking the X-ray structure of the glycyltyrosine complex of carboxypeptidase A[38] as the basis for the model, hippurylphenylalanine was introduced into the active site maintaining the positions of the aryl ring, the C-terminal carboxylate, and the carbonyl group of the scissile peptide bond. The Phe inhibitor (**14**) is simply the cyclopropane analogue of the substrate and once the latter has been satisfactorily fitted into the active site, it remains only to place the cyclopropane ring in an appropriate conformation. This was done with regard to the known conformational preference for the interaction of a carbonyl group and a cyclopropane ring[1,39]. Together, these considerations led to the model shown in Figure 10 for the interesting case of the Pro inhibitor (**15**). The immediate suggestion of this model was that the most likely nucleophile to attack the cyclopropane ring was the carboxylate of Glu-270.

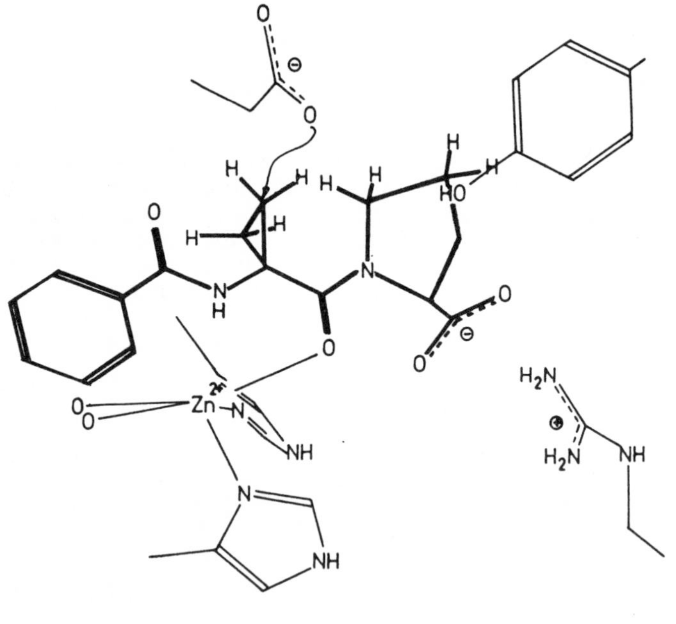

Figure 10 Model of active site of carboxypeptidase A with the inhibitor (**15**) introduced in place of the crystallographic substrate, glycyltyrosine.

A further significant suggestion was that the lack of substrate activity of the Pro inhibitor was due to the inaccessibility of the carbonyl group to attack by nucleophiles; the three and five membered rings form a protective barrage. In effect, the Pro inhibitor was behaving not so much as an amide but as a ketone. This molecular graphics investigation does not, of course, prove the mechanism of inhibition but it was important in suggesting further experiments, in particular, the possibility of using nonpeptide inhibitors[40].

The simplest ketonic inhibitors to examine were the methyl and benzyl ketones (18) and (19); the former is highly truncated and, if the mechanism proposed is sound, may be expected to be a general inhibitor of many peptidases at sufficiently high concentratations. We found that the methyl ketone (18) had very similar kinetic properties to the Pro inhibitor (15). The benzyl ketone was a slightly more effective compound presumably because it has some affinity for the hydrophobic binding pocket of the enzyme.[40]

The extensions of the original inhibition observation on carboxypeptidase A are not yet exhausted. Recalling the unusual hydrolysis reaction of bicycloheptan[4,1,0]-7-methanol (1) catalysed by HLADH mentioned earlier, the possibility that the secondary alcohols derived from the ketones (20) and (21) might also be inhibitors is interesting to investigate.

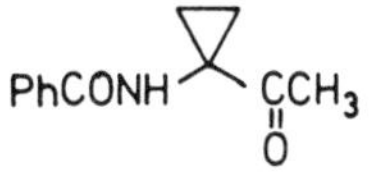

PhCONH—CCH₃
18

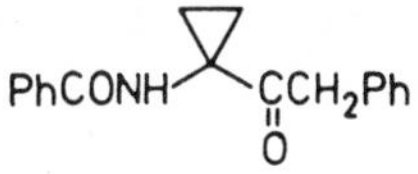

PhCONH—CCH₂Ph
19

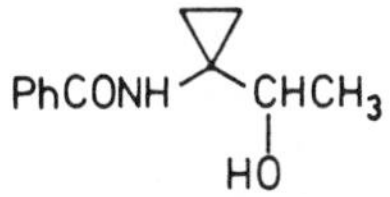

PhCONH—CHCH₃
20

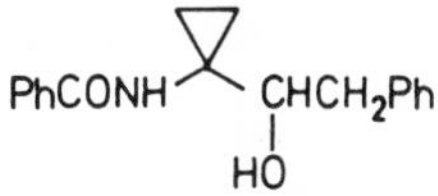

PhCONH—CHCH₂Ph
21

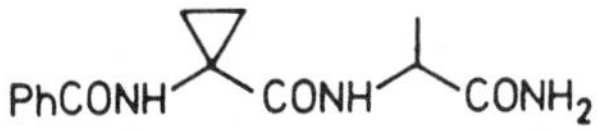

PhCONH—CONH—CONH₂
22

This idea is given enhanced significance in view of the use of secondary alcohols as transition state analogues in peptidase inhibitors[41]. Preliminary experiments have confirmed this expectation and the secondary alcohol (21) was found to be the most reactive, about twice as active as any of the other peptide analogues with a half life of 90 min. at a concentration of 1 mM for the inhibition of carboxypeptidase A (3.9 x 10^{-7} M) at 25°C.

It is notable that none of the compounds showed significant inhibition of metallopeptidases with different substrate specificities such as the carboxypeptidase B-type enzyme enkephalin convertase or of endopeptidases such as endopeptidase 24.11[42].

Serine, Cysteine, and Acid Peptidases

Although the most probable type of peptidase to be inhibited by our strategy was exemplified by carboxyxpeptidase A based upon the experience with alcohol dehydrogenase, it was obviously of interest to determine the extent to which the strategy could be applied. We have carried out some preliminary screening experiments on chymotrypsin, papain, and pepsin as representatives of the serine, cysteine, and acid peptidases respectively using the available ketones (18), and (19) and the dipeptide amide (22). Since none of these compounds contains groups appropriate to primary recognition sites, high activity was not expected; all that was sought in this screen was an indication of time-dependent inhibition or irreversibility.

The results obtained so far suggest that the ketones inhibit chymotrypsin weakly but more needs to be done to determine the nature of the interaction. The dipeptide (22) is, not surprisingly, a poor substrate for this enzyme but it inhibits in a time-dependent manner. Inhibition was, however, reversed by gel filtration. The most positive indication of inhibition was with papain. Although the ketones were not time-dependent inhibitors, the dipeptide (5.5 mM) inhibited papain (0.34 mM) with a half life of 50 min. at 20°C. Activity was not recovered on gel filtration. It is, however, possible that the active site thiol could have added to the carbonyl group of the ketones as has been observed for aldehydes[44] and this possibility will have to be considered in further experiments. Pepsin, although not believed to act through nucleophilic catalysis, has a potential nucleophile available at the active site in the form of the carboxylate anion of aspartate-32 which is known to react with epoxides[45]. It is also a relative of renin, another significant target for drugs aimed at the control of blood pressure[46]. We found that the dipeptide (22) was an inhibitor that reacted in a time-dependent manner with pepsin but that the ketones (18) and (19) were unreactive. The detailed kinetics of this reaction have not yet been investigated.

There is therefore a little evidence to suggest that all of the four main classes of peptidase are susceptible to the enzyme-activated cyclopropane strategy for inhibitor design. It is now necessary to learn how to tailor the selectivity to attack enzymes of chemotherapeutic significance. This selectivity can clearly be sought by empirical synthesis or by molecular modelling. However before it is reasonable to attempt extended molecular modelling studies, it is important to establish the mechanisms of the reactions described beyond the present merely conceptual understanding.

7 CONCLUSIONS

The extent to which the concept of cyclopropanes in enzyme inhibition can be stretched has more than equalled our expectations from the time when the first suggestion of the approach was made in a discussion with an able graduate student. In the coming months we shall be paying closer attention to the mechanism of the reactions that we have discovered. In the coming years we hope to develop compounds built from this or related strategies with the ability to irreversibly inhibit enzymes of chemotherapeutic importance with high specificity.

ACKNOWLEDGEMENTS

I thank the many colleagues who have contributed to the work described; their names are cited in the references. Financial support from SERC, MRC, Wellcome Research, Burroughs Wellcome, and the Smith and Nephew Foundation is also acknowledged.

REFERENCES

1. C.J. Suckling, <u>Angew.Chem.Int.Edn.Engl.</u>, in press.

2. J.P. McCormick and D.L. Barton, J.Chem.Soc.,
    Chem.Commun., 1975, 303.
3. M. Julia, S. Julia, and R. Guegan, Bull.Chem.Soc.
    Fr., 1960, 1072; S.F. Brady, M.I. Ilton, and W.S.
    Johnson, J.Am.Chem.Soc., 1968, **90**, 2822.
4. S.Danishevsky, Accts.Chem.Res., 1979, **12**, 66; R.V.
    Stevens, Pure Appl.Chem., 1979, **51**, 1317.
5. for recent examples R.B. Silverman and P.A. Zieske,
    Biochem.Biophys.Res.Commun., 1986, **135**, 154; J.S.
    Wiseman, J.S. Nichols, and M. Kolpak, J.Biol.Chem.,
    1982, **257**, 6328; B. Sherry and R.H. Abeles,
    Biochemistry, 1985, **24**, 2594.
6. P. Dowd, C. Kaufman, and R.H. Abeles, J.Am.Chem.
    Soc,. 1984, **106**, 2703; J.S. Wiseman, G. Tayrien,
    and R.H. Abeles, Biochemistry, 1980, **19**, 4222.
7. K. Bloch, Accts.Chem.Res., 1969, **2**, 1971.
8. for example E.J. Corey, M. D´Alarco and S.P.T.
    Matsuda, Tetrahedron Lett., 1986, **27**, 3585; P.R.
    Ortiz de Montellano, B.A. Mico, J.M. Mathews, K.L.
    Kunze, G.T. Miwa, and A.Y.H. Lu, Arch. Biochem.
    Biophys., 1981, **210**, 717.
9. for example R.J. Auchus and D.F. Covey,
    Biochemistry, 1986, **25**, 7288; B.W. Metcalf, C.L.
    Wright, J.P. Burkhardt, and J.O. Johnson, J.Am.
    Chem.Soc., 1981, **103**, 3221.
10. for example I.A. McDonald, J.M. Lacoste, P. Bey, J.
    Wagner, M. Zreika, and M.G. Palfreyman, Bioorg.
    Chem., 1986, **14**, 103; R.B. Silverman and B.J.
    Invergo, Biochemistry, 1986, **25**, 6817.
11. W. Boisvert, K.S. Cheung, S.A. Lerner, and M.
    Johnston, J.Biol.Chem., 1986, **261**, 7871; C. Walsh,
    R. Badet, E. Daub, N. Esaki, and N. Galakatos, in
    Third SCI-RSC Medicinal Chemistry Symposium, ed.
    R.W. Lambert, Royal Society of Chemistry, London,
    1986, p. 193.
12. I.MacInnes, D.C. Nonhebel, S.T. Orszulik, C.J.
    Suckling, and R. Wrigglesworth, J.Chem.Soc.,Perkin
    Trans.1, 1983, 2771.
13. D. Laurie, E. Lucas, D.C. Nonhebel, and C.J.
    Suckling, unpublished results reviewed in ref. 1.
14. I. MacInnes, D.E. Schorstein, C.J. Suckling, and R.
    Wrigglesworth, J.Chem.Soc.,Perkin Trans.1, 1981,
    1103.
15. R.J. Breckenridge and C.J. Suckling, Tetrahedron,
    1986, **42**, 5665.
16. I. MacInnes, D.C. Nonhebel, S.T. Orszulik, and C.J.
    Suckling, J.Chem.Soc.,Perkin Trans.1, 1983, 2277;
    D. Laurie, E. Lucas, D.C. Nonhebel, C.J. Suckling,
    and J.C. Walton, Tetrahedron, 1986, **42**, 1035.
17. P.R. Ortiz de Montellano, K.L. Kunze, H.S. Beilan,

and C. Wheeler, Biochemistry, 1982, **21**, 1331; K.L.
Kunze, B.L.K. Mangold, C. Wheeler, H.S. Beilan, and
P.R. Ortiz de Montellano, J.Biol.Chem., 1983, **258**,
4202; P.R. Ortiz de Montellano, B.L.K. Mangold, C.
Wheeler, K.L. Kunze, and N.O. Reich, ibid., 4208.

18. K.E. Suckling, C.G. Smellie, I.E. Ibrahim, D.C.
    Nonhebel, and C.J. Suckling, FEBS Letters, 1982,
    145, 179.

19. K.E. Suckling, C.R. Wolf, L. Brown, D.C. Nonhebel,
    and C.J. Suckling, Biochem.J., 1985, **232**, 199.

20. J.D. Houghton, S.E. Beddows, K.E. Suckling, L.
    Brown, and C.J. Suckling, Tetrahedron Lett., 1986,
    27, 4655.

21. P.R. Ortiz de Montellano and R.A. Stearns, J.Am.
    Chem.Soc., 1987, **109**, 3415.

22. P.F. Fitzpatrick and J.E. Villafranca, J.Am.Chem.
    Soc., 1985, **107**, 5022.

23. T.L. Macdonald, K. Zirvi, L.T. Burka, P. Peyman,
    and F.P. Guengerich, J.Am.Chem.Soc., 1982, **104**, 2050;
    R.P. Hanzlik and R.H. Tullman, ibid, 2048.

24. T.W. Kensler and D.A. Cooney, Adv.Pharmacol.
    Chemother., 1981, **18**, 273.

25. I.G. Buntain, J.C. Courtney, W. Fraser, C.J.
    Suckling, and H.C.S. Wood, unpublished results.

26. I.G. Buntain, C.J. Suckling, and H.C.S. Wood,
    J.Chem. Soc.,Chem.Commun., 1985, 242.

27. W. Gutteridge, personal communication.

28. ´Inhibition of folate metabolism in chemotherapy´,
    ed. G.H. Hitchings, Springer Verlag, Berlin, 1983.

29. J.T. Bolin, D.J. Filman, D.A. Mathews, and R.C.
    Hamlin, J.Biol.Chem., 1982, **257**, 13650; D.J.
    Filman, D.A. Mathews, and J. Kraut, ibid., 13663.

30. G.C.K. Roberts in ´Chemistry and Biology of
    Pteridines´, ed. J. Blair, de Gruyter, Berlin,
    1983, p. 197.

31. S.R. Stone and J.F. Morrison, Biochemistry, 1984,
    23, 2753.

32. J.E. Villafranca, E.E. Howell, D.H. Voet, M.S.
    Strobel, R.C. Ogden, J.N. Abelson, and J. Kraut,
    Science, 1983, **222**, 782.

33. J. Haddow, C.J. Suckling, and H.C.S. Wood, J.Chem.
    Soc.,Chem.Commun., 1987, 478.

34. V.Cody, J.Mol.Graphics, 1986, **4**, 69; C.D.
    Selassie, Z.-X. Fang, R.R. Li, C. Hansch, T. Klein,
    R. Langridge, and B.T. Kaufmann, J.Med.Chem., 1986,
    29, 621.

35. M.A. Ondetti and D.W. Cushman, Annu.Rev.Biochem.,
    1982, **51**, 283.

36. S.K. Ner, Ph.D. Thesis, University of Strathclyde,
    1986.

37. S.K. Ner, C.J. Suckling, A.R. Bell, and R. Wrigglesworth, J.Chem.Soc.,Chem.Commun, 1987, 480.
38. D.C. Ress and W.N. Lipscomb, Proc.Nat.Acad.Sci.USA, 1983, **80**, 7151.
39. A. DeMeijere, Angew.Chem.Int.Edn.Engl., 1979, **18**, 809.
40. E. Lucas, L. Rees, and C.J. Suckling, unpublished results.
41. for example E.M. Gordon, J.D. Godfrey, J. Pluseac, D. VanLangen, and S. Natarajan, Biochem.Biophys. Res. Commun., 1985, **126**, 419; J.G. Dann, D.K. Stammers, C.J. Harris, R.J. Arrowsmith, D.E. Davies, G.W. Hardy, and J.A. Morton, ibid., 1986, **134**, 71.
42. S.H. Snyder, personal communication; A.J. Turner, personal communication. Thanks to both gentlemen.
43. J.C. Powers, J.W. Harper, K. Hemmi, A. Yasutake, and H. Hori, in Third SCI-RSC Medicinal Chemistry Symposium, ed. R.W. Lambert, Royal Society of Chemistry, London, 1986, p.241.
44. M.P. Gamscik, J.P.G. Malthouse, W.V. Primrose, N.E. Mackenzie, A.S.F. Boyd, R.A. Russell and A.I. Scott, J.Am.Chem.Soc., 1983, **105**, 6324.
45. B.M. Dunn, and A.L. Fink, Biochemistry, 1984, **23**, 5241; D.H. Rich, M.S. Beratowicz, and P.G. Schmidt, J.Am.Chem.Soc., 1982, **104**, 3535; P.G. Schmidt, M.W. Holloday, F.G. Salituro and D.H. Rich, Biochem.Biophys.Res.Commun., 1985, **129**, 597; A.K. Newmark and J.R. Knowles, J.Am.Chem.Soc., 1975, **97**, 3557.
46. J. Boger, in Third SCI-RSC Medicinal Chemistry Symposium, ed. R.W. Lambert, Royal Society of Chemistry, London, 1986, p. 271.

Amino-sugar Derivatives and Related Compounds as Glycosidase Inhibitors

G. W. J. Fleet

THE DYSON PERRINS LABORATORY, UNIVERSITY OF OXFORD,
SOUTH PARKS ROAD, OXFORD OX1 3QY, UK

Replacement of the pyranose oxygen of a sugar by nitrogen leads to a class of specific and powerful glycosidase inhibitors. Thus the naturally occurring nojirimycin (1),[1,2] nojirimycin B (2),[3] and galactostatin (3)[4] are inhibitors of glucosidases, mannosidases[5] and galactosidases[6] respectively. Removal

$$(1)$$
nojirimycin

$$(2)$$
nojirimycin B

$$(3)$$
galactostatin

$$(4)$$
deoxynojirimycin

$$(5)$$
deoxymannojirimycin

$$(6)$$
iminogalactitol

of the anomeric hydroxyl group gives a more chemically stable class of compound such as (4), (5) and (6) with powerful and specific glycosidase inhibitory properties.[7,8] Several naturally occurring azafuranose derivatives have also been shown to be enzyme inhibitors.[9,10] It has also been found that synthetic analogues of other aza-sugars provide opportunities for the design of specific inhibitors for other

glycosidases; thus the azapyranose analogues of N-
acetylglucosamine (7) and of fucose (9) are powerful and
specific inhibitors of a number of β-N-
acetylglucosaminidases[11] and α-L-fucosidases[12]
respectively. Also, 1,4-dideoxy-1,4-imino-D-mannitol
(DIM) (10), the azafuranose analogue of mannose, is an
inhibitor of several α-mannosidases.[13,14]

(7)   (9)   (10)

Bicyclic compounds containing polyoxygenated piperidines
and pyrrolidines such as castanospermine and swainsonine
are also glycosidase inhibitors. Additionally several of
these compounds have been indicated to possess potential
as chemotherapeutic agents. Thus a derivative of
homonojirimycin may be a candidate for an orally active
agent for the treatment of diabetes mellitus[15] and
swainsonine may have promise in aspects of cancer
chemotherapy;[16,17] castanospermine has been reported as
inhibiting the growth of HIV.[18] Also, studies on the
effects of castanospermine on insulin receptor
biosynthesis have provided evidence that glucose removal
from core oligosaccharides represents an important
signal in the translocation and rate of processing of
the insulin receptor to the plasma membrane.[19]
    The synthetic problem for the preparation of
polyhydroxylated piperidines (including bicyclic

mannose   DEOXYMANNOJIRIMYCIN   glucose
                         (11)

compounds such as castanospermine in which the majority
of the hydroxyl groups are in the piperidine ring) may
be illustrated by considering the synthesis of
deoxymannojirimycin (11) from sugars; the preparation
from D-mannose requires a double inversion of
configuration with introduction of nitrogen at C-5
together with joining the nitrogen function at C-5, a
procedure which in general is unattractive. In contrast,
a synthesis from glucose requires the introduction of
nitrogen at C-2 with a single inversion of configuration
together with joining this nitrogen function to C-6 of
glucose; this approach is attractive from the point of
view of considering the protecting group strategy in
that a system can be set up in both the anomeric
position and either C-5 (as a pyranose) or C-4 (as a
furanose).

CONNECTION OF 2-6 PYRANOSE

    In syntheses involving the joining of C-2 and C-6
of glucose via pyranoside intermediates, the nitrogen
may first be introduced at C-2 to form (12), followed by
nucleophilic displacement by the nitrogen of a
sulphonate leaving group on C-6 to give a [2.2.2]
bicyclic intermediate (13); alternatively, the nitrogen
may first be introduced onto C-6 (14) and the same
bicyclic intermediate (13) produced by subsequent
displacement by nitrogen of the leaving group at C-2.
Subsequent acid catalysed hydrolysis of the acetal to
the lactol (15) followed by borohydride reduction and
removal of the protecting groups by hydrogenolysis gives
deoxymannojirimycin (11).[20] In this route the slowness
of both the formation of the bicyclic system and of its
subsequent hydrolysis makes this an unattractive
procedure for the preparation of significant amounts of
deoxymannojirimycin.

CONNECTION OF 2-6 FURANOSE.

In the connection of C-2 and C-6 of glucose by nitrogen using a protected furanoside of glucose, again nitrogen may be first introduced either at C-2 or C-6. Diacetone glucose may readily be converted into the 2-azidomannose derivative (16) which cyclises efficiently to give the [3.2.1] bicyclic derivative (17);[21],[22] in contrast to the inconveniently slow hydrolysis of the [2.2.2] system, this furanoside (17) undergoes a fast and clean acid hydrolysis to give the lactol which on sodium borohydride reduction and hydrogenolytic removal of the protecting groups affords deoxymannojirimycin.[23] Although the overall yields in this synthesis are high, the preparation of the α-mannoazide (16) requires a troublesome separation of anomers; this can be avoided by introduction of nitrogen at C-6 first to form (18) so that the formation of the C-2 nitrogen bond occurs intramolecularly with either of the epimeric furanosides. Thus diacetone glucose is converted to the azidodiol (19), which on sequential benzylation, methanolysis and triflation gives the azidotriflate (20); reduction of (20) allows the crucial cyclisation to form the bicyclic intermediate (21) efficiently on a 50 g scale (Scheme 1).[24]

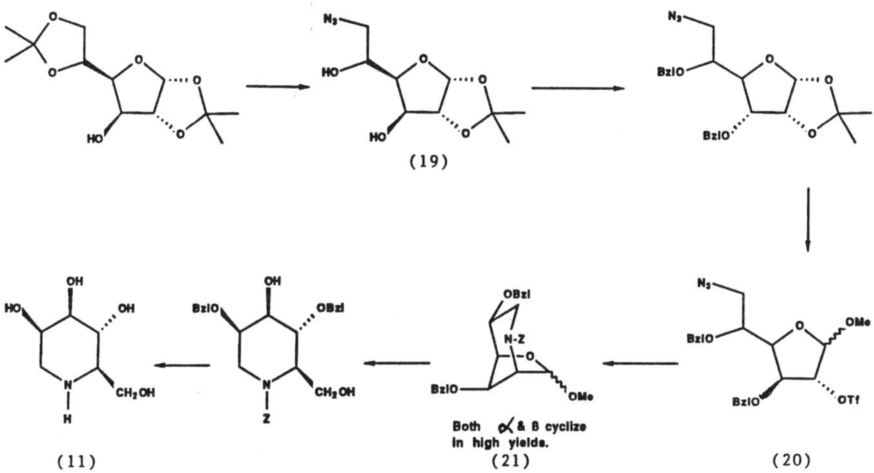

(19)

(11) (21) (20)

Both α & β cyclize in high yields.

SCHEME 1

As well as being an efficient intermediate for the synthesis of deoxymannojirimycin (11), the bicyclic amine (17), with only the original C-5 hydroxyl group of the sugar free, is a powerful divergent intermediate for the synthesis of a whole range of azapyranoses. Thus inversion of the free hydroxyl group gives deoxynojirimycin (4), removal gives fagomine (22),[25] and replacement by nitrogen gives the N-acetylglucosamine (NAG) (8) and N-acetylmannosamine (NAM) (23) analogues. The lactol formed by hydrolysis of the bicyclic intermediates may be oxidised to give trihydroxypipecolic acids, such as the glucuronic acid analogue BR1 (24) and the mannuronic acid equivalent (25); also, the aldehyde function of the lactol may be used for chain extension (for example by Ganem's procedure)[26] to give bicyclic compounds such as castanospermine (26) and epicastanospermine (27).[27]

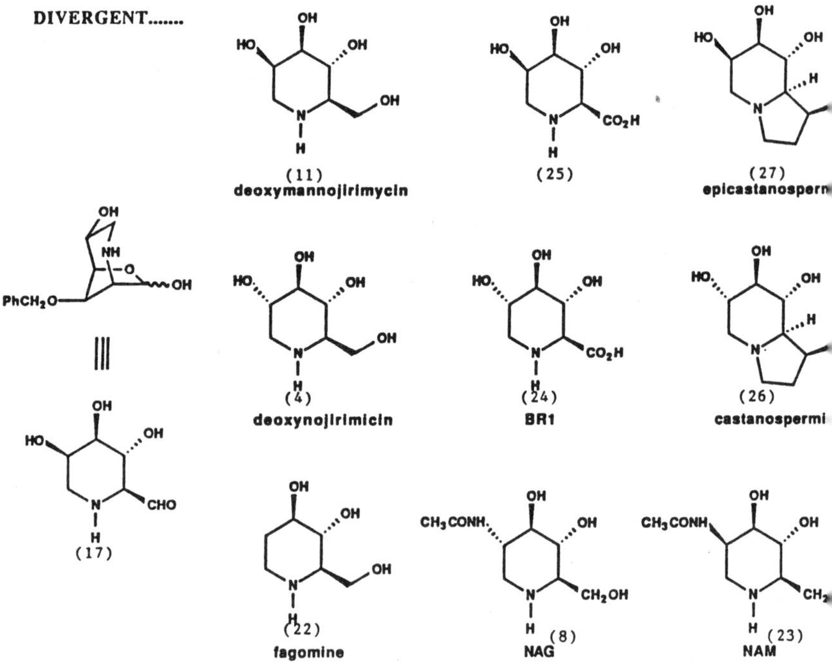

DIVERGENT........

(11)
deoxymannojirimycin

(25)

(27)
epicastanosperm

(17)

(4)
deoxynojirimicin

(24)
BR1

(26)
castanospermi

(22)
fagomine

(8)
NAG

(23)
NAM

The iminofucitol (9), an exceptionally powerful inhibitor of a number of mammalian α-L-fucosidases, is an analogue of an L-sugar and so may be prepared from the closure from C-1 to C-5 of a D-glucose with a single inversion of configuration at C-5. Thus methyl α-D-glucopyranoside may be converted to a protected form of the altritol dimesylate (28) which with substituted amines produces N-alkylated derivatives (29) of the iminofucitol; thus benzylamine forms (29:R=benzyl) which on hydrogenation gives (9). Treatment of (28) with methyl 6-aminohexanoate gives (29:R=(CH$_2$)COOMe) suitable for the preparation of an affinity column.

CONNECTION OF 1-5.

(29)   (28)   glucose

(31)   (30)

Alternatively, the iminofucitol (9) may prepared by way of catalytic reduction of the azidolactone (30) to form the lactam (31) which on further reduction gives the required product (9). Thus, diacetone glucose is converted to the allose derivative (32) with inversion of configuration at C-3 (Scheme 2); removal of the side chain acetonide and selective tosylation of the primary hydroxyl group gives (33) which with lithium aluminum hydride and subsequent introduction of azide by prior

mesylation at C-5 allows the formation of (34).
Hydrolysis of the acetonide followed by bromine water
oxidation of the resulting lactol gives the lactone (35)
which on inversion of the unprotected C-2 hydroxyl
function gives (36). Hydrogenation of the azide (36)

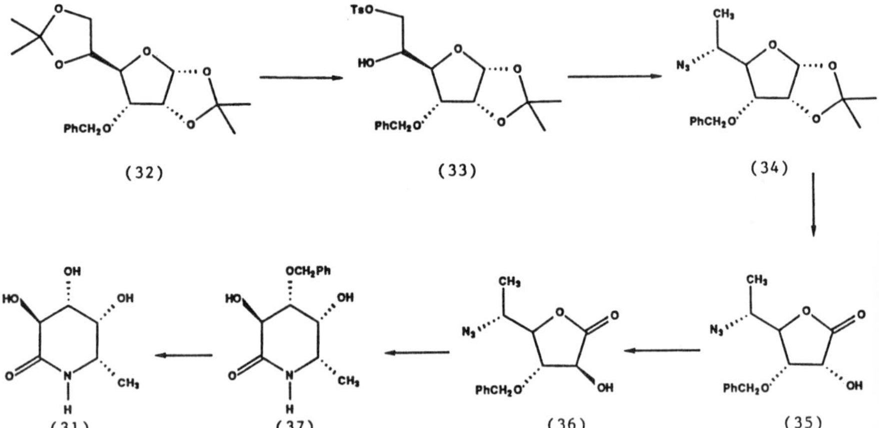

(32)   (33)   (34)

(31)   (37)   (36)   (35)

SCHEME 2

gives the lactam (37) which on hydrogenolysis gives the
required fuconolactam (31).[29]

     Three possible strategies available for the
synthesis of hydroxylated pyrrolidines from hexoses are
the connection by nitrogen of C-1 to C-4, or C-2 to C-5,
or of C-3 to C-6 (Scheme 3). We have used the first of
these strategies in the synthesis of 1,4-dideoxy-1,4-
imino-D-mannitol (DIM) (10) and swainsonine (41) from 4-
aminomannose derivatives, and of 1,4-dideoxy-1,4-imino-
L-arabinitol (L-AB1) (51) from xylose. In contrast,
connection of C-2 to C-5 has been utilised in the
synthesis of 1,4-dideoxy-1,4-imino-D-arabinitol (AB1)
(52) from xylose and of 2R,3R,4R,5R-2,5-dihydroxymethyl-
3,4-dihydroxy-pyrrolidine (DMDP) from glucose. Formation
of the nitrogen bonded between C-3 and C-6 allows the
synthesis from glucose of 1,4-dideoxy-1,4-imino-D-
lyxitol (a powerful α-galactosidase inhibitor) (54), and
of bicyclic compounds such as the epi-swainsonine (55).

5 RING STRATEGIES

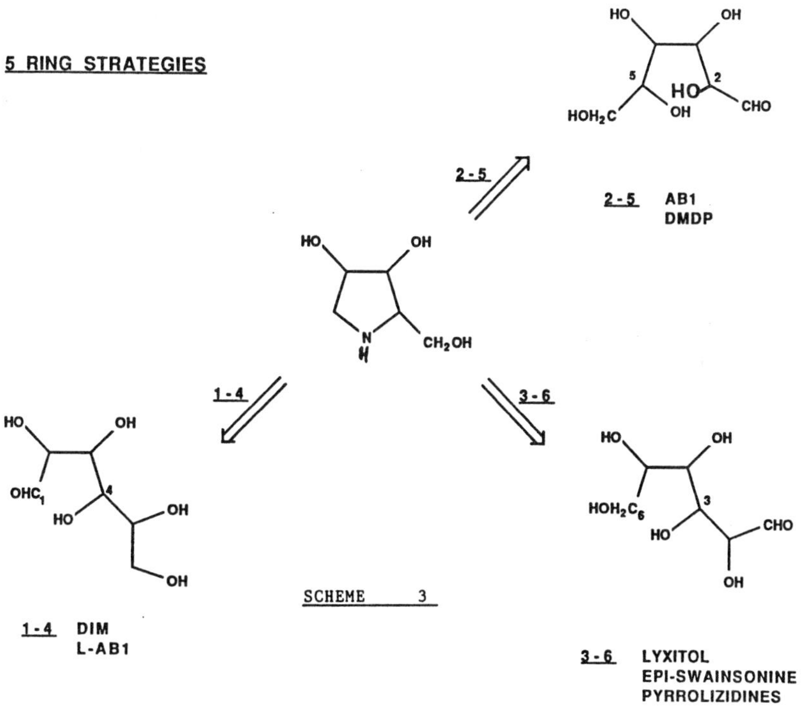

SCHEME 3

2 - 5 AB1
DMDP

1 - 4 DIM
L-AB1

3 - 6 LYXITOL
EPI-SWAINSONINE
PYRROLIZIDINES

The connection of C-1 to C-4 is the most commonly employed strategy and can be illustrated by the hydrogenation of the 4-azidomannose derivative (38) to give reduction of the azide to an amine, followed by hydrogenolysis and intramolecular reductive amination of the resulting amino lactol (39) to give the mannosidase inhibitor, 1,4-dideoxy-1,4-imino-D-mannitol (10). The azidomannose (38) is readily prepared in an overall yield of about 50% from benzyl α-D-mannopyranoside; a two carbon extension from the primary hydroxyl group in (38) by oxidation followed by a Wittig reaction with the stabilised ylid, formylmethylenetriphenylphosphorane, gives an azidoaldehyde which on hydrogenation with three equivalents of hydrogen in the presence of a palladium catalyst gives 4-aminomannopyranoside (40), formed by reduction of the carbon-carbon double bond, reduction of the azide to an amine and subsequent intramolecular

reductive amination. Further hydrogenation of (40) leads
to cleavage of the anomeric benzyl group followed a
second intramolecular reductive amination of the
resulting lactol to give swainsonine (41).[30,31]

(38)   (39)

(40)   (41)

swainsonine   DIM

(10)

Commonly available lactones allow easy protection
of all carbons other than C-1 and C-4, permitting
relatively short syntheses of 1,4-dideoxy-1,4-
iminohexitols. For example, the protected
galactonolactone (42) may similarly be converted to the
dimesylate (43) which with benzylamine yields (44) as a
protected form of the iminoglucitol (45).

(42)   (43)   (44)   (45)

Similarly, the diacetonide of D-gulonolactone (46) with sodium borohydride followed by esterification with methanesulphonyl chloride gives the dimesylate of the diol (47) which with benzylamine forms (48); removal of all the protecting groups gives the iminoallitol (49). Partial hydrolysis of the diacetonide (48) followed by periodate oxidation and sodium borohydride reduction allowed the synthesis of 1,4-dideoxy-1,4-imino-L-ribitol (50).[32]

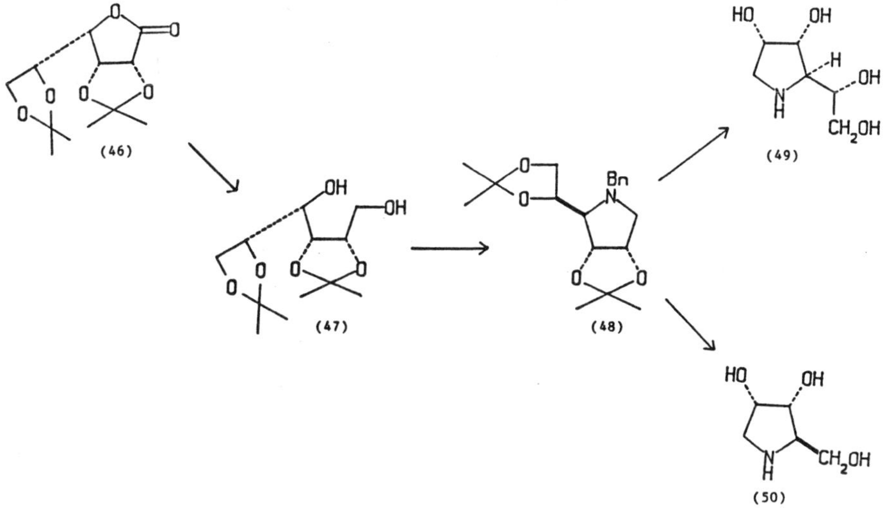

The enantiomers of 1,4-dideoxy-1,4-iminoarabinitol are both powerful α-glucosidase inhibitors.[9] The L-enantiomer of 1,4-dideoxy-1,4-iminoarabinitol (51) may be prepared by the connection of C-1 and C-4 of D-xylose together by nitrogen with inversion of configuration at C-4, whereas the D-isomer (52) arises from connection of C-2 of xylose with C-5 (Scheme 4).[33]

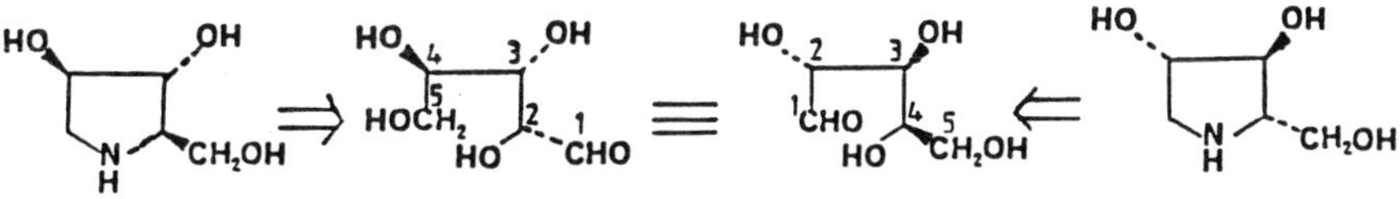

(52) AB1   D-xylose   (51) L-AB1

SCHEME 4

The value of the bicyclic iminoglucofuranose (53),
which can be prepared by a short and efficient sequence
from diacetone glucose involving ring closure of
nitrogen between C-3 and C-6 of glucose, as a divergent
intermediate for the synthesis of polyhydroxylated
pyrrolidines has been demonstrated by its conversion to
a number of mono- and bicyclic amines (Scheme 5); thus
it may be converted to the α-galactosidase inhibitor,
1,4-dideoxy-1,4-imino-D-lyxitol (54), the iminogulitol
(56) and the dihydroxyproline (57). Introduction of a
two carbon extension onto C-1 followed by ring closure
onto nitrogen allows the synthesis of the epi-
swainsonine (55), whereas two carbon extension at C-2
followed by cyclisation leads to pyrrolidizines (58) and
(59).[34,35]

SCHEME   5

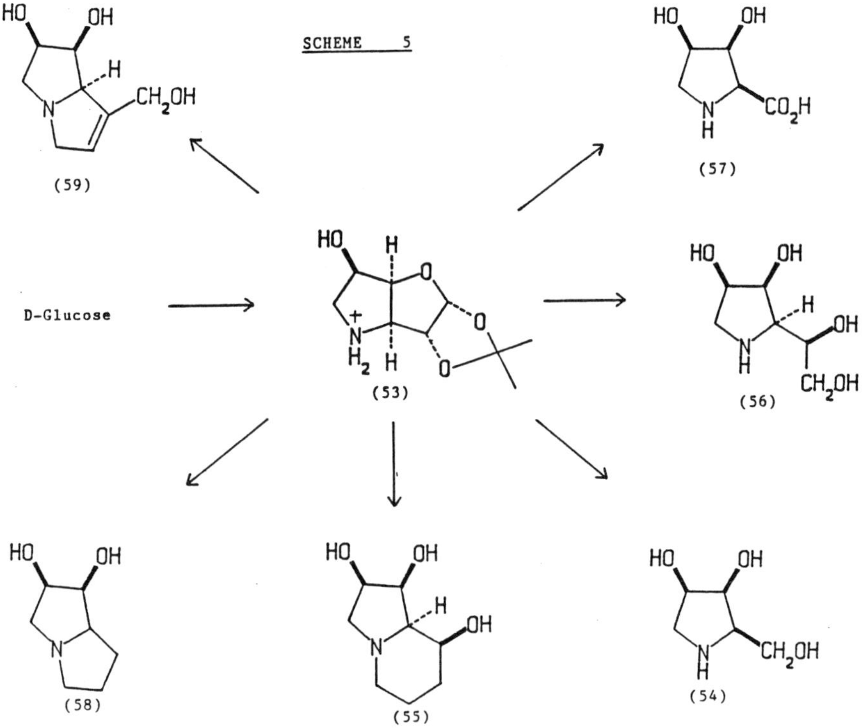

(59)

(57)

D-Glucose

(53)

(56)

(58)

(55)

(54)

I am extremely grateful to Jong Chan Son, Paul Smith,
Max Gough, Sigtor Petersen, Nigel Ramsden, Tony Shaw,
David Witty, Bharat Bashyal and Son Namgoong who have

worked on this project at Oxford, and also to Dr. Linda Fellows of the Royal Botanic Gardens at Kew for many discussions; support from SERC and Monsanto is gratefully acknowledged.

REFERENCES

1. N. Ishida, K. Kumagi, T. Niida, T. Tsuruoka and H, Yumoto, J. Antibiot. Ser. A, 1967, 20, 66.
2. S. Inouye, T. Tsuruoka, T. Ito and T. Niida, Tetrahedron, 1968, 23, 2125.
3. T. Niwa, T. Tsuruoka, H. Goi, Y. Kodama, J. Itoh, S. Inouye, Y. Yamada, T. Niida, M. Nobe, and Y. Ogawa, J. Antibiot., 1984, 37, 1579.
4. Y. Miake and E. Ebata, J. Antibiot., 1987, 40, 122.
5. G. Legler and E. Julich, Carbohydr. Res., 1984, 128, 61.
6. G. Legler and S. Pohl, Carbohydr. Res., 1986, 155, 119-131.
7. L. E. Fellows, Pestic. Sci., 1986, 17, 602.
8. S. V. Evans, L. E. Fellows, T. K. M. Shing and G. W. J. Fleet, Phytochemistry, 1985, 24, 1953.
9. G. W. J. Fleet, S. J. Nicholas, P. W. Smith, S. V. Evans, L. E. Fellows and R. J. Nash, Tetrahedron Lett., 1985, 26, 3127.
10. A. M. Scofield, L. E. Fellows, R. J. Nash and G. W. J. Fleet, Life Sci., 1986, 39, 645.
11. G. W. J. Fleet, P. W. Smith, R. J. Nash, L. E. Fellows, R. B. Parekh and T. W. Rademacher, Chem. Lett., 1986, 1051.
12. G. W. J. Fleet, A. N. Shaw, S. V. Evans and L. E. Fellows, J. Chem. Soc., Chem. Commun., 1985, 841.
13. G. W. J. Fleet, P. W. Smith, S. V. Evans and L. E. Fellows, J. Chem. Soc., Chem. Commun., 1984, 1240.
14. G. Palamartczky, M. Mitchell, P. W. Smith, G. W. J. Fleet and A. D. Elbein, Arch. Biochem. Biophys., 1985, 242, 35.
15. B. L. Rhinehart, K. M. Robinson, P. S. Liu, A. J. Payne, M. E. Wheatly and S. R. Wagner, J. Pharmacl. Exptl. Therapeutic., 1987, 241, 915.
16. J. W. Dennis, S. Laferte, C. Waghorne, M. L. Breitman and R. S. Kerbel, Science, 1987, 236, 582.
17. P. B. Ahrens and H. Ankel, J. Biol. Chem., 1987, 262, 7575.
18. B. D. Walker, M. Kowalski, W. Goh, L. Rohrschneider, W. Haseltine and J. Sodroski, Abs. 3rd Int. Conference on AIDS, Washington DC, June 1987.
19. R. F. Arakaki, J. A. Hedo, E. Collier, and P. Gorden, J. Biol. Chem., 1987, 262, 11886.

20. G. W. J. Fleet, M. J. Gough and T. K. M. Shing, Tetrahedron Lett., 1984, 25, 4029.

21. G. W. J. Fleet and P. W. Smith, Tetrahedron Lett., 1985, 26, 1469.

22. G. W. J. Fleet and P. W. Smith, Tetrahedron, 1987, 43, 971.

23. G. W. J. Fleet, L. E. Fellows and P. W. Smith, Tetrahedron, 1987, 43, 979.

24. G. W. J. Fleet and D. J. Witty, in preparation.

25. S. V. Evans, A. R. Hayman, L. E. Fellows, T. K. M. Shing, G. W. J. Fleet and A. E. Derome, Tetrahedron Lett., 1985, 26, 1465.

26. R. B. Bernotas and B. Ganem, Tetrahedron Lett., 1984, 25, 165.

27. R. J. Molyneux, J. N. Roitman, G. Dunnheim, T. Szumilo and A. D. Elbein, Arch. Biochem. Biophys., 1987, 251, 450.

28. G. W. J. Fleet, A. N. Shaw and S. Peterson, unpublished results.

29. G. W. J. Fleet and N. G. Ramsden, unpublished results.

30. G. W. J. Fleet, M. J. Gough and P. W. Smith, Tetrahedron Lett., 1984, 25, 1853.

31. G. W. J. Fleet, B. P. Bashyal, M. J. Gough and P. W. Smith, Tetrahedron, 1987, 43, 3083.

32. G. W. J. Fleet and J. C. Son, unpublished results.

33. G. W. J. Fleet and P. W. Smith, Tetrahedron, 1986, 42, 5685.

34. G. W. J. Fleet, G. N. Austin, P. D. Baird, J. M. Peach, P. W. Smith and D. J. Watkin, Tetrahedron, 1987, 43, 3095.

35. G. W. J. Fleet, J. A. Seijas and V. P. Vazquez, unpublished results.

The Mechanism of Action and Selectivity of the Antiherpetic Drug Acyclovir

G. B. Elion

WELLCOME RESEARCH LABORATORIES, BURROUGHS WELLCOME CO., 3030 CORNWALLIS ROAD, RESEARCH TRIANGLE PARK, NORTH CAROLINA 27709, USA

1   INTRODUCTION

The clinical success of acyclovir in the treatment of herpes simplex virus and varicella zoster virus infections has reawakened the interest of medicinal chemists in antiviral research after many years of discouragement and frustration. It is now almost ten years since the first publications of the antiviral activity and mechanism of action of acyclovir (ACV), 9-(2-hydroxyethoxymethyl)guanine (originally called acycloguanosine).[1,2] Since that time, this body of information has been enlarged and periodically reviewed.[3-6] However, interesting new information continues to be uncovered.

   This review will consolidate the current state of knowledge of the mechanism of action and selectivity of ACV and will follow the story as it unfolded histori-cally. Acyclovir (1) is a nucleoside analog in which the sugar moiety has been replaced by an acyclic side chain. It closely resembles 2'-deoxyguanosine except that the 2' and 3' carbons and the 3'-hydroxyl group of the deoxyriboside (shown in dotted lines) are missing. It was one of a series of acyclic analogs synthesized because of the finding that some nucleoside metabolizing enzymes, e.g. adenosine deaminase, could accept these compounds as substrates or be inhibited by them. The discovery that the arabinosides of adenine, 2,6-diamino-purine and guanine could act as inhibitors of DNA viruses gave impetus to the testing of purine nucleoside analogs as antiviral agents.

(1)

2   ACTIVATION OF ACYCLOVIR

There were two reasons for the excitement engendered by
ACV. First, it was highly potent in its activity
against herpes simplex viruses types 1 and 2 (HSV-1,
HSV-2) and varicella zoster virus (VZV).[1-3,7-9] Second,
it was highly selective for the virus-infected cells,
being essentially inert on uninfected cells. Thus,
there was a 3000-fold difference between the
IC_{50} = 0.1 μM for HSV-1 and the IC_{50} = 300 μM for the
host Vero cell in plaque-reduction assays *in vitro*. Such
selectivity had never been described previously for
compounds which inhibited DNA viruses and it therefore
demanded investigation.

The first clue to the reason for the selectivity of
ACV came from studies of the fate of radioactive ACV in
HSV-1 infected Vero cells, compared with that in uninfected cells.[1] From high pressure liquid chromatography
of the cell extracts it was apparent that ACV was
unchanged in uninfected Vero cells but was converted to
three new compounds, which were identified as the mono-,
di-, and triphosphates of ACV, in HSV-infected cells.
The question then was: "Which enzymes are induced by
viral infection which are responsible for this
metabolism?" The highly specific step in this series of
reactions turned out to be the first one, the phosphorylation of ACV to its monophosphate (ACVMP).

Although the production of HSV-specified thymidine
kinase (HSV-TK) in HSV-infected cells had been previously reported, it seemed unlikely at first that this
was the enzyme responsible for the phosphorylation of a
guanine nucleoside analog. However, all attempts to
separate the enzymatic phosphorylating activity for ACV

from the HSV-TK failed.[1,10] Confirmatory evidence that
HSV-TK was indeed the responsible enzyme came from the
prevention of ACV phosphorylation by excess thymidine
and by the inability of a temperature-sensitive
thymidine kinase-deficient mutant of HSV to phosphorylate ACV. In addition, cells biochemically transformed
by transfection with the portion of HSV DNA containing
the genetic information for HSV-TK acquired the ability
to phosphorylate ACV.[11] The unusual specificity of this
enzyme was intriguing and led to an extensive investigation of its ability to bind and to phosphorylate
purine and pyrimidine nucleoside analogs.[12] The enzyme
had fairly broad specificity with respect to the ali-
phatic side chain which had replaced the sugar, provided
there was a terminal hydroxyl group on the substituent
for phosphorylation. The requirements for the heterocyclic base moiety were more restricted, and among the
purines only guanine analogs were good substrates.
Surprisingly, since the natural substrates of HSV-TK are
thymidine and deoxycytidine, pyrimidine acyclic nucleosides are poor substrates and are inactive as antiherpetic agents.

While the studies on the structure-activity
relationships of the substrates of HSV-TK from HSV-1
were going on, work continued on the mechanism of action
of ACV. Specific viral TK's capable of phosphorylating
ACV were also found in cells infected with HSV-2 and
VZV, and large amounts of ACVTP were produced in such
cells on exposure to ACV.[9,13] The second step in the
activation of ACV, the conversion of ACVMP to ACVDP, was
found to be carried out by cellular guanylate kinase,[14]
while the final step to ACVTP could be accomplished by a
variety of cellular enzymes, particularly phosphoglycerate kinase.[15] Some compounds which serve as substrates for HSV-TK nevertheless have no antiviral
activity.[11] This suggests that their monophosphates are
not transformed to triphosphates by the cellular enzymes
or, alternatively, that their triphosphates are inactive
on viral DNA polymerase.

The formation of ACVTP in HSV-infected cells was
found to begin soon after viral infection, a peak
concentration being reached eight hours after infection.[16] Since viral DNA synthesis begins early in the
infective cycle, the presence of ACVTP during that
period is important. If ACV is added to cells eight
hours after HSV infection, it has much less effect on
viral DNA synthesis. The intracellular concentration of
ACVTP in infected cells dropped after eight hours, with

a $t_{1/2}=4$ hours, even when the cells remained exposed to
ACV in the medium. If cells containing ACVTP were
washed and resuspended in fresh medium without ACV, the
level of intracellular ACVTP dropped 50% during the
first 1.5 hours, but the breakdown then slowed, and the
ACVTP level reached a plateau at about 15% of the
initial concentration after six hours. The ACV resulting from the breakdown of ACVTP diffused back into the
medium.

3   EFFECTS ON DNA POLYMERASE

It had become apparent that ACVTP was the active
metabolite of ACV and the logical assumption was that
the locus of its action must be the viral DNA polymerase. The investigations which followed compared the
effects of chemically synthesized ACVTP on purified DNA
polymerases from several strains of HSV-1 and HSV-2, as
well as various cellular α DNA polymerases, *in vitro* in
cell-free systems.[1,17] These experiments utilized activated calf thymus DNA as template, although synthetic
templates were also investigated.

The early studies revealed that there was a pronounced quantitative difference between the inhibitory
effects of ACVTP on the viral DNA polymerases and the
cellular enzymes. Depending on the strain of HSV used,
the inhibition constants (K_i values) were 10 to 40 times
lower for the viral DNA polymerases. Thus, the selectivity of ACV for herpes virus-infected cells over normal
cells was further enhanced. Not only was much more
ACVTP formed in HSV-infected cells, but this product was
more inhibitory to the viral DNA polymerase.[1,17]

One interesting facet of the inhibition of viral DNA
polymerase by ACVTP *in vitro* was that the inhibited reaction slowed down within minutes and seemed to come to a
halt long before the substrates or template were
depleted.[17] Moreover, it appeared that ACVTP could act
as a substrate for the viral DNA polymerase, but not, to
any extent, for the cellular polymerase. These studies
were extended and refined over the ensuing years and it
has been shown that, whereas in the initial binding of
ACVTP to the viral DNA polymerase, dGTP is competitive
with ACVTP, a change takes place, presumably conformational, in the enzyme-template-ACVTP complex, so that
the enzyme is not released for further action.[18] Thus,
the HSV DNA polymerase is progressively inactivated.
Activity at this stage cannot be restored by adding dGTP

but only by adding fresh enzyme.   The cellular α DNA
polymerase is not inactivated by ACVTP.

The competition of ACVTP and dGTP for binding to the
viral DNA polymerase suggested the possibility that a
reduction in the dGTP pool size might result in a
potentiation of the effects of ACVTP on the DNA
polymerase. This was accomplished by the use of a
ribonucleotide reductase inhibitor, a 2-acetylpyridine
thiosemicarbazone BW A723U.[19,20] The synergism by this
reductase inhibitor on the antiviral activity of ACV
appears to result not only from the reduction of dGTP
levels but also from the elevation of ACVTP concentrations. The reason for the latter elevation is unclear
at present. However, the large increase in the ACVTP to
dGTP ratio accounts for the increased inhibitory effect
on the HSV/DNA polymerase.

Before complete inactivation of the viral DNA
polymerase occurs, the enzyme is capable of incorporating ACVMP into the DNA template.[3,17,21] However,
such an action is chain-terminating since the ACV lacks
a 3'-OH group on which to elongate the DNA chain. The
incorporation of ACV into growing chains of new viral
DNA was shown to result in small fragments of DNA and
absence of long-chain viral DNA synthesis.[22]

The effect of ACV on viral protein synthesis is
mainly on the late proteins of the γ class which are
being synthesized eight hours after infection. This is
probably due to the inhibition of viral progeny DNA
synthesis by ACV. In addition, the synthesis of four of
the major viral glycosylated polypeptides was markedly
inhibited; this could affect the infectivity of newly
made virus.[23]

The mechanism of action of ACV on VZV is very
similar to that on HSV, with a VZV-specified TK and a
specific viral DNA polymerase sensitive to ACVTP.[9,24]
For the other two clinically important herpes viruses,
HCMV and EBV, the picture is different. Neither of
these viruses induces a specific TK which phosphorylates
ACV. Nevertheless, a small amount of ACVTP is formed in
cells infected by HCMV[13] and EBV,[25] presumably by way of
the 5'-nucleotidase which phosphorylates ACV in normal
cells to a small extent.[26] The DNA polymerase of EBV
is, fortunately, exquisitely sensitive to ACVTP
($K_i=0.015$ μM) so that the replication of this virus is
inhibited by ACV at clinically achievable levels.[6,27,28]
The situation is somewhat similar with HCMV, except that

different strains of HCMV differ greatly in the sensitivity of their DNA polymerases to ACVTP. Hence, the
sensitivity of HCMV strains to ACV can vary widely, e.g.
$IC_{50}=40$ μM to >200 μM.[8,24]

4   RESISTANCE

As might be predicted from knowledge of the mechanism of
action of ACV, acquired resistance of the virus to the
drug can arise in three different ways. Since the HSV-
or VZV-specified thymidine kinases are so important in
the activation of ACV, a deletion of this enzyme[29-31] or
a change in its substrate specificity[32,33] would result
in resistance. Both of these mechanisms have been found
to occur clinically, although the altered enzyme appears
to be rare. Finally, an altered viral DNA polymerase
could be less sensitive to the effects of ACVTP.[29,30]
This type of resistance has been documented in the
laboratory but does not appear to occur clinically.
Fortunately, clinical resistance of HSV to ACV has been
relatively infrequent and mainly confined to immunocompromised, chronically infected individuals on
prolonged treatment.[34-36]

5   GANCICLOVIR

Because of the clinical need for a potent inhibitor of
HCMV, especially in immunosuppressed patients, the
search for such a compound has continued. A close
analog of ACV is ganciclovir (GCV) (previously called
BW B759U, DHPG, 2'-NDG, and BIOLF-62 . This compound,
9-(1,3-dihydroxy-2-propoxymethyl)gua ine, contains two
hydroxymethyl groups in the acyclic side chain.

     GCV is approximately as active as ACV against HSV-1,
HSV-2 and VZV, somewhat more active than ACV against
EBV, but much more inhibitory than ACV against HCMV (for
references see review).[6] A search for the explanation
for this high activity of GCV revealed that GCV is
phosphorylated in HCMV-infected cells to a much greater
extent than is ACV. The exact nature of the phosphorylating enzyme induced by HCMV is still not clear.
The enzyme appears not to be a thymidine kinase, nor is
it identical with the deoxyguanosine kinase which has
been reported.[37,38] In any event, the high levels of
GCVTP achieved and maintained for a long time in HCMV-
infected cells are sufficient to inhibit HCMV DNA
polymerase at levels of GCV in the medium as low as

0.8 µM.[37] GCV can be incorporated into both viral and cellular DNA and is not chain-terminating, since a hydroxyl group similar to the 3'-OH group of deoxyguanosine is present on the side chain.[39]

6 CONCLUSION

The studies on mechanism of action and selectivity of ACV on herpes viruses have been a continuing source of scientific excitement and reward. The discovery that a compound non-toxic to normal cells required a specific virally-induced enzyme for activation and that the ultimate product of this activation (ACVTP) could preferentially inhibit and inactivate viral DNA polymerases was a dream of selectivity come true. It demonstrated once again that active drugs can be powerful tools for unraveling the mysteries of nature, and for promoting the discovery of new medically important drugs.

REFERENCES

1. G.B. Elion, P.A. Furman, J.A. Fyfe, P. de Miranda, L. Beauchamp and H.J. Schaeffer, Proc. Natl. Acad. Sci. USA, 1977, 74, 5716.

2. H.J. Schaeffer, L. Beauchamp, P. de Miranda, G.B. Elion, D.J. Bauer and P. Collins, Nature, 1978, 272, 583.

3. G.B. Elion, Am. J. Med., 1982, 73(1A), 7.

4. G.B. Elion, J. Antimicrob. Chemother., 1983, 12 Suppl. B, 9.

5. G.B. Elion, "Antiviral Drugs and Interferon", Y. Becker, ed., Martinus Nijhoff, Boston, 1984, Chapter 5, p. 71.

6. G.B. Elion, "Antiviral Chemotherapy: New Directions for Clinical Application and Research", J. Mills and L. Corey, eds., Elsevier, New York, 1986, p. 118.

7. P. Collins and D.J. Bauer, J. Antimicrob. Chemother., 1979, 5, 431.

8. C.S. Crumpacker, L.E. Schnipper, J.A. Zaria and M.J. Levin, J. Antimicrob. Chemother., 1979, 5, 431.

9. K.K. Biron and G.B. Elion, Antimicrob. Agents Chemother., 1980, 18, 443.

10. J.A. Fyfe, P.M. Keller, P.A. Furman, R.L. Miller and
    G.B. Elion, J. Biol. Chem., 1978, 253, 8721.

11. P.A. Furman, P.V. McGuirt, P.M. Keller, J.A. Fyfe and
    G.B. Elion, Virology, 1980, 102, 420.

12. P.M. Keller, L. Beauchamp, C.M. Lubbers, P.A. Furman,
    H.J. Schaeffer and G.B. Elion, Biochem. Pharmacol., 1981, 30,
    3071.

13. M. St. Clair, P.A. Furman, C.M. Lubbers and G.B. Elion,
    Antimicrob. Agents Chemother., 1980, 18, 741.

14. W.H. Miller and R.L. Miller, J. Biol. Chem., 1980, 255, 7204.

15. W.H. Miller and R.L. Miller, Biochem. Pharmacol, 1982, 31,
    3879.

16. P.A. Furman, P. de Miranda, M.H. St. Clair and G.B. Elion,
    Antimicrob. Agents Chemother., 1981, 20, 518.

17. P.A. Furman, M.H. St. Clair, J.A. Fyfe, J.L. Rideout,
    P.M. Keller and G.B. Elion, J. Virol., 1979, 32, 72.

18. P.A. Furman, M.H. St. Clair and T. Spector, J. Biol. Chem.,
    1984, 259, 9575.

19. T. Spector, D.R. Averett, D.J. Nelson, C.H. Lambe,
    R.W. Morrison, M.H. St. Clair and P.A. Furman, Proc. Natl.
    Acad. Sci. USA, 1985, 82, 4254.

20. T. Spector, Pharmac. Ther., 1985, 31, 295.

21. D. Derse, Y.-C. Cheng, P.A. Furman, M.H. St. Clair and
    G.B. Elion, J. Biol. Chem., 1981, 256, 11447.

22. P.V. McGuirt, J.E. Shaw, G.B. Elion and P.A. Furman,
    Antimicrob. Agents Chemother., 1984, 25, 507.

23. P.A. Furman and P.V. McGuirt, Antimicrob. Agents Chemother.,
    1983, 23, 332.

24. K.K. Biron, P.J. Stenbuck and J.B. Sorrell, "Herpesvirus",
    F. Rapp, ed., Alan R. Liss, New York, 1984, p. 677.

25. B.M. Colby, P.A. Furman, J.E. Shaw, G.B. Elion and
    J.C. Pagano, J. Virol., 1981, 38, 606.

26. P.M. Keller, S.A. McKee and J.A. Fyfe, J. Biol. Chem., 1985,
    260, 8664.

27. B. Colby, J.E. Shaw, G.B. Elion and J.S. Pagano, <u>J. Virol.</u>, 1980, <u>34</u>, 560.

28. E. Pagano and A.K. Datta, <u>Am. J. Med.</u>, 1982, <u>73(1A)</u>, 18.

29. D.M. Coen and P.A. Schaffer, <u>Proc. Natl. Acad. Sci. USA</u>, 1980, <u>77</u>, 2265.

30. L.E. Schnipper and C.S. Crumpacker, <u>Proc. Natl. Acad. Sci. USA</u>, 1980, <u>77</u>, 2270.

31. D.M. Coen, P.A. Schaffer, P.A. Furman, P.M. Keller and M.H. St. Clair, <u>Am. J. Med.</u>, 1982, <u>73(1A)</u>, 351.

32. G. Darby, H.J. Field and S.A. Salisbury, <u>Nature</u>, 1981, <u>289</u>, 81.

33. B.A. Larder, Y.-C. Cheng and G. Darby, <u>J. Gen. Virol.</u>, 1983, <u>63</u>, 523.

34. S. Nusinoff Lehrman, E.L. Hill, J.F. Rooney, M.N. Ellis, D.W. Barry and S.E. Straus, <u>J. Antimicrob. Chemother.</u>, 1986, 18 Suppl. B, 85.

35. P. Collins, and N.M. Oliver, <u>J. Antimicrob. Chemother.</u>, 1986, 18 Suppl. B, 103.

36. J.C. Wade, C. McLaren and J.D. Meyers, <u>J. Infect. Dis.</u>, 1983, <u>148</u>, 1077.

37. K.K. Biron, S.C. Stanat, J.B. Sorrell, J.A. Fyfe, P.M. Keller, C.U. Lambe and D.J. Nelson, <u>Proc. Natl. Acad. Sci. USA</u>, 1985, <u>82</u>, 2473.

38. K.K. Biron, J.A. Fyfe, S.C. Stanat, L.K. Leslie, J.B. Sorrell, C.U. Lambe and D.M. Coen, <u>Proc. Natl. Acad. Sci. USA</u>, 1986, <u>83</u>, 8769.

39. Y.-C. Cheng, S.P. Grill, G.E. Dutschman, K. Nakayama and K.F. Bastow, <u>J. Biol. Chem.</u>, 1983, <u>258</u>, 12460.

Synthesis of Some Antiviral Carbocyclic Nucleosides

S. M. Roberts*

DEPARTMENT OF CHEMISTRY, UNIVERSITY OF EXETER, EXETER,
DEVON EX4 4QD, UK

K. Biggadike, A. D. Borthwick, and B. E. Kirk

DEPARTMENT OF MICROBIOLOGICAL CHEMISTRY, GLAXO GROUP
RESEARCH, GREENFORD, MIDDLESEX UB6 0HE, UK

At the present time the search for effective antiviral agents is one of the most challenging areas of
investigation involving medicinal chemists. The gross
inconvenience due to influenza infections, the morbidity
caused by certain herpes infections and the terrifying
AIDS epidemic emphasise the urgent need to achieve
significant progress in various areas of anti-viral researc'
Infections caused by herpes simplex viruses HSV-1 and HSV-2
are widespread and, of the diseases associated with these
viruses, recurrent genital herpes is a major concern.

Conceptually the task of finding an anti-viral agent
is considerably more difficult than that of finding an
effective anti-bacterial agent or a useful anti-fungal
agent. Bacterial cells differ considerably from mammalian
cells; for example they are prokaryotic and the cell-
walls are made up of peculiar peptidoglycan units. Fungal
cells are eukaryotic but other features, for example, the
steroid components within the cell-wall are significantly
different from the host cell. Viruses replicate within
the host cells taking over the machinery of these cells
for the good of the virus often with detriment to the
host. Thus viruses present fewer obvious starting points
from which to initiate searches for new chemotherapeutic
agents; obviously a more detailed examination of the
process of virus infection is needed before targeted
medicinal chemistry can be envisaged.

The life-cycle of a typical virus consists of adsorpt
ion of the virus to the host cell, penetration into the
cell, synthesis of virus specific macromolecules, virus
assembly within the cell and release of the organism from
the cell. Theoretically anti-viral agents could be

designed to interfere with any part of this cycle.[1]
Interference with the manufacture of viral macromolecules
is one of the easier and is probably, at present, the most
popular approach to anti-viral chemotherapy.

The herpes viruses HSV-1 and HSV-2 are viruses that
carry the genome on double-stranded DNA.   Transcription
and translation of this DNA provides virally coded enzymes
such as thymidine kinase and DNA polymerase that are used
for the synthesis of macromolecules vital to the virus.
Such virally coded enzymes will differ, to a lesser or
greater extent, from the equivalent host cell enzymes and
exploitation of these differences is a valid approach to
anti-herpes chemotherapy.   Suppose, for example, a
nucleoside analogue is only phosphorylated by a virally
coded kinase and is not phosphorylated by any of the host
cell kinases.   The derived monophosphate could be taken
to the corresponding triphosphate by virally coded or host-
cell enzymes and incorporated into the nascent DNA chain.
If the nucleoside analogue lacks a 3'-hydroxyl group, chain
extension could not take place and DNA synthesis could not
be completed.   The interference with DNA synthesis would
only take place in virally infected cells and viral replication would be inhibited.   Suppose, on the other hand, a
nucleoside analogue was a specific inhibitor of virally
coded DNA polymerase.   Only viral DNA synthesis would be
affected.   Once again the virus would not be able to
replicate.   The best of both worlds would be the result
of specific phosphorylation of a compound by the viral
kinase to give (after further phosphorylation) a triphosphate which was a specific inhibitor of virally coded DNA
polymerase (Figure 1).

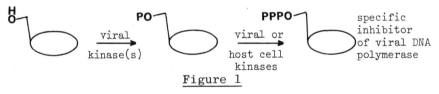

Figure 1

A number of compounds have been shown to inhibit
herpes virus replication by interference with DNA synthesis
in the above manner.   Acyclovir (ACV) (1) is a compound
which has been shown to be effective in the clinic against
HSV-1 and HSV-2.   The compound is selectively phosphorylated by herpes-coded thymidine kinase and acts as a
terminator of DNA synthesis.[2]   Phosphonoformate (PFA) (2)
is a selective inhibitor of herpes coded DNA polymerase;[3]
unfortunately the compound has been shown to be too toxic
for use in the clinic.   Bromovinyldeoxyuridine (BVDU) (3)

is a very potent agent against HSV-1 infections but is
relatively inactive against HSV-2.   This compound is
selectively phosphorylated by HSV-1 kinase(s) and after
further phosphorylation, disturbs viral DNA replication.[4]
The inactivity of BVDU against HSV-2 seems likely to limit
the usefulness of this compound in the clinic.   Finally,
the fluorine containing compounds 2'-fluoro-5-iodoarabino-
furanosyl uracil (FIAU) (5) and 2'-fluoro-5-methylarabino-
furanosyl uracil (FMAU) (6) have been shown to be very
powerful inhibitors of the replication of HSV-1 and HSV-2.
The mechanism of action again depends on the selectivity

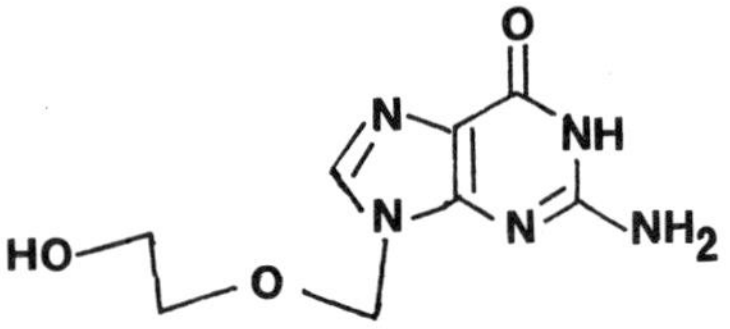

(1)

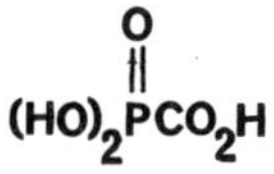

(2)

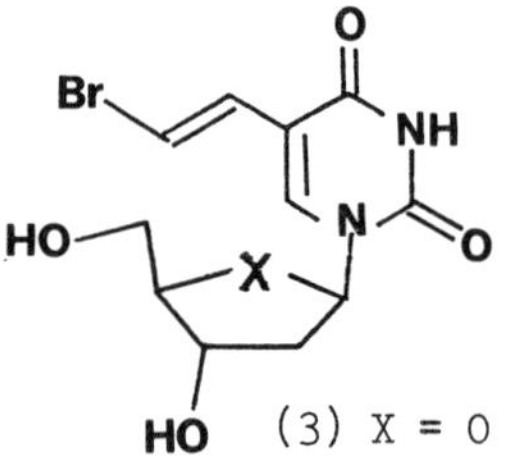

(3) X = O

(4) X = CH$_2$

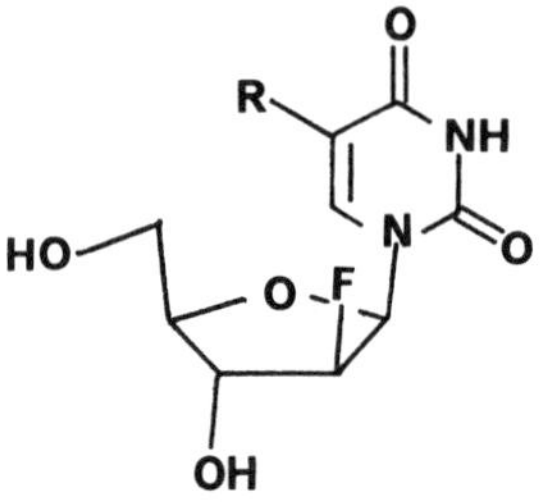

(5) R = I

(6) R = Me

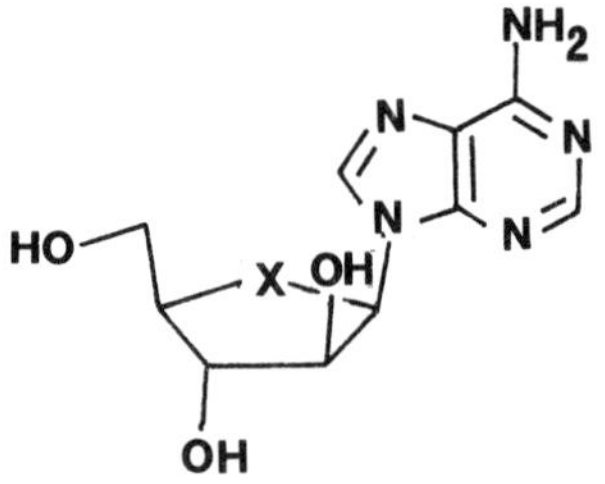

(7) X = O

(8) X = CH$_2$

of the initial phosphorylation of the 5'-hydroxyl group by
virus-induced thymidine kinase and selective disruption of
DNA synthesis in virally infected cells.[5]   Despite the
powerful anti-herpes activities of these and related
fluorinated compounds, toxicity problems may prevent their
widespread use in the clinic.[6]

    Anti-viral agents possessing a sugar moiety as part
structure are susceptible to degradation *in vivo* due to th
action of nucleoside phosphorylase enzymes that catalyse
cleavage of the heterocyclic unit from the tetrahydrofuran
ring.[7]   Not surprisingly, therefore, a considerable amoun
of attention has been directed towards the replacement of
the sugar unit with an appropriately substituted cyclo-

pentane ring system.[8] For example carbocyclic-BVDU (4) has been synthesised; the compound is more stable to degradation *in vivo* and displays good anti-viral activity against HSV-1.[9] Similarly arabinofuranosyl adenine (ara-A) (7) and the corresponding carbocyclic compound, cyclaridine (8)[10] have been prepared and the latter compound shows considerable anti-herpes activity. The long-established anti-herpes agent 5-iodo 2'-deoxyuridine (9) and the carbocyclic compound (10) represent another pair of comparable compounds.[11] Recently carbocyclic 2'-deoxyguanosine (11) has been reported to show a good level of biological activity against both HSV-1 and HSV-2.[12]

(9) X = O

(10) X = CH_2

(11)

(12) X = CHF, α-F

(13) X = CHF, β-F

When comparing compounds such as the sugar derivative (9) and the corresponding cyclopentane derivative (10) it is clear that the replacement of the oxygen atom by a methylene group is quite a reasonable switch from a stereochemical point of view but it does cause major changes from the electronic point of view, *i.e.* the inductive effect of the electronegative oxygen atom on the rest of the system is removed. Blackburn has argued that the fluoromethylene unit is iso-steric and iso-electronic with an oxygen atom[13] and with this in mind we undertook the preparation of compounds (12) and (13). Immediately one is faced with an interesting situation of having to construct, in a stereocontrolled fashion, a tetra-substituted cyclopentane ring system with four different atoms (C, F, N, O) attached to the five-membered ring!

The preparation of compound (12) started with cyclopentadiene (Scheme 1). Alkylation followed by hydroboration using the requisite dipinanylborane (a procedure closely analogous to one used previously in the Hofmann-La Roche laboratories[14]) furnished the homochiral alcohol (14) (Scheme 1). Sharpless epoxidation and protection of the free hydroxyl group gave the oxirane (15) which was attacked regioselectively by azide ion to afford the azido-alcohol (16). The system is now set up for the intro-

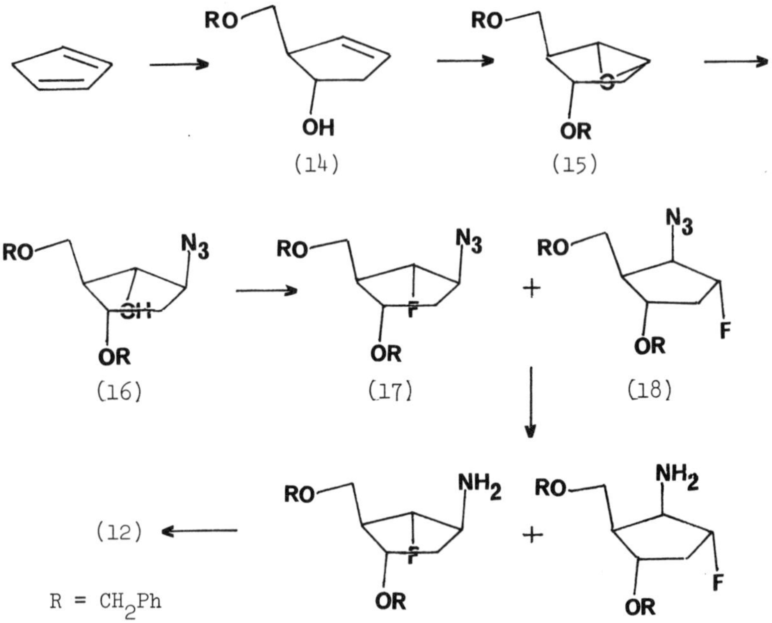

(14)   (15)

(16)   (17)   +   (18)

(12) ◄───   (19)   +   (20)

R = CH$_2$Ph

Scheme 1

duction of the fluorine atom using diethylaminosulphur
trifluoride (DAST) a reagent introduced almost simultane-
ously by Middleton[15] and a Russian group[16] for the
conversion of alcohols into mono-fluoro compounds and for
the conversion of aldehydes and ketones into the corresponding di-fluoro compounds.   Treatment of the compound (16)
with DAST gave the fluoro-compounds (17) and (18) in the
ratio 2:3.   Two features of this reaction are worthy of
mention.   First the fluorine atom is introduced into
compound (17) with retention of configuration at the
participating carbon centre.   Secondly the major product
results from a 1,2-shift of the azide group.   Both of
these facets of the reaction have precedent in the literature.   While DAST reactions normally proceed with
inversion of configuration, the influence of a neighbourin
functional group can lead to overall retention.[17]   The
shift of an azide group during a DAST reaction has been
used to good effect by Nicolaou in carbohydrate chemistry.
The conversion (16) → (17) + (18) presumably proceeds as
shown in Figure 2.   The mixture of compounds (17) and
(18) was reduced to give a separable mixture of the amines
(19), (20).   The amine (19) was converted into carbocycli

1-(2'-deoxy-6'α-fluororibofuranosyl)-5-iodouracil (12) in four steps in 30% yield.[19]

Figure 2

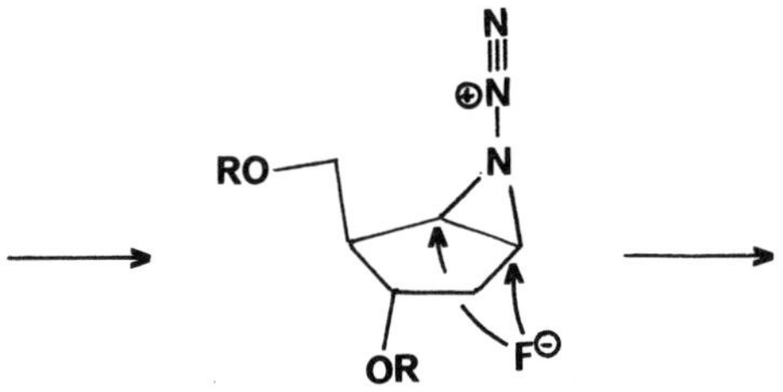

The alcohol (16) was converted into the corresponding triflate and treated with tetrabutylammonium fluoride to give the fluoro-azide (21). This compound was converted into the required iodouracil derivative (13) in five steps in 25% yield (Scheme 2).

Scheme 2

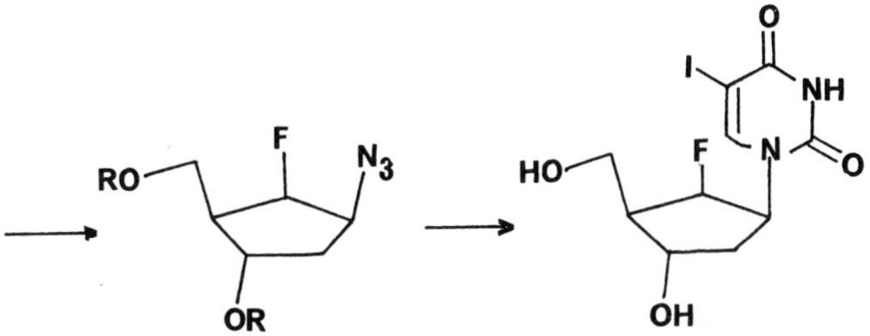

The 6'α-fluoro-compound (12) was highly active against HSV-1 infected cells in the microtitre assay (about ½ x the activity of acyclovir) while the 6'β-fluoro-compound (13) was two orders of magnitude less active.

The level of biological activity associated with compound (12) encouraged us to consider the preparation of other carbocyclic nucleosides containing a fluorine atom. In particular, the well-documented, highly potent anti-herpes activity of FIAU (5) and FMAU (6) suggested the corresponding carbocyclic compounds (26), (27) would be interesting target structures.

The known aminotriol (22)[20] was protected using 2,4-dinitrofluorobenzene followed by dichlorotetraisopropyl-

disiloxane to give the alcohol (23) (Scheme 3).   Treatment
of this alcohol with DAST gave two products, the desired
fluoro-compound (24) (74%) and the unexpected product (25)
(18%).   The formation of the latter compound is discussed
later.   The success of the DAST reaction for the conversion
of the alcohol (23) into the fluoro-compound (24) is very
dependent on the choice of protecting group for the amino-
function.   A strongly electron-withdrawing protecting
group is required to minimize electron density on the
nitrogen atom in order to reduce the participation of this
atom in the reaction.   Thus a protecting group such as
trityl is useless;   a DAST reaction on the corresponding
substrate gave rise to the aziridine (28).

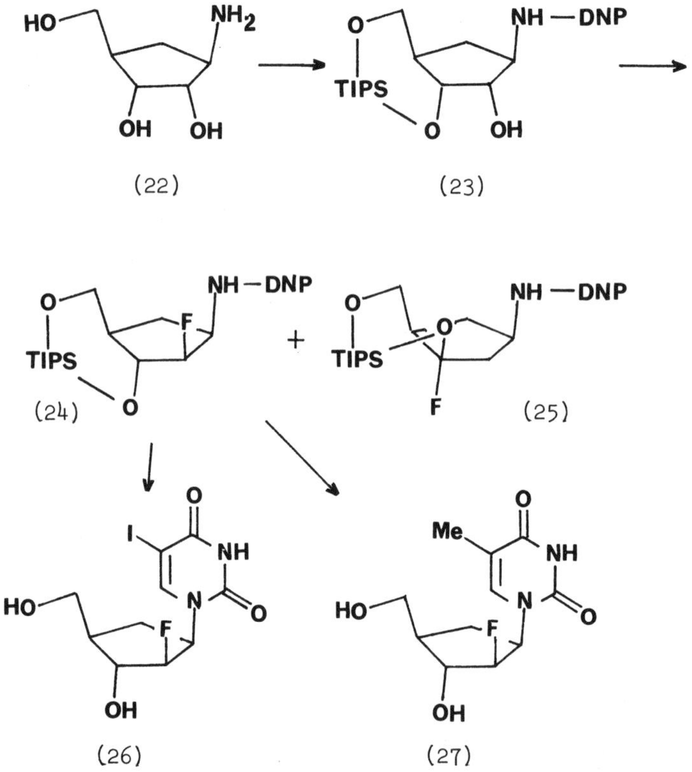

Scheme 3

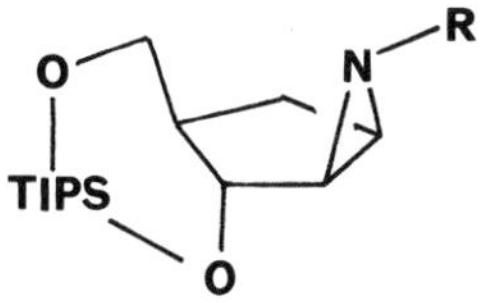

(28)   R = CPh$_3$
(29)   R = DNP

The survival of the oxybis(di-isopropylsilyl) (TIPS) protecting group during the DAST reaction is also worthy of note.[21]   The TIPS protecting group is rapidly removed by fluoride ion (*e.g.* tetrabutylammonium fluoride in tetrahydrofuran) indicating that if fluoride ion is generated in the DAST reaction (as seems likely at first sight - Figure 3) it must be unable to migrate from the area close to the C-2' position.

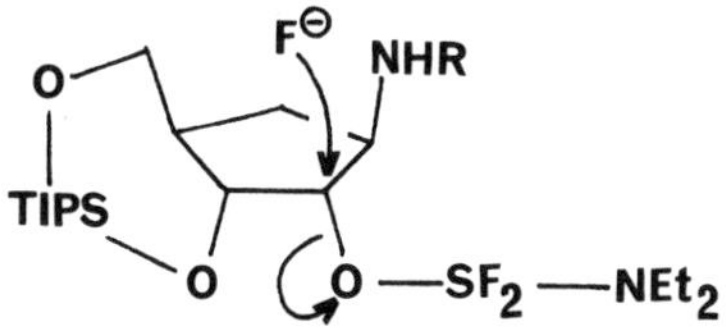

Figure 3

In the event, compound (24) could be produced on a multi-gram scale, and was converted using conventional chemistry (five steps, 17% overall yield) into the required compound, (±)-carbocyclic-FIAU (26) and, by another sequence of reactions (four steps, 18% yield), into the second target compound (±)-carbocyclic-FMAU (27).[22]

It was also of interest to prepare the carbocyclic analogue of FMAU having the fluorine atom in the configuration *trans* to the pyrimidinedione ring.   To this end, the configuration of the hydroxyl group in compound (23) was inverted using a three step sequence involving mesylation, SN$_2$ displacement of the mesylate group with acetate ion and hydrolysis (68% overall yield) to afford the required alcohol (30) (Scheme 4).   Interestingly, an attempted Mitsunobu reaction on compound (23) with methyl p-toluenesulphonate as the attendant nucleophile, furnished the aziridine (29);   it seems that in the Mitsunobu reaction the carbenium ion is so well developed

at C2' that even the dinitrophenylamino-group cannot
resist the temptation to get involved.

Treatment of the alcohol (30) with DAST gave a three-
component mixture containing roughly equal quantities of
the required fluoro-compound (31), the protected fluoro-
hydrin (27) and the imine (32).   The imine (32) is
derived by a process of dehydration and [1,3]-migration of
a hydrogen atom.   The compound (25) is presumably formed
by a [1,2]-migration of a hydrogen atom to form an oxa-
carbenium ion which is attacked by fluoride ion from the
less hindered face (Figure 4).   The rapid conversion of

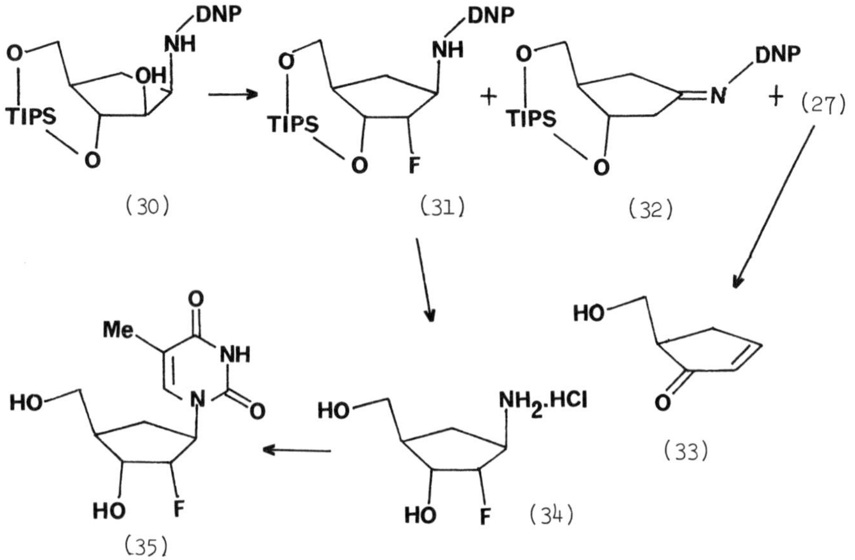

Scheme 4

the compound (25) into 5-hydroxymethylcyclopent-2-enone
(33)[23] using tetrabutylammonium fluoride in tetrahydrofura
serves to emphasise that 'free' fluoride ion cannot be
produced in a DAST reaction.

The crude fluoro-compound (31) was deprotected to giv
the hydrochloride (34) [17% yield from compound (30)] and
the latter compound was transformed into the desired produ
(35) (34%).

Since the sugar derivative (39) had been shown to
possess anti-viral activity,[24] we felt that it would be
prudent to prepare the corresponding carbocyclic compound

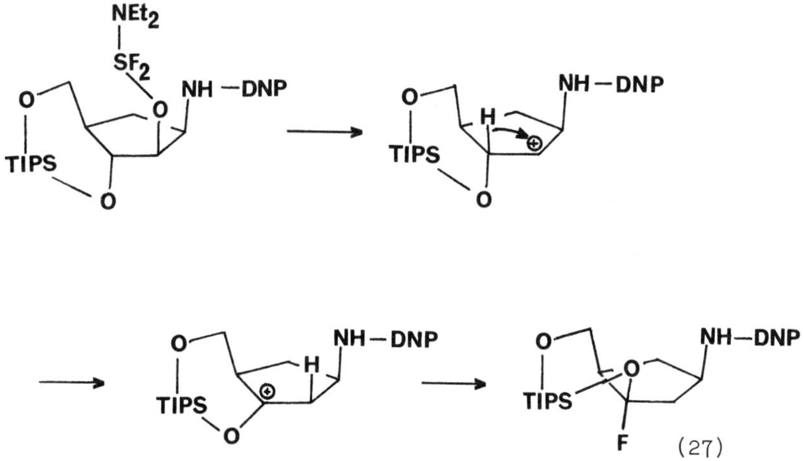

Figure 4

(40). The ketone (36) was a key intermediate and this compound was produced from the alcohol (23) by a Swern oxidation (Scheme 5). Treatment of the ketone with DAST followed by desilylation using $F^{\ominus}$ gave two identifiable products, the trifluoro compound (37) and the required difluoro compound (38) (*ca.* 20%). The conversion of a

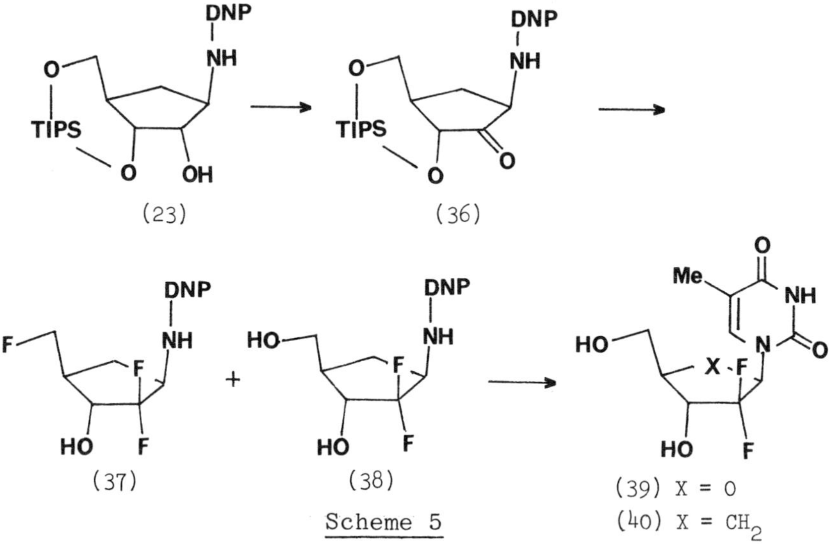

Scheme 5

ketone into the corresponding difluoromethylene compound
does require relatively harsh conditions (prolonged reaction times and higher temperatures) and under these more
severe conditions the TIPS group is partially lost and the
exposed primary hydroxyl group is converted into the alkyl
fluoride.   The diol (38) was converted into the difluoro-
carbocyclic nucleoside (40) in 35% yield.

      The considerable effort involved in the preparation
of compounds (26), (27), (35), and (40) was completely and
utterly unrewarded in terms of anti-viral activity of these
nucleoside analogues.   Only carbocyclic-FMAU showed
activity in the HSV-1 plaque reduction assay and even then
the activity was three orders of magnitude less than that
observed for FMAU.

      Undaunted by these disappointing results, it was
decided that we should extend the series of 2'-F-carbo-
cyclic nucleosides to include selected purine derivatives
(41).   Accordingly, the fluoro-compound (24) was depro-

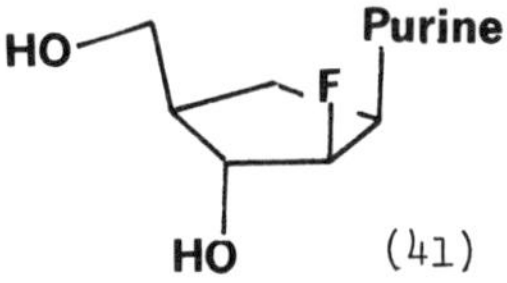

(41)

tected and reacted with 2-amino-4,6-dichloropyrimidine to
give the diamine (42) (Scheme 6).   Reaction with *p*-chloro-
phenyldiazonium chloride and reduction of the resulting
diazo compound with zinc in acetic acid gave the triamine

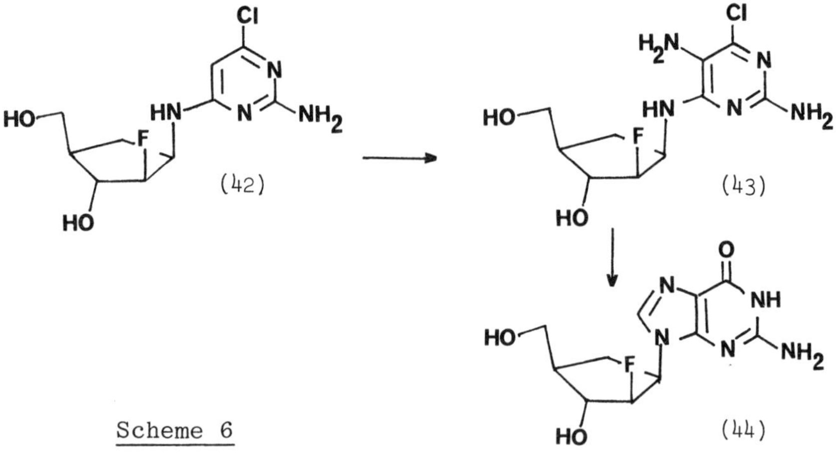

Scheme 6

(43) which was treated with triethyl orthoformate followed by hydrochloric acid to give the guanine derivative, carbocyclic 9(2'-fluoro-arabinofuranosyl)guanine (44). Delightfully this carbocyclic nucleoside showed quite remarkable biological activity against both HSV-1 and HSV-2 in infected cells. For example the compound is more active than FMAU and ACV in the *in vitro* assay against HSV-1 infected cells. In stark contrast the compound showed no effect on uninfected cells even at very much higher concentrations. In short, the compound displayed an unprecedented level of highly selective anti-herpes activity.[25]

Besides the issue of whether there would be closely related analogues showing similar exquisite biological activity, four points were immediately raised. First, the method of preparation of compound (44) involves a long linear synthesis: could a shorter more convergent synthesis be devised using cheap and readily available starting materials? Secondly, the compound (44) was prepared as a racemic modification: was the biological activity due to one enantiomer? Thirdly, carbocyclic nucleosides often mimic the biological activity of the corresponding sugar: would 9(2'-fluoroarabinofuranosyl)guanine show anti-herpes activity? Finally, what was the mechanism of action of the new anti-herpes compound? All four of these points were addressed.

A convergent synthesis of the carbocyclic nucleoside (44) was achieved by conversion of the ether (45)[26] into the epoxide (46) using *tert*-butylhydroperoxide and reaction conditions recommended by Sharpless. Reaction of the derived oxirane (47) with potassium hydrogendifluoride in ethylene glycol at 150-160 °C gave the fluorodiol (48) (the trityl protecting group is lost under the reaction conditions). The tosylate (49) was obtained in three steps from the fluorodiol (48). Coupling of the tosylate (49) with 2-amino-6-chloropurine was achieved using potassium carbonate in dimethylsulphoxide at 80 °C and the product (50) was hydrolysed to give the desired carbocyclic nucleoside (44) (Scheme 7).

All attempts to resolve intermediates in both of the established routes to the compound (44) proved fruitless. Hence we decided to resolve the end-product by means of some enzyme-catalysed reactions. Thus the racemic compound (44) was phosphorylated using HSV-1 thymidine kinase (TK) (Scheme 8). The monophosphate (51) was isolated by ion exchange h.p.l.c.; the enzyme catalysed phosphorylation process showed low enantioselectivity.

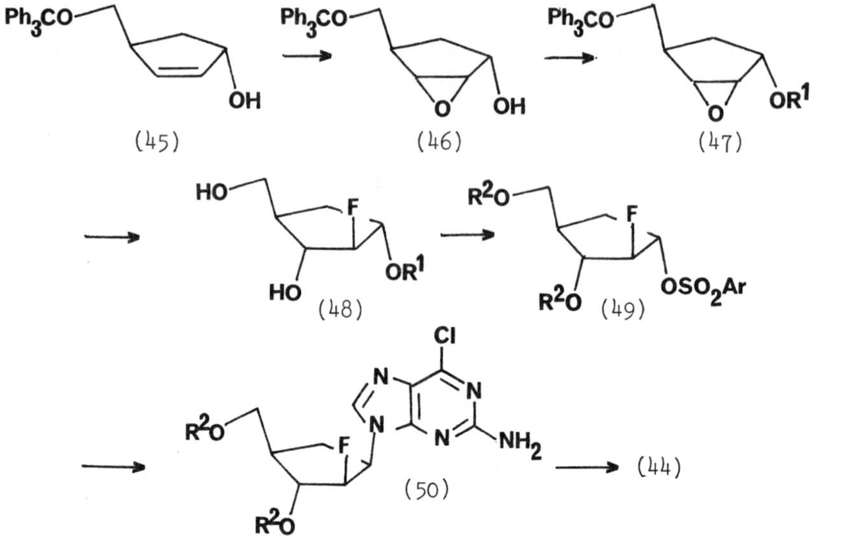

(45) → (46) → (47)

(48) → (49)

(50) → (44)

$R^1 = CH_2Ph$
$R^2 = CH_2OMe$

Scheme 7

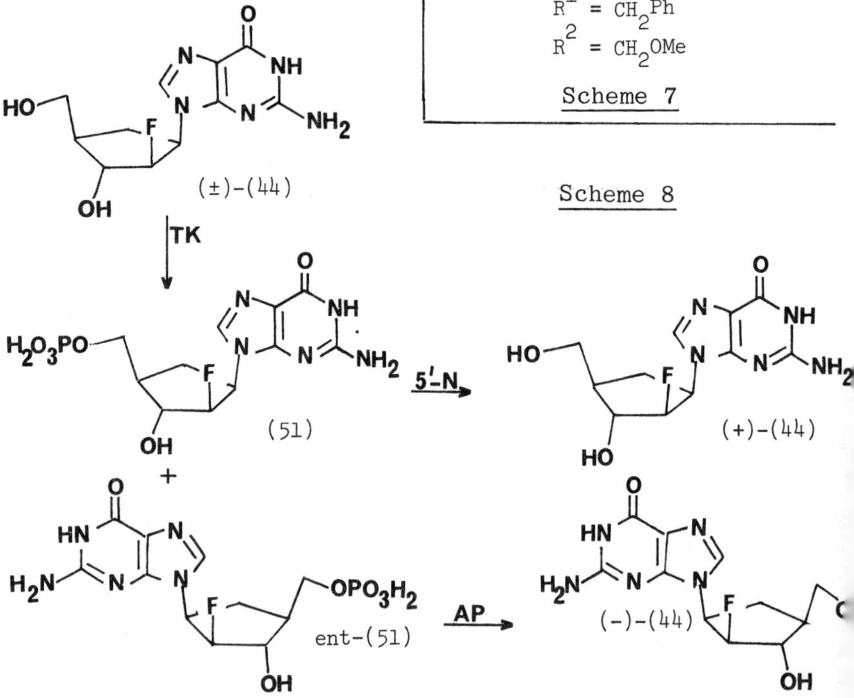

(±)-(44)

Scheme 8

TK

(51)

$\xrightarrow{5'\text{-N}}$ (+)-(44)

+

ent-(51) $\xrightarrow{AP}$ (−)-(44)

The monophosphate (51) was incubated with 5'-nucleotidase
(5'-N) (EC 3.1.3.5) from *Crotalus atrox* venom. The resultant mixture was separated to give recovered monophosphate
(51) and optically active carbocyclic nucleoside (44)
$\{[\alpha]_D + 48^0$ (c, 3.57, $H_2O)\}$. This reaction is enantio-
selective and presumably the nucleotide corresponding to
the natural compound (guanosine 5'-phosphate) is hydrolysed
more readily.[27] The monophosphate recovered from the
nucleotidase treatment was hydrolysed using the enzyme
alkaline phosphatase (AP) to give the laevorotatory enantiomer of the carbocyclic nucleoside (44) $\{[\alpha]_D -68^0$ (c, 1.92,
$H_2O)\}$. The dextrorotatory enantiomer (+)-(44) was twice
as active against herpes-infected cells when compared to
the racemic material. The laevorotatory enantiomer was at
least 100-fold less active in the same test system.

The synthesis of 9-(2'-fluoroarabinofuranosyl)guanine
(56) involved coupling the bromo-compound (52) and the
silylated chloropurine (53) followed by deprotection and
separation of the required amine (54) (27%) from the epimer
(55) (11%) (Scheme 9). Hydrolysis of the compound (54)
afforded the nucleoside (56) $\{[\alpha]_D^{24}$ 41.6^0 (c, 0.31 methanol)$\}$.
Very recently a somewhat less efficient synthesis of the
same nucleoside was reported by Montgomery, Secrist, *et al.*[28]
We found that the sugar displayed only modest activity (4%
of the activity of ACV) against cells infected with HSV.
Comparing the biological results obtained for the compounds
(44) and (56) shows that the carbocyclic nucleoside is

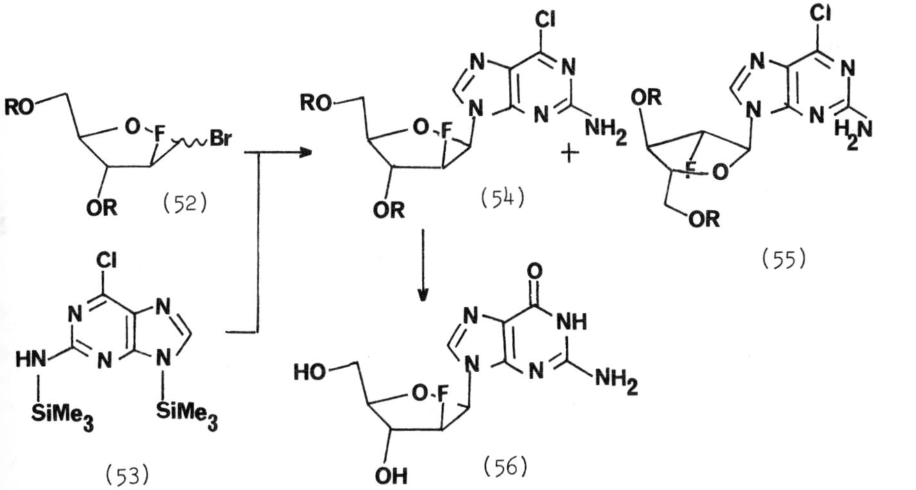

Scheme 9

considerably more active than the corresponding sugar, a
most surprising and unusual observation.

The mechanism of action of the anti-viral carbocyclic
nucleoside (44) must remain speculative at this stage. The
nucleoside is a substrate for HSV-thymidine kinase and,
after this first phosphorylation, further reaction to give
the triphosphate is probably very facile. The nucleoside
triphosphate may then act as a potent inhibitor of herpes-
derived DNA polymerase or alternatively the nucleoside
surrogate may be incorporated into the growing DNA chain
ultimately to provide a defective gene. Studies concerning the biochemistry of the novel anti-viral agent will be
reported elsewhere.

In summary, the synthesis of selected fluorine containing carbocyclic nucleosides has resulted in the discovery of the most potent and selective inhibitor of HSV-1
and HSV-2 infected cells *in vitro*. Further biological
data including the results of *in vivo* testing will be
communicated in due course.

Acknowledgements: this work was carried out in the
Chemical Research Laboratories, Glaxo Group Research,
Greenford, U.K. Dr. Hebblethwaite, Miss Exall and
Messrs. Evans and Stephenson were involved in the initial
studies involving the DAST reactions. We also thank
Miss Butt and Mr. Youds for other important contributions.
Biological results were obtained under the supervision of
Mr. Knight and Dr. Coates. Drs. Cameron and Booth and
members of their team were involved in the use of the
enzymes for the resolution of the nucleoside (44).

References

1. G.J. Galasso, *J. Antimicrobial Chemother.*, 1984, *14A*, 127;
    R.T. Walker, *ibid.*, 1984, *14A*, 119; W.H. Prusoff, T.-S. Lin,
    and M. Zucker, *Anti-viral Res.*, 1986, *6*, 311.

2. G.B. Elion in 'Acyclovir: An Update', Wiley and Sons, 1985, p.7;
    ACV-triphosphate is also an inhibitor of HSV-1 DNA polymerase.

3. B. Eriksson and B. Öberg in 'Anti-viral Drugs and Interferon:
    The Molecular Basis of Their Activity', Martinus Nijhoff,
    Boston, 1984, Chapter 9, p. 127.

4. E. De Clercq, *Biochem. Pharmacol.*, 1984, *33*, 2159.

5. D.L. Swallow and G.L. Kampfner, *Brit. Med. Bull.*, 1985, *41*, 322
    and references therein.

6. T.-L. Su, K.A. Watanabe, R.F. Shinazi, and J.J. Fox, *J. Med. Chem.*, 1986, *29*, 151; C. McLaren, M.S. Chen, R.H. Barbhaiya, R.A. Buroker, and F.B. Oleson, in 'Pharmacological and Clinical Approaches to Herpes Viruses and Virus Chemotherapy', R. Kono (ed.), *Excepta Medica*, Amsterdam, 1985, p. 57.

7. See, for example, Z. Veres, A. Szabolcs, I. Szinai, G. Denes, and A. Jeney, *Biochem. Pharmacol.*, 1986, *35*, 1057; Y.F. Shealy, C.A. O'Dell, G. Arnett, and W.M. Shannon, *J. Med. Chem.*, 1986, *29*, 79.

8. For a useful review on carbocyclic nucleosides see V.E. Marquez and M.I. Lim, *Med. Chem. Rev.*, 1986, *6*, 1.

9. E. De Clercq, J. Balzarini, R. Bernaerts, P. Herdewijn, and A. Verbruggen, *Biochem. Biophys. Res. Commun.*, 1985, *126*, 397; P. Herdewijn, E. De Clercq, J. Balzarini, and H. Vanderhaeghe, *J. Med. Chem.*, 1985, *28*, 550; P. Ravenscroft, R.F. Newton, D.I.C. Scopes and C. Williamson, *Tetrahedron Lett.*, 1986, *27*, 747.

10. R. Vince and S.J. Daluge, *J. Med. Chem.*, 1977, *20*, 612.

11. G. Streissle, A. Paessens, and H. Oediger, *Adv. Virus Res.*, 1985, *30*, 83.

12. Y.F. Shealy, C.A. O'Dell, W.M. Shannon, and G. Arnett, *J. Med. Chem.*, 1984, *27*, 1416.

13. G.M. Blackburn and D.E. Kent, *J. Chem. Soc., Perkin Trans. I*, 1986, 913 and references therein.

14. J.J. Partridge, N.K. Chadha, and M.R. Uskokovic, *J. Am. Chem. Soc.*, 1973, *95*, 532.

15. W.J. Middleton, *J. Org. Chem.*, 1975, *40*, 574.

16. L.N. Markovskii, V.E. Pashinik, and A.V. Kirsanov, *Synthesis*, 1973, 787; *idem.*, *Z. Org. Khim.*, 1976, *12*, 965.

17. S. Castillon, A. Dessinges, R. Faghih, G. Lukacs, A. Olesker, and T.T. Thang, *J. Org. Chem.*, 1985, *50*, 4913.

18. K.C. Nicolaou, T. Ladduwahetty, J.L. Randall, and A. Chucholowski, *J. Am. Chem. Soc.*, 1986, *108*, 2466.

19. Preliminary communication, K. Biggadike, A.D. Borthwick, A.M. Exall, B.E. Kirk, S.M. Roberts, P. Youds, A.M.Z. Slawin, and D.J. Williams, *J. Chem. Soc., Chem. Commun.*, 1987, 255.

20. R.C. Cermak and R. Vince, *Tetrahedron Lett.*, 1981, *22*, 2331.

21. Loss of trialkylsilyl protecting groups during DAST reactions has been observed, P.J. Card, *J. Carbohydrate Chem.*, 1985, *4*, 451.

22. Preliminary communication, K. Biggadike, A.D. Borthwick, D. Evans, A.M. Exall, B.E. Kirk, S.M. Roberts, L. Stephenson, P. Youds,

A.M.Z. Slawin, and D.J. Williams, *J. Chem. Soc.*, *Chem. Commun.*, 1987, 251.

23. *Cf.* F. Hansske and M.J. Robins, *J. Am. Chem. Soc.*, 1983, *105*, 6736.

24. Eli Lilley and Co., U.K. Pat. Appl. G.B., 2,136,425A, 1984; F. Wohlrab, A.T. Jamieson, J. Hay, R. Mengel, and W. Guschlbauer, *Biochim. Biophys. Acta*, 1985, *824*, 233.

25. Full details of the *in vitro* biological activity of this compound will be published elsewhere.

26. H. Paulsen and U. Maass, *Chem. Ber.*, 1981, *114*, 346.

27. *Cf.* P. Herdewijn, J. Balzarini, E. De Clercq, and H. Vanderhaeghe, *J. Med. Chem.*, 1985, *28*, 1385.

28. J.A. Montgomery, A.T. Shortnacy, D.A. Carson, and J.A. Secrist III, *J. Med. Chem.*, 1986, *29*, 2389.

The Development of Agents Active against Herpes Viruses and Human Immunodeficiency Virus

J. L. Kelley

WELLCOME RESEARCH LABORATORIES, BURROUGHS WELLCOME CO., 3030 CORNWALLIS ROAD, RESEARCH TRIANGLE PARK, NORTH CAROLINA 27709, USA

1 INTRODUCTION

Infectious viral diseases are a serious medical problem, as evidenced by the high incidence of herpes virus infections and the high rate of mortality associated with acquired immunodeficiency syndrome (AIDS), whose etiological agent is the human immunodeficiency virus (HIV). Significant progress has been made in the development of chemotherapeutic agents for the treatment of these two virus infections within the past decade. Several cyclic nucleosides and acyclic nucleosides are available for the treatment of infections caused by the herpes family of DNA viruses. And very recently 3'-azido-3'-deoxythymidine (zidovudine, AZT) was shown to be clinically efficacious against the retrovirus HIV. In this review I will discuss the development of agents that have *in vitro* activity against the herpes family of viruses and emphasize the structure-activity relationships of the acyclic nucleosides such as acyclovir. I will also discuss some of the agents, such as zidovudine and the 2',3'-dideoxynucleosides, that have activity against HIV.

2 AGENTS ACTIVE AGAINST HERPES VIRUSES

An iodinated analogue of 2'-deoxyuridine, 5-iodo-2'-deoxyuridine (idoxuridine, IUDR), was the first clinically effective antiherpetic agent.[1] The compound was synthesized in 1959 by Prusoff[2] as part of an anticancer program and was reported later to have activity against herpes simplex virus (HSV) *in vitro*.[3] Although idoxuridine has long been approved for use in the topical treatment of herpes simplex keratitis, a sight-threatening eye infection, it is not useful as a systemic antiherpetic drug due to its toxicity.[4]

A closely related pyrimidine nucleoside, trifluridine (5-trifluoromethyl-2'-deoxyuridine, TFT), has become a standard drug for treatment of herpetic keratitis, especially in patients unresponsive or hypersensitive to idoxuridine.[5] It was first synthesized by Heidelberger[6] as a potential antineoplastic agent. Although its efficacy in the treatment of herpes keratitis in rabbits was reported in 1964,[7] it was not until 1980 that trifluridine was approved for use in humans in the United States. As with idoxuridine, trifluridine's use is limited to topical application because of its systemic toxicity.[8]

The antiherpetic purine nucleoside, vidarabine [9-(1-β-*D*-arabinofuranosyl)adenine, ara-A], also emerged from the search for anticancer agents. Ara-A was synthesized originally by Baker and co-workers[9] and was soon reported to possess antiherpes activity *in vitro*.[10] Vidarabine has been used for the treatment of ocular herpes diseases[11] and has approval for the systemic treatment of HSV encephalitis.[12]

These three nucleosides are examples of first-generation antiviral agents; although they are potent antiherpetic agents, they do not exhibit a high degree of selective activity.[13,14] They actually function as prodrugs, since they must be activated by enzymatic conversion to the triphosphate metabolites. This metabolism occurs both in herpes-infected and uninfected cells. Subsequent incorporation of the triphosphate metabolite into the DNA of uninfected cells limits their utility for systemic administration.[15,16]

(*E*)-5-(2-Bromovinyl)-2'-deoxyuridine (BVDU) is representative of the second generation antiherpetic agents, which exhibit more selective activity against the herpes viruses. BVDU has potent activity against herpes simplex virus type 1 (HSV-1) and varicella-zoster virus (VZV), but it does not affect host cell growth unless drug concentrations are very high.[17] This selective toxicity is due to phosphorylation of BVDU by a virus-induced thymidine kinase, which is accompanied by little or no metabolic activation in uninfected cells. Although BVDU was effective in the treatment of

lethal herpetic infections in animal models[18] its
clinical usefulness is in question.[19]

A series of 2'-fluoro-5-substituted-arabino-
furanosylcytosines and uracils, which has very potent
and selective antiherpetic activity *in vitro*, was reported
by Watanabe et al.[20] FIAC (2'-fluoro-5-iodo-
1-β-*D*-arabinofuranosylcytosine), which is representative
of this group of fluorinated pyrimidines, is phosphorylated in virus-infected cells by the viral thymidine
kinase. The subsequently formed FIAC-triphosphate is
more inhibitory to the herpetic DNA polymerase than to
the cellular polymerase, resulting in a potent and
selective antiherpetic effect. The clinical usefulness
of these compounds remains in question.[21]

Not all potentially useful antiherpetic agents are
nucleosides. Phosphonoformic acid (foscarnet, PFA) is
an inhibitor of HSV replication both *in vitro* and *in vivo*.[22]
It is a selective inhibitor of HSV-induced DNA polymerase and appears to have a beneficial effect against
cutaneous herpes virus infections in humans when applied
topically.[23]

Medicinal chemists have tried a variety of
approaches to improve the clinical efficacy of these
first- and second-generation antiviral agents. For
example, vidarabine, which suffers from poor water
solubility, was chemically converted to its
5'-monophosphate, ara-AMP. Ara-AMP has improved
solubility properties but is rapidly metabolized to
vidarabine *in vivo*.[24] Another problem with vidarabine is
its rapid metabolism to the less active hypoxanthine
derivative. This occurs because vidarabine is a good
substrate for adenosine deaminase.[25] Attempts to
develop analogues that are adenosine deaminase resistant
led to the carbocyclic analogue cyclaradine,[26] which is
not a substrate for adenosine deaminase and has antiherpetic activity *in vitro* and *in vivo*.[27] Inhibition of
adenosine deaminase by use of *erythro*-9-(2-hydroxy-3-
nonyl)adenine (EHNA) in conjunction with vidarabine has
been used successfully to enhance the antiviral as well
as the antitumor properties of vidarabine.[28,29] Furthermore, EHNA was reported to have *in vitro* antiviral
activity against HSV.[30]

EHNA emerged from a systematic effort by Schaeffer
in the 1960s to study the structural requirements for
substrate activity and to develop inhibitors of
adenosine deaminase.[31] EHNA was the apical product of

this instructional study in enzyme inhibitor development.[32]   It is a potent inhibitor of adenosine deaminase
with a K_i = 6.5 nM.[33] This work also showed that the
cyclic carbohydrate moiety of adenosine is not necessary
for an adenine derivative to exhibit potent inhibitory
properties towards adenosine deaminase.[34]

Because a number of nucleoside analogues have shown
good activity as antiviral agents, a program was
initiated in the Wellcome Research Laboratories to
synthesize nucleoside analogues in which the cyclic
carbohydrate moiety was replaced by a non-cyclic side
chain. This work led to the discovery and development
of the potent antiherpetic drug acyclovir
(9-(2-hydroxyethoxymethyl)guanine, ACV).[35,36] This
guanosine analogue, which contains a side chain in which
the 2'- and 3'-carbons of the ribose moiety have been
excised, mimics a portion of the natural sugar moiety of
2'-deoxyguanosine. Acyclovir possesses potent activity
against viruses of the herpes group, especially against
HSV-1, HSV-2, VZV and Epstein-Barr virus (EBV).

The level of *in vitro* antiherpetic activity of
acyclovir relative to several first- and second-
generation agents was published by De Clercq et al.[37]
The ID_{50}s, which are in µg/ml, are the average values
for eleven strains of HSV-1 and seven strains of HSV-2
(Table 1). Against HSV-1 acyclovir has an ID_{50} =
0.04 µg/ml, which is 0.18 µM. It is 175-fold more
active than ara-A (vidarabine), 17-fold more active than
TFT (trifluridine) and 3-fold more active than IUDR
(idoxuridine). This group found FIAC and BVDU to be
about 2.5- and 5-fold, respectively, more active than
acyclovir. A similar profile of relative potencies
prevails against HSV-2 except for BVDU, which is
125-fold less active against HSV-2 than against HSV-1.
This divergence for BVDU is probably due to the
decreased ability of the HSV-2-induced thymidine kinase
to phosphorylate BVDU beyond the monophosphate level.[38]
Acyclovir has equivalent activity *in vitro* against both
types of HSV.

There is a wealth of information in the literature
on the structure-activity relationships of acyclic
nucleosides and antiviral activity, but it is broadly
disseminated in many publications. Several efforts have
been made to consolidate this information.[39,40] A large
variety of structural modifications of acyclovir has
been reported. Table 2 shows ten 2-substituted
analogues of acyclovir and their activity against HSV-1

Table 1. Potencies of Antiherpes Compounds in PRK Cells Against
Eleven Strains of HSV-1 and Seven Strains of HSV-2[a]

Compound	Average ID_{50} ($\mu g/ml$) for	
	HSV-1	HSV-2
PFA	13	8
ara-A	7	5
TFT	0.7	0.7
IUDR	0.13	0.3
acyclovir	0.04	0.04
FIAC	0.017	0.05
BVDU	0.008	1

[a]From reference 37.

in cell culture.[41-43] Acyclovir (1) has an IC_{50} =
0.1 μM. Replacement of the 2-amino substituent with
hydrogen (2) was reported to give no detectable activity
at 400 μM.[42] The 2-methylamino analogue (3) has an IC_{50}
= 19.5 μM, but the dimethylamino (4), 2-oxo (5) and
2-methylthio (6) analogues are over 2500-fold less
active.[41] The 2-methyl (7), 2-phenyl (8), $N(1)$-methyl
(9) and 2-aza (10) analogues of acyclovir are also
inactive.[43]

A variety of 8-substituted analogues of acyclovir
were reported by M.J. Robins.[42] The activity of
ten 8-substituted compounds against HSV-1 and HSV-2 are
tabulated in Table 3. The IC_{50}s are in $\mu g/ml$ with
acyclovir (1) active at 0.05 $\mu g/ml$. Substitution with
an 8-methyl (11), 8-bromo (13), 8-iodo (14) or 8-amino
(16) substituent gives analogues that are only 7- to 14-
fold less active than acyclovir. The 8-chloro (12)
analogue has anomalously low activity against both types
of HSV. The 8-oxo (15) and substituted 8-amino analogues (17-19) are substantially less active than acyclovir with 75- to 6000-fold higher IC_{50}s.

Several variations at the 6-position of acyclovir
were reported.[41,42,44] The 6-amino and 6-chloro
analogues have IC_{50}s 170-fold and 40-fold higher than
for acyclovir, and the 6-H compound when tested at
100 μM showed no detectable activity.[45] However, the
significance of these IC_{50}s is in question, because

Table 2. HSV-1 Activity of 2-Substituted Analogues of Acyclovir in Cell Culture

Compound	R	IC_{50}, μM	Reference
1	NH$_2$ (acyclovir)	0.1	41
2	H	>400	42
3	NHCH$_3$	19.5	41
4	N(CH$_3$)$_2$	>250	41
5	OH (oxo)	>250	41
6	SCH$_3$	>250	41
7	CH$_3$	inactive	43
8	C$_6$H$_5$	inactive	43
9	NH$_2$ (*N*(1)-CH$_3$)	inactive	43
10	(2-aza)	inactive	43

these compounds can be enzymatically converted to acyclovir. Indeed, the 2,6-diaminopurine analogue, BW A134U, is a prodrug for acyclovir due to its *in vivo* conversion to acyclovir by adenosine deaminase.[46] BW A134U is better absorbed from the gut, which results in higher plasma levels of acyclovir.[47] The 6-chloro analogue is also a substrate for adenosine deaminase. This enzyme was used by M.J. Robins to convert 2-amino-6-chloro-9-(2-hydroxyethoxymethyl)-9<u>H</u>-purine to acyclovir in an efficient chemical synthesis commencing with dioxolane.[48] Adenosine deaminase is present in the cell culture system, so the *in vitro* activity values may reflect enzymatically formed acyclovir. The 6-hydrogen analogue, BW A515U, is under evaluation as a prodrug for acyclovir. It is readily oxidized by xanthine oxidase to acyclovir both *in vitro* and *in vivo*.[44] Oral administration of BW A515U gives plasma levels and urinary excretions of acyclovir some 5- to 6-fold higher than those observed after similar administration of acyclovir; therefore, BW A515U may be useful in the clinic as a prodrug for acyclovir.

Table 3. HSV Activity of 8-Substituted Analogues of Acyclovir In Cell Culture[a]

Compound	R	IC$_{50}$, µg/ml	
		HSV-1	HSV-2
1	H (acyclovir)	0.05	0.04
11	CH3	0.5	0.5
12	Cl	70	>400
13	Br	0.5	0.5
14	I	0.4	0.4
15	OH (oxo)	15	25
16	NH2	0.7	0.3
17	NHCH3	7	3
18	N(CH3)2	30	250
19	N(CH2)5	70	150

[a] From reference 42.

These data show that a 2-amino-6-oxo substitution pattern in the purine ring of acyclovir is needed for optimal activity against herpes simplex viruses. A variety of changes in the heterocyclic ring has been reported, but these data only further substantiate the uniqueness of the purine ring system.[49] Variation of the purine ring amongst pyrrolo[2,3-d]pyrimidine, pyrazolo[3,4-d]pyrimidine and triazolo[4,5-d]pyrimidine ring systems gave only weakly active compounds at best.

A variety of side chain variations of acyclovir have been reported. Substitution of a methylene for the side chain ether oxygen of acyclovir (IC$_{50}$ = 0.1 µM) results in a 23-fold loss in *in vitro* activity against HSV-1, whereas the sulfur analogue is quite active with an IC$_{50}$ = 0.5 µM.[41] Several other linear side chain variations result in compounds with decreased activity.[41,50] Branched side chain analogues are depicted in Table 4. [41,51,52] Substitution on the

Table 4. HSV-1 Activity of Branched Side Chain Analogues of
Acyclovir In Cell Culture

Compound	R	IC$_{50}$, µM[a]	Reference
1	CH$_2$OCH$_2$CH$_2$OH (acyclovir)	0.1	41
20	CH(CH$_3$)OCH$_2$CH$_2$OH	>250	41
21	CH(CH$_2$OH)OCH$_2$CH$_2$OH	inactive	51
22	CH$_2$OCH(CH$_3$)CH$_2$OH	2.3	41
23	CH$_2$OCH(CH$_2$OH)CH$_2$OH	0.3[b]	52

[a]Cl strain except for ganciclovir. [b]Average IC$_{50}$ of four strains;
F, MGH 06, MGH 10, Shealey. The average IC$_{50}$ for acyclovir was
0.5 µM.

1'-carbon with methyl (**20**) or hydroxymethyl (**21**)
causes substantial decrease or complete loss of
activity. A 23-fold loss of activity occurs upon
substitution of a methyl on the 3'-carbon (**22**).
However, if the 3'-carbon substituent is a hydroxymethyl
group (**23**), a compound with *in vitro* potency comparable to
acyclovir (**1**) is obtained.[52] In 1982 no fewer than four
laboratories reported the potent
activity of compound **23**, variously referred to as
DHPG[53], BW B759U[13,54], 2'NDG[55] and BIOLF-62.[56] Compound
23 was recently assigned the generic name ganciclovir.

        Ganciclovir has potent *in vitro* activity against
herpes simplex virus (HSV). This activity is equivalent
to or up to 3-fold better than acyclovir depending on
the virus strain.[52] Smee, et al. reported IC$_{50}$s for
ganciclovir against HSV-1 ranging from 0.2 to 0.5 µM
under conditions where the acyclovir IC$_{50}$ ranges from
0.3 to 0.8 µM. Against HSV-2 the IC$_{50}$s range from 0.3
to 1.8 µM for ganciclovir and from 0.5 to 2.3 µM for
acyclovir.[52] A more striking aspect of ganciclovir's
antiherpetic activity is its potent activity against
human cytomegalovirus (HCMV).[57,58] It is active against

HCMV with IC_{50}s ranging from 1.0 to 7.0 μM under conditions where acyclovir is substantially less active with IC_{50}s ranging from 39 to 98 μM. This difference in activity is apparently due to the small amount of phosphorylation of acyclovir that occurs in HCMV-infected cells.[59] In contrast, ganciclovir undergoes extensive phosphorylation, which results in high levels of ganciclovir-triphosphate in HCMV-infected cells.[60] Ganciclovir is also active against VZV and EBV with IC_{50}s ranging from 1.5 to 5.9 μM for VZV[61] and 0.05 μM for EBV.[62]

A variety of side chain analogues of ganciclovir were reported, but all are less active.[41] Replacement of the hydroxymethyl substituent of ganciclovir with side groups such as hydroxyethyl, methoxymethyl, azidomethyl, aminomethyl and halogenomethyl results in substantial losses in activity.[63] The effect of changing the bridge length between guanine and the symmetrical 1,3-dihydroxypropyl substituent was studied by several groups.[63-65] Substitution of a methylene for the side chain ether oxygen of ganciclovir (IC_{50} = 0.2 μM) gives an analogue with an IC_{50} = 0.5 μM against HSV-1.[64] The side chain sulfur analogue is also very active against HSV-1 (IC_{50} = 0.9 μM), but its IC_{50}s against HSV-2 and HCMV were only 8 and 80 μM, respectively.[65] Several analogues with other types of bridges are less active.[63,65]

Whereas ganciclovir has the symmetrical 1,3-dihydroxypropyl side chain, a non-symmetrical 2,3-dihydroxypropyl side chain analogue was reported.[66] This analogue, (*S*)-iNDG ((*S*)-9-[2,3-dihydroxy-1-propoxy)methyl]guanine), is active against HSV-1 and HSV-2, but it is not active against HCMV or VZV at 25 μg/ml. The *R* enantiomer of iNDG is about 25-fold less active than the *S* enantiomer, which is active against HSV-1 and HSV-2 at concentrations comparable to ganciclovir and acyclovir.

Another dihydroxy acyclic nucleoside with selective antiherpetic activity is buciclovir (9-(3,4-dihydroxybutyl)guanine).[67] Although buciclovir does not contain the side chain ether oxygen, it has the requisite hydroxyl functions. The compound is phosphorylated in virus-infected cells resulting in a selective antiherpetic effect at concentrations where no toxic effect is seen on normal cell metabolism. The antiherpetic activity of a variety of 9-(hydroxyalkyl)-guanine analogues was reported.[68,69] Under conditions

where acyclovir has an IC_{50} = 0.3 μM, the parent
4-hydroxylbutyl analogue of buciclovir is 10-fold less
active than acyclovir with an IC_{50} = 3 μM. The four
atom chain length is critical for activity in this
linear hydroxyalkyl series, because the 3-hydroxypropyl
and 5-hydroxypentyl homologues are inactive. For
buciclovir there is a preference for the *R* configuration
of the 3-hydroxyl by about 5-fold in potency. Buciclovir
is only weakly active against HCMV[70] and HSV-1 defective or deficient in thymidine kinase.[69]

An interesting branched chain variation in this
series is represented by a compound that has a
hydroxymethyl substituent on the 2'-carbon of 9-(4-
hydroxybutyl)guanine. This compound, which has the code
name 2HM-HBG, has good anti-HSV-1 activity and was
recently reported to be very active against VZV.[71] Its
IC_{50}s against three strains of TK+ VZV are comparable to
those of BVDU and 10- to 20-fold lower than the IC_{50}s
for acyclovir and ganciclovir. However, its clinical
usefulness is in question due to clastogenic effects in
human lymphocytes and poor activity in animals against
HSV.[72]

The structure-activity data discussed so far have
dealt largely with purine acyclic nucleosides. Indeed,
although acyclovir has potent anti-HSV activity, substitution of the 2-hydroxyethoxymethyl side chain on a
pyrimidine base results in compounds with no significant antiviral activity.[41,42,49,73] Little or no significant anti-HSV activity was observed for pyrimidine
acyclic nucleosides containing the symmetrical
dihydroxypropyl side chain of ganciclovir.[74] However,
an interesting exception to this rule recently emerged
(Table 5).[75]

Beauchamp   et al.   have observed that the cytosine
acyclic analogue **28** (BW A1117U), although only weakly
active against HSV-1 and HSV-2, has activity against
HCMV that is comparable to the activity of ganciclovir.
Furthermore, BW A1117U is very active against EBV with
an IC_{50} of about 0.07 μM for inhibition of EBV DNA
replication. Although BW A1117U was mutagenic in vitro,
it represents a novel lead amongst pyrimidine acyclic
nucleosides with a spectrum of antiherpetic activity not
heretofore observed.

The various purine acyclic nucleosides have similar
mechanisms of action. For each agent, be it acyclovir,

Table 5. Antiviral Activity of Pyrimidine Analogues of Ganciclovir
in Cell Culture

			IC$_{50}$, μM			
Compound	R4	R5	HSV-1	HCMV	EBV[a]	Reference
24	OH	H	>100	>250	-[b]	74,75
25	OH	CH$_3$	>100	>250	100/206/	74,75
26	OH	F	>100			74
27	OH	Br	>100	>100	33/610/35	74
28[c]	NH$_2$	H	140	2-14	~0.07	74,75
29	NH$_2$	F	>100	14	22/206/	74,75
30	NH$_2$	I	-[d]	>100	43/577/35	75
31	NH$_2$	CH$_3$	-[d]	>100	S[e]	75
23	ganciclovir		0.1	3.4	0.05	75

[a]Number of viral genome copies per cell/control value/acyclovir
value (when run) at 50 μM or IC$_{50}$. [b]Inactive at 50 μM. [c]BW
A1117U. [d]Inactive in plaque inhibition assay at 50 μg per disc.
[e]Stimulated EBV induction by 70% at 50 μM.

ganciclovir, (S)-iNDG or buciclovir, selective toxicity
in virus-infected cells is related to the agent being a
substrate for virus-encoded thymidine kinase but not a
substrate for the normal cellular kinase.[76] Initial
monophosphorylation is followed by conversion to the
diphosphate derivative. This second step is catalyzed
by a cellular guanosine monophosphate kinase.[77] Several
cellular enzymes may contribute to formation of the
acyclic nucleoside triphosphate. For acyclovir, the
cellular phosphoglycerate kinase is the most active and
most prevalent enzyme that contributes to the final
metabolic step.[78] The triphosphate of the acyclic
nucleoside inhibits the viral DNA polymerase. For
acyclovir, the triphosphate is a more potent inhibitor

of viral DNA polymerase than of the host cell α or β DNA
polymerases. In addition, with acyclovir, rapid
inactivation of the viral DNA polymerase occurs as the
reaction proceeds.[79] Incorporation of acyclovir into
the DNA primer-template results in chain termination,
since the acyclovir side chain is monofunctional.[80] This
acyclovir-enzyme template complex is not displaceable,
and irreversible inactivation occurs.

According to this mechanism, a prerequisite for a
compound to have antiherpetic activity is that it be a
substrate for the viral thymidine kinase. Being a
substrate for thymidine kinase does not, however, ensure
that a compound will be antiviral, because there are
several steps involved in the lethal activation process
and each has its own structure-activity requirements.
Keller et al. reported the thymidine kinase and
antiviral properties of several analogues of
acyclovir.[41] The rate of phosphorylation of acyclovir
was 36% that of thymidine. The $N(2)$-methyl and
$N,N(2)$-dimethyl analogues of acyclovir have better
affinity for the enzyme and are efficiently phosphorylated, but they are much weaker as antiherpetic agents.
This also holds for the 2-methylthio analogue of
acyclovir, which is efficiently phosphorylated but has
substantially decreased antiviral activity. Amongst
acyclovir analogues with purine ring modifications,
there is no direct, positive correlation between the
ability to be phosphorylated and antiherpetic activity.

This lack of a correlation is also observed for a
set of side chain analogues.[41] The one carbon
((3-hydroxypropoxy)methyl side chain) and two carbon
(4-hydroxybutyloxy)methyl side chain) homologues of
acyclovir are better substrates for the thymidine
kinase, but they have weak activity against HSV-1. With
the side chain sulfur analogue and the methylene
analogue there is some correlation. However, the
(2-aminoethoxy)methyl analogue is quite an exception; it
is not significantly phosphorylated, but it has antiherpetic activity with an $ED_{50} = 8$ μM. This compound
may be antiviral by another mechanism of action or be
deaminated to acyclovir in small amounts.[50] Ganciclovir
is a very efficient thymidine kinase substrate, and this
is amply reflected in its potent antiherpetic
activity.[52] Thus we see that there is not a clear
correlation between substrate properties towards
thymidine kinase and antiherpetic activity. Although
substrate properties are essential for antiherpetic
activity, they are not sufficient to assure activity due

to the multiple steps in the mechanism by which these acyclic nucleosides exert their antiviral activity.

A consequence of the mechanism of action of the acyclic nucleosides is that viruses that do not induce a thymidine kinase are usually not susceptible to inhibition by agents that must be phosphorylated to be active. Several laboratories have attempted to circumvent dependence on the viral kinase for drug activation by developing agents that contain a phosphate or phosphate-like group.[81-84] The phosphonate analogues of acyclovir and ganciclovir have been reported.[81,83] These phosphorus compounds are not esters but contain an isosteric substitution of a methylene for the phosphate ester oxygen. The phosphonates of acyclovir and ganciclovir are weak or inactive against HSV-1 and HSV-2. However, the phosphonate of ganciclovir has *in vitro* activity against human CMV (IC_{50} = 10 μM) that is comparable to that of ganciclovir. Its toxicity towards Vero cells is low. Perhaps most interesting is the *in vivo* activity of this phosphonate against murine cytomegalovirus (MCMV). Its protective effect against a MCMV infection in mice was comparable to that of ganciclovir.[82] These as well as other results show that charged, phosphorus derivatives of acyclic nucleosides can have interesting *in vitro* antiviral activity that is accompanied by *in vivo* efficacy.

Within the past decade very significant advances have been made towards developing agents with clinical utility in treating herpes virus infections. We have advanced from toxic nucleosides limited in use for topical application to a nontoxic acyclic nucleoside that can be used systemically to treat several different herpetic infections. Structure-activity relationships, which are fairly well understood, point to a high degree of specificity for a 9-substituted guanine moiety with some leeway for various side chains.

3 AGENTS ACTIVE AGAINST HUMAN IMMUNODEFICIENCY VIRUS (HIV)

Human immunodeficiency virus (HIV) is the etiologic agent for the acquired immunodeficiency syndrome (AIDS). AIDS, which was first recognized in 1981 in the United States, is seen with increasing frequency throughout the world. HIV belongs to the family of RNA viruses known as retroviruses. The genetic information flows from RNA to DNA, a process that is mediated by a virus-induced reverse transcriptase. HIV has the ability to replicate in helper/inducer (T4+ or CD4+) T cells, which results in destruction of these cells that are critical for normal function of the immune system. The immuno-deficiency of AIDS predisposes patients to life-threatening opportunistic infections.[85-88]

In the past few years substantial interest has emerged to identify and develop agents with activity against HIV and AIDS. Several compounds have been reported to have activity against HIV or AIDS. These include suramin, a hexasodium salt of naphthalene-trisulfonic acid; HPA-23, a polyanion of ammonium 21-tungsto-9-antimoniate; phosphonoformate, a systemic antiherpetic agent; AL-721, a mixture of lipids that purportedly alters HIV membrane properties; ribavirin, an *in vitro* broad spectrum antiviral agent; 3'-azido-3'-deoxythymidine (AZT, zidovudine); and 2',3'-dideoxy-cytidine (ddC).[87-89]

Suramin was the first compound to be recognized as a selective inhibitor of HIV, but it has a narrow *in vitro* safety margin. HPA-23 has been claimed to be active in patients, but it does not have *in vitro* activity against HIV and is toxic to host cells. Phosphonoformate is a good reverse transcriptase inhibitor, but quite high concentrations are needed against the virus. Ribavirin has weak activity against HIV and is toxic to host cells at concentrations as low as 10 μg/ml. Both AZT and ddC are very good inhibitors of HIV

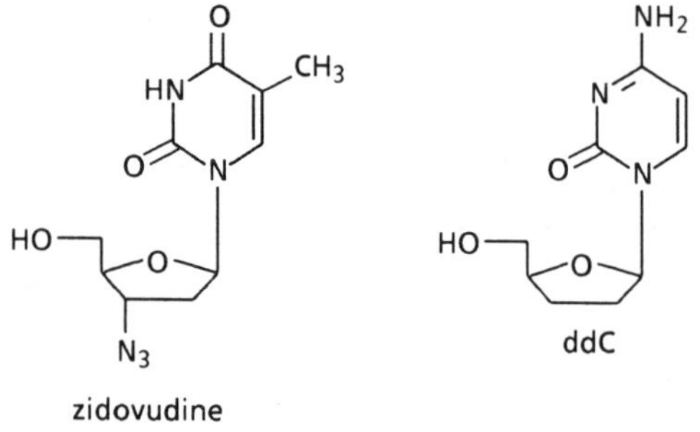

zidovudine   ddC

Figure 1.   Structures of 3'-Azido-3'-deoxythymidine (AZT, zidovudine, Retrovir®) and 2',3'-Dideoxycytidine (ddC).

with MICs around 1 μM and are more selective than other inhibitors of HIV.[87-90]

AZT was first reported in 1964 by Horwitz[91] and was subsequently resynthesized in the Wellcome Research Laboratories. The compound was recently under investigation in the Wellcome Research Laboratories as an antimicrobial agent, because it has potent, bactericidal *in vitro* activity against various members of the family *Enterobacteriaceae*.[92] Its potential use as an anti-HIV agent was demonstrated *in vitro*.[93] AZT is a potent inhibitor of HIV replication. A concentration of 1 μM is sufficient to protect ATH8 cells against the cytopathic effects of HIV *in vitro*.[93] AZT is not active in its own right but must be metabolized to the triphosphate. The triphosphate of AZT is an effective inhibitor of the HIV reverse transcriptase.[94] It is a clinically efficacious antiviral agent, which was recently approved for use in the treatment of certain patients with AIDS and AIDS-related complex under the trade name Retrovir®.[95,96]

Several 2',3'-dideoxynucleosides are reported to have good *in vitro* activity against HIV. One of the most potent inhibitors is 2',3'-dideoxycytidine (ddC), which protects ATH8 cells against the cytopathogenicity of HIV at 0.5 μM.[97] This agent, which appears to have a mechanism similar to AZT,[99] has been under development as a potential agent for treating HIV infections.[85] Other 2',3'-dideoxynucleosides are substantially less active than ddC against HIV.[97,98] The 2',3'-dideoxy-purine nucleoside analogues of adenosine, guanosine and inosine confer complete protection at 10 μM.[97] Thus, the purines are about 20-fold less active than the cytosine analogue ddC. The 5-methyl and 5-bromo analogues of ddC are inactive at 10 μM, but the 5-fluoro analogue has *in vitro* activity against HIV equivalent to ddC.[98] The thymidine analogue is only weakly active at 100 μM under these assay conditions. However, Baba et al. recently reported quite good activity for 3'-deoxythymidine in an assay using MT-4 cells.[100] Thus, the relative potency of the thymidine analogue appears to vary with assay conditions and may be attributable to differences in metabolism.

The effect of a few structural modifications of AZT have been reported.[101] Under assay conditions where AZT gives 78% protection at 1 μM with slight toxicity, the carbocyclic analogue has little or no protective effect at 7 μM. The 3'-deoxy-3'-fluoro

analogue is much less active than AZT and exhibits
substantial toxicity towards ATH8 cells. A few
analogues in which the base moiety is varied have been
reported.[101] The adenosine analogue has appreciable
activity at 5 μM with no toxicity. However, the
cytosine analogue has very poor activity at concentrations as high as 20 μM. Although comparable data for
the guanosine analogue is not available in this
reference,[101] 3'-azido-2',3'-dideoxyguanosine was
reported by others to protect MT-4 T-lymphocytes against
the cytopathic effects of HIV at 1 μg/ml.[102]

AZT is a potent inhibitor of HIV replication *in vitro*
and exhibits little effect on the growth of most
mammalian cells.[94] It is a selective agent because its
site of action is the HIV reverse transcriptase, a
target not present in normal cells. AZT is phosphorylated to the monophosphate by the cytosolic thymidine
kinase. The monophosphate is further phosphorylated by
the host-cell thymidylate kinase, and other cellular
enzymes generate AZT-triphosphate. The AZT-triphosphate
is about 100-fold more active as an inhibitor of HIV
reverse transcriptase than of the cellular DNA
polymerases. Thus, although AZT is nonselectively
phosphorylated, the resultant triphosphate selectively
inhibits the HIV DNA polymerase, a reverse transcriptase.[94] The compound may have a dual mechanism. AZT-
triphosphate was shown to be a substrate for the reverse
transcriptase and to be incorporated into a DNA oligomer
which resulted in chain termination.[103] Further
elongation of the viral DNA was terminated and viral
replication was aborted.

4   SUMMARY

This is an exciting era for antiviral chemotherapy.[104]
Substantial progress has been made during the past
decade in the development of agents for the treatment of
herpes virus infections. In a very short time, the
cause of AIDS was identified and an antiviral agent with
clinical utility was discovered and developed. From
first-generation, nonselective antiherpetic agents have
emerged compounds with selective activity. Acyclovir is
the agent of clinical choice for use against several
herpes virus infections. The successes of acyclovir
fueled the search for other selective agents, which led
to ganciclovir, a clinically useful cytomegalovirus
agent. The properties of acyclovir may be further
modified through the use of prodrugs. The successes of

antiherpetic drug development set the stage for the fast discovery and development of AZT, an anti-HIV agent with clinical utility against AIDS and AIDS-related complex. Thus, very significant progress has been made in the development of antiherpetic and anti-HIV agents. It is evident from recent reports that other gems remain to be mined in this fertile field of research on antiviral nucleoside analogues.

ACKNOWLEDGEMENT

I extend a special thanks to Ms. P. Hursey for her invaluable help with the preparation of the manuscript. I also acknowledge the assistance of Ms. Cozart, Ms. S. Paris and Ms. D. Alston in the final step of manuscript preparation, and thank Drs. J. Rideout, G. Elion, Ms. L. Beauchamp, and Mr. A. Jones for proofreading the manuscript.

REFERENCES

1. H.E. Kaufman, Proc. Soc. Exp. Biol. Med., 1962, 109, 251.

2. W.H. Prusoff, Biochem. Biophys. Acta, 1959, 32, 295.

3. E.C. Herrman Jr., Proc. Soc. Exp. Biol. Med., 1961, 107, 142.

4. W.M. Shannon, in "Antiviral Agents and Viral Diseases of Man", Second Edition, G.J. Galasso, T.C. Merigan and R.A. Buchanan, eds., Raven Press, New York, 1984, p. 60.

5. G.J. Galasso, Antiviral Res., 1981, 1, 73.

6. C. Heidelberger, D. Parsons and D.C. Remy, J. Amer. Chem. Soc., 1962, 84, 3597.

7. H.E. Kaufman and C. Heidelberger, Science, 1964, 145, 585.

8. C. Heidelberger and D.H. King, Pharmacol. Ther., 1979, 6, 427.

9. W.W. Lee, A. Benitez, L. Goodman and B.R. Baker, J. Amer. Chem. Soc., 1960, 82, 2648.

10. M. Privat de Garilhe and J. de Rudder, C.R. Acad. Sci. [D] (Paris), 1964, 259, 2725.

11. J. Colin, J. Fr. Opthalmol., 1981, 4, 525.

12. R.J. Whitley and C.A. Alford, Hosp. Pract., 1981, 16, 109.

13. H.J. Schaeffer, in "Nucleosides, Nucleotides, and Their
     Biological Applications", J.L. Rideout, D.W. Henry, and
     L.M. Beacham III, eds., Academic Press, New York, 1983, p.1.

14. G.B. Elion, in "Antiviral Chemotherapy: New Directions for
     Clinical Application and Research", J. Mills and L. Corey,
     eds., Elsevier, New York, 1986, p. 118.

15. P.H. Fischer and W.H. Prusoff, Handb. Exp. Pharmacol., 1982,
     61, 95.

16. N.H. Park and D. Pavan-Langston, Handb. Exp. Pharmacol., 1982,
     61, 117.

17. E. De Clercq, J. Descamps, P. DeSomer, P.J. Barr, A.S. Jones
     and R.T. Walker, Proc. Natl. Acad. Sci. USA, 1979, 76, 2947.

18. W.M. Shannon, in "Antiviral Agents and Viral Diseases of Man",
     Second Edition, G.J. Galasso, T.C. Merigan and R.A. Buchanan,
     eds., Raven Press, New York, 1984, p.68.

19. Bromovinyldeoxyuridine. In: PHAR, Pharmaprojects [Data
     base]. Geneva, Switz: DATA-STAR, Radio-Suisse, [1987 July];
     accession no. 003346.

20. W.M. Shannon, in "Antiviral Agents and Viral Diseases of Man",
     Second Edition, G.J. Galasso, T.C. Merigan and R.A. Buchanan,
     eds., Raven Press, New York, 1984, p. 69.

21. C. McLaren, M.S. Chen, R.H. Barbhaiya, R.A. Buroker and
     F.B. Oleson, in "Pharmacological and Clinical Approaches to
     Herpes Viruses and Virus Chemotherapy", R. Kono et al.,
     eds., Excerpta Medica, Amsterdam, 1985, p. 57.

22. E.R. Kern, J.T. Richards, J.C. Overall, Jr. and L.A. Glasgow,
     Antiviral Res., 1981, 1, 225.

23. B. Oberg, Pharmacol. Ther., 1983, 19, 387.

24. J.K. Preiksaitis, B. Lank, P.K. Ng, L. Brox, G.A. LePage and
     D.L.J. Tyrrell, J. Infect. Dis., 1981, 144, 358.

25. J.C. Drach, Ann. Rep. Med. Chem., 1980, 15, 149.

26. R. Vince, S. Daluge and H. Lee, W.M. Shannon, G. Arnett,
     T.W. Schafer, T.L. Nagabhushan, P. Reichert and H. Tsai,
     Science, 1983, 221, 1405.

27. J. Schwartz, M. Ostrander, N.J. Butkiewicz, M. Lieberman,
    C. Lin, J. Lim, and G.H. Miller, <u>Antimicrob. Agents</u>
    <u>Chemother.</u>, 1987, <u>31</u>, 21.

28. T.W. North and S.S. Cohen, <u>Pharmacol. Ther.</u>, 1979, <u>4</u>, 81.

29. W.M. Shannon and F.M. Schabel, Jr., <u>Pharmacol. Ther.</u>, 1980,
    <u>11</u>, 263.

30. T.W. North and S.S. Cohen, <u>Proc. Natl. Acad. Sci. USA</u>, 1978,
    <u>75</u>, 4684.

31. H.J. Schaeffer in "Topics in Medicinal Chemistry",
    J.L. Rabinowitz and R.M. Myerson, John Wiley & Sons, New York,
    1970, p. 1.

32. H.J. Schaeffer and C.F. Schwender, <u>J. Med. Chem.</u>, 1974, <u>17</u>, 6.

33. R.P. Agarwal, T. Spector and R.E. Parks, Jr., <u>Biochem.</u>
    <u>Pharmac.</u>, 1977, <u>26</u>, 359.

34. H.J. Schaeffer, S. Gurwara, R. Vince and S. Bittner, <u>J. Med.</u>
    <u>Chem.</u>, 1971, <u>14</u>, 367.

35. G.B. Elion, P.A. Furman, J.A. Fyfe, P. de Miranda,
    L. Beauchamp, and H.J. Schaeffer, <u>Proc. Natl. Acad. Sci. USA</u>,
    1977, <u>74</u>, 5716.

36. H.J. Schaeffer, L. Beauchamp, P. de Miranda, G.B. Elion,
    D.J. Bauer, and P. Collins, <u>Nature (London)</u>, 1978, <u>272</u>, 583.

37. E. De Clercq, J. Descamps, G. Verhelst, R.T. Walker,
    A.S. Jones, P.F. Torrence, and D. Shugar, <u>J. Infect. Dis.</u>,
    1980, <u>141</u>, 563.

38. J.A. Fyfe, <u>Mol. Pharmacol.</u>, 1982, <u>21</u>, 432.

39. C.K. Chu and S.J. Cutler, <u>J. Heterocyclic Chem.</u>, 1986, <u>23</u>,
    289.

40. R.J. Remy and J.A. Secrist III, <u>Nucleosides and Nucleotides</u>,
    1985, <u>4</u>, 411.

41. P.M. Keller, J.A. Fyfe, L. Beauchamp, C.M. Lubbers,
    P.A. Furman, H.J. Schaeffer and G.B. Elion, <u>Biochem. Pharmac.</u>,
    1981, <u>30</u>, 3071.

42. M.J. Robins, P.W. Hatfield, J. Balzarini and E. De Clercq, <u>J.</u>
    <u>Med. Chem.</u>, 1984, <u>27</u>, 1486.

43. A. Parkin and M.R. Harnden, J. Heterocyclic Chem.,1982, 19, 33.

44. T.A. Krenitsky, W.W. Hall, P. de Miranda, L.M. Beauchamp, H.J. Schaeffer and P.D. Whiteman, Proc. Natl. Acad. Sci. USA, 1984, 81, 3209.

45. P. Collins, Wellcome Research Laboratories, unpublished.

46. T. Spector, T.E. Jones and L.M. Beacham, III, Biochem. Pharma., 1983, 32, 2505.

47. H.C. Krasny, S.H.T. Liao and S.S. Good, Clin. Pharmac. Ther., 1983, 33, 256.

48. M.J. Robins and P.W. Hatfield, Can. J. Chem., 1982, 60, 547.

49. L.M. Beauchamp, B.L. Dolmatch, H.J. Schaeffer, P. Collins, D.J. Bauer, P.M. Keller and J.A. Fyfe, J. Med. Chem., 1985, 28, 982.

50. J.L. Kelley, M.P. Krochmal and H.J. Schaeffer, J. Med. Chem., 1981, 24, 1528.

51. D.P.C. McGee and J.C. Martin, Can. J. Chem., 1986, 64, 1885.

52. D.F. Smee, J.C. Martin, J.P.H. Verheyden and T.R. Matthews, Antimicrob. Agents Chemother., 1983, 23, 676.

53. J.C. Martin, C.A. Dvorak, D.F. Smee, T.R. Matthews and J.P.H. Verheyden, J. Med. Chem., 1983, 26, 759.

54. E.A. Rollinson and G. White, Antimicrob. Agents Chemother., 1983, 24, 221.

55. W.T. Ashton, J.D. Karkas, A.K. Field and R.L. Tolman, Biochem. Biophys. Res, Commun., 1982, 108, 1716.

56. K.O. Smith, K.S. Galloway, W.L. Kennell, K.K. Ogilvie and B.K. Radatus, Antimicrob. Agents Chemother., 1982, 22, 55.

57. Y.-C. Cheng, E.-S. Huang, J.-C. Lin, E.-C. Mar, J.S. Pagano, G.E. Dutschman and S.P. Grill, Proc. Natl. Acad. Sci. USA, 1983, 80, 2767.

58. V.R. Freitas, D.F. Smee, M. Chernow, R. Boehme, and T.R. Matthews, Antimicrob. Agents Chemother., 1985, 28, 240.

59. M.H. St. Clair, P.A. Furman, C.M. Lubbers and G.B. Elion, Antimicrob. Agents Chemother., 1980, 18, 741.

60. K.K. Biron, S.C. Stanat, J.B. Sorrell, J.A. Fyfe, P.M. Keller, C.U. Lambe and D.J. Nelson, <u>Proc. Natl. Acad. Sci., USA</u>, 1985, <u>82</u>, 2473.

61. K.K. Biron, P.J. Stenbuck, J.B. Sorrell in "Herpesvirus", F. Rapp, ed., Alan R. Liss, Inc., New York, 1984, p. 677.

62. J.C. Lin, D.J. Nelson, C.U. Lambe, E.I. Choi and J.S. Pagano in "Herpes Viruses and Virus Chemotherapy; Pharmacological and Clinical Approaches," R. Kono, ed., Elsevier Publishers, Amsterdam, 1985, p. 225.

63. J.C. Martin, D.P.C. McGee, G.A. Jeffrey, D.W. Hobbs, D.F. Smee, T.R. Matthews and J.P.H. Verheyden, <u>J. Med. Chem.</u>, 1986, <u>29</u>, 1384.

64. M.A. Tippie, J.C. Martin, D.F. Smee, T.R. Matthews and J.P.H. Verheyden, <u>Nucleosides & Nucleotides</u>, 1984, <u>3</u>, 525.

65. D.P.C. McGee, J.C. Martin, D.F. Smee, T.R. Matthews and J.P.H. Verheyden, <u>J. Med. Chem.</u>, 1985, <u>28</u>, 1242.

66. W.T. Ashton, L.F. Canning, G.F. Reynolds, R.L. Tolman, J.D. Karkas, R. Liou, M.-E.M. Davies, C.M. DeWitt, H.C. Perry and A.K. Field, <u>J. Med. Chem.</u>, 1985, <u>28</u>, 926.

67. A. Larsson, B. Oberg, S. Alenius, C.-E. Hagberg, N.-G. Johansson, B. Lindborg and G. Stening, <u>Antimicrob. Agents Chemother.</u>, 1983, <u>23</u>, 664.

68. A.-C. Ericson, A. Larsson, F.Y. Aoki, W.-A. Yisak, N.-G. Johansson, B. Oberg and R. Datema, <u>Antimicrob. Agents Chemother.</u>, 1985, <u>27</u>, 753.

69. R. Datema, N.G. Johannson and B. Oberg, <u>Chemica Scripta</u>, 1986, <u>26</u>, 49.

70. B. Wahren, A. Larsson, U. Ruden, A. Sundqvist and E. Solver, <u>Antimicrob. Agents Chemother.</u>, 1987, <u>31</u>, 317.

71. G. Abele, A. Karlstrom, J. Harmenberg, S. Shigeta, A. Larsson, B. Lindborg and B. Wahren, <u>Antimicrob. Agents Chemother.</u>, 1987, <u>31</u>, 76.

72. A. Larsson, K. Stenberg, A.C. Ericson, U. Haglund, W.-A. Yisak, N.G. Johansson, B. Oberg and R. Datema, <u>Antimicrob. Agents Chemother.</u>, 1986, <u>30</u>, 598.

73. J.L. Kelley, J.E. Kelsey, W.R. Hall, M.P. Krochmal and
    H.J. Schaeffer, J. Med. Chem., 1981, 24, 753.

74. J.C. Martin, G.A. Jeffrey, D.P.C. McGee, M.A. Tippie,
    D.F. Smee, T.R. Matthews and J.P.H. Verheyden, J. Med.
    Chem., 1985, 28, 358.

75. L. Beauchamp, B.L. Serling, J.E. Kelsey, K.K. Biron,
    P. Collins, J. Selway, J.-C. Lin and H.J. Schaeffer, J. Med.
    Chem., in press (1987).

76. J.A. Fyfe, P.M. Keller, P.A. Furman, R.L. Miller and
    G.B. Elion, J. Biol. Chem., 1978, 253, 8721.

77. W.H. Miller and R.L. Miller, J. Biol. Chem., 1980, 255, 7204.

78. W.H. Miller and R.L. Miller, Biochem. Pharmacol., 1982, 31,
    3879.

79. P.A. Furman, M.H. St. Clair and T. Spector, J. Biol. Chem.,
    1984, 259, 9575.

80. P.A. Furman, M.H. St. Clair, J.A. Fyfe, J.L. Rideout,
    P.M. Keller and G.B. Elion, J. Virol., 1979, 32, 72.

81. E.J. Prisbe, J.C. Martin, D.P.C. McGee, M.F. Barker,
    D.F. Smee, A.E. Duke, T.R. Matthews and J.P.H. Verheyden,
    J. Med. Chem., 1986, 29, 671.

82. A.E. Duke, D.F. Smee, M. Chernow, R. Boehme and
    T.R. Matthews, Antiviral Res., 1986, 6, 299.

83. W. Streicher, G. Werner and B. Rosenwirth, Chemica Scripta,
    1986, 26, 179.

84. E. De Clercq, A. Holy, I. Rosenberg, T. Sakuma, J. Balzarini
    and P.C. Maudgal, Nature, 1985, 323, 464.

85. H. Mitsuya and S. Broder, Nature, 1987, 325, 773.

86. R. Yarchoan and S. Broder, N. Engl. J. Med., 1987, 316, 557.

87. L. Resnick, The Mount Sinai J. Med., 1986, 53, 648.

88. E. De Clercq, J. Med. Chem., 1986, 29, 1561.

89. T.-S. Lin, R.F. Schinazi, M.S. Chen, E. Kinney-Thomas and
    W.H. Prusoff, Biochem. Pharmacol., 1987, 36, 311.

90. H. Mitsuya, M. Popovic, R. Yarchoan, S. Matsushita, R.C. Gallo and S. Brodor, <u>Science</u>, 1984, <u>226</u>, 172.

91. J.P. Horwitz, J. Chua and M. Noel, <u>J. Org. Chem.</u>, 1964, <u>29</u>, 2076.

92. L.P. Elwell, R. Ferone, G.A. Freeman, J.A. Fyfe, J.A. Hill, P.H. Ray, C.A. Richards, S.C. Singer, V.B. Knick, J.L. Rideout and T.P. Zimmerman, <u>Antimicrob. Agents Chemother.</u>, 1987, <u>31</u>, 274.

93. H. Mitsuya, K.J. Weinhold, P.A. Furman, M.H. St. Clair, S.N. Lehrman, R.C. Gallo, D. Bolognesi, D.W. Barry and S. Broder, <u>Proc. Natl. Acad. Sci. USA</u>, 1985, <u>82</u>, 7096.

94. P.A. Furman, J.A. Fyfe, M.H. St. Clair, K. Weinhold, J.L. Rideout, G.A. Freeman, S.N. Lehrman, D.P. Bolognesi, S. Broder, H. Mitsuya and D.W. Barry, <u>Proc. Natl. Acad. Sci. USA</u>, 1986, <u>83</u>, 8333.

95. R. Yarchoan, K.J. Weinhold, H.K. Lyerly, E. Gelmann, R.M. Blum, G.M. Shearer, H. Mitsuya, J.M. Collins, C.E. Myers, R.W. Klecker, P.D. Markham, D.T. Durack, S.N. Lehrman, D.W. Barry, M.A. Fischl, R.C. Gallo, D.P. Bolognesi and S. Broder, <u>Lancet</u>, 1986, 575.

96. R. Yarchoan, P. Brouwers, A.R. Spitzer, J. Grafman, B. Safai, C.F. Perno, S.M. Larson, G. Berg, M.A. Fischl, A. Wichman, R.V. Thomas, A. Brunetti, P.J. Schmidt, C.E. Myers and S. Broder, <u>Lancet</u>, 1987, 132.

97. H. Mitsuya and S. Broder, <u>Proc. Natl. Acad. Sci. USA</u>, 1986, <u>83</u>, 1911.

98. C.-H. Kim, V.E. Marquez, S. Broder, H. Mitsuya and J.S. Driscoll, <u>J. Med. Chem.</u>, 1987, <u>30</u>, 862.

99. D.A. Cooney, M. Dalal, H. Mitsuya, J.B. McMahon, M. Nadkarni, J. Balzarini, S. Broder and D.G. Johns, <u>Biochem. Pharmacol.</u>, 1986, <u>35</u>, 2065.

100. M. Baba, R. Pauwels, P. Herdewijn, E. De Clercq, J. Desmyter and M. Vandeputte, <u>Biochem. Biophys. Res. Comm.</u>, 1987, <u>142</u>, 128.

101. H. Mitsuya, M. Matsukura and S. Broder in "AIDS Modern Concepts and Therapeutic Challenges," S. Broder, ed., Marcel Dekker, Inc., 1986, p. 303.

102. H. Hartmann, G. Hunsmann and F. Eckstein, <u>Lancet</u>, 1987, 40.

103. M.H. St. Clair, C.A. Richards, K.J. Weinhold, W.H. Miller, A. Langlois and P.A. Furman, submitted for publication·

104. M.C. Nahata, <u>Drug Intell. Clin. Pharm.</u>, 1987, <u>21</u>, 399.

New Purine Derivatives with Selective Antiviral Activity

M. R. Harnden,* S. Bailey, M. R. Boyd, M. Cole, R. L. Jarvest, and P. G. Wyatt

BEECHAM PHARMACEUTICALS RESEARCH DIVISION, BIOSCIENCES RESEARCH CENTRE, GREAT BURGH, YEW TREE BOTTOM ROAD, EPSOM, SURREY KT18 5XQ, UK

Since the discovery of the potent and selective anti-herpesvirus activity of acyclovir,[1-3] many groups have prepared and evaluated additional acyclic analogues of nucleosides.

Replacement of the guanine base by other purines, such as adenine, hypoxanthine, xanthine, 2,6-diaminopurine and 8-azaguanine, has in each case resulted in greatly diminished or no antiviral activity. Analogous N1-substituted derivatives of the pyrimidines uracil, thymine and cytosine also generally have little activity. However, many N9-substituted guanines have now been synthesized and a number of them possess selective anti-herpesvirus activity.[4] Of these, the compounds shown overleaf (Figure 1) are of major interest and have been extensively investigated.

These compounds have been evaluated in many cell culture tests by different investigators, often using different strains of virus and different cells. A range of virus inhibitory concentrations is therefore available for each compound, and in Table 1 only an overview of their relative activities is presented.

Figure 1

Acyclonucleosides with Anti-herpesvirus Activity

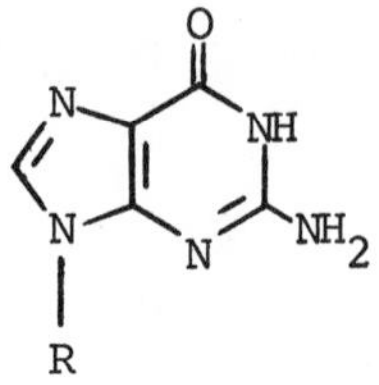

	$\underline{R}$
ACV	$HOCH_2CH_2OCH_2-$
DHPG	$(HOCH_2)_2CHOCH_2-$
BRL 39123	$(HOCH_2)_2CHCH_2CH_2-$
($\underline{R}$)-DHBG	$HOCH_2CCH_2CH_2-$ $HO\ H$
($\underline{S}$)-iNDG	$HOCH_2CCH_2OCH_2-$ $HO\ H$

Table 1

Relative Antiviral Activities

of Acyclonucleosides in Cell Culture

	HSV-1	HSV-2	VZV	EBV	CMV
ACV	+++	+++	++	++	+
DHPG	++++	++++	+++	+++	+++
BRL 39123	+++	+++	++	++	+
($\underline{R}$)-DHBG	++	++	+	++	NA
($\underline{S}$)-iNDG	+++	+++	NA		NA

NA = Inactive

In many systems, DHPG (ganciclovir) is more active than ACV (acyclovir) and it is the only acyclonucleoside that has useful activity against human cytomegalovirus. Unfortunately, ganciclovir has proved to be more toxic than acyclovir[5,6] and consequently it is not now being developed for use in the treatment of herpes simplex and varicella zoster virus infections. Cytomegalovirus infections are, however, a major clinical problem in the immunocompromised, and initial studies using ganciclovir in AIDS patients with cytomegalovirus retinitis and pneumonitis have been encouraging.[7]

(R̲)-DHBG (buciclovir)[8] and (S̲)-iNDG[9] have also been extensively investigated. Both of these compounds are highly selective inhibitors of herpes simplex virus replication with a biochemical mode of action similar to that of acyclovir. Although in tests in herpes simplex virus-infected cell cultures and animals each has sometimes shown activity as good as, or better than, that of acyclovir, buciclovir has reduced activity against varicella zoster virus and (S̲)-iNDG has none. Overall, neither compound appears to possess significant advantages in comparison with acyclovir.

In 1983 we prepared BRL 39123, the carba analogue of ganciclovir, and noted its antiviral properties.[10] Our synthesis (Figure 2) involved alkylation of 2-amino-6-chloropurine with the acetonide of 4-hydroxy-3-hydroxymethylbutyl bromide. The N9-substituted purine was obtained almost exclusively and upon acid hydrolysis this provided BRL 39123 in about 50% overall yield.[11]

Figure 2

Synthesis of

9-(4-Hydroxy-3-hydroxymethylbut-1-yl)guanine (BRL 39123)

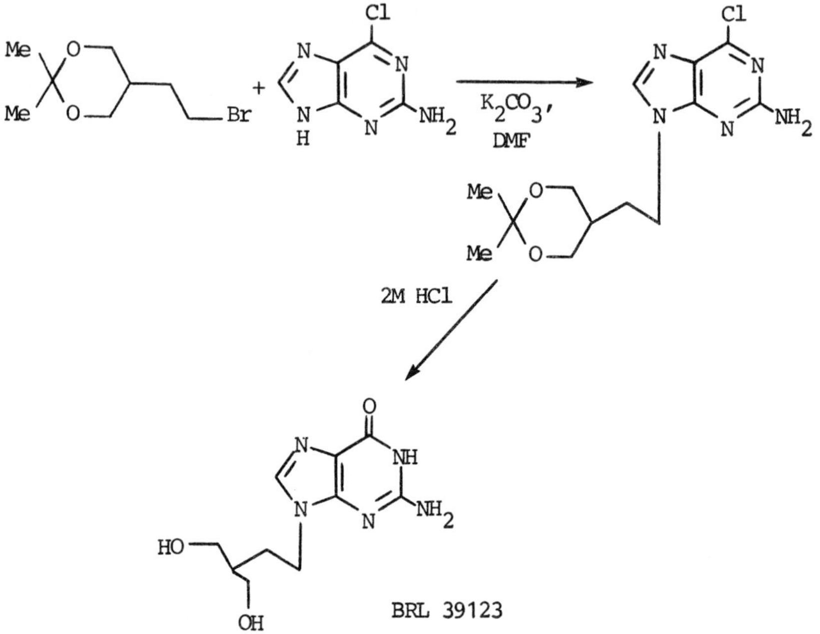

BRL 39123

In initial plaque reduction tests against clinical isolates of herpesviruses (Table 2), BRL 39123 was slightly less active than acyclovir against herpes simplex type 1 viruses, 2-3 fold less active against the type 2 viruses and slightly more active against varicella zoster virus strains. Uninfected mammalian cell growth was not affected in the presence of concentrations of BRL 39123 that were very much higher than those that were antiviral.

Table 2

Activity of BRL 39123 and Acyclovir Against Clinical
Strains of Herpesviruses[a] and Cell Growth Inhibition

Mean IC_{50} (µg/ml)

	HSV-1 (16)	HSV-2 (13)	VZV (5)	Uninfected MRC-5[b] cell replication
BRL 39123	0.3	1.5	3.1	>100
ACV	0.2	0.6	3.8	>100

[a] Plaque reduction tests in MRC-5 (human fibroblast cells)

( ) Number of strains tested

[b] Inhibition of growth of uninfected cells over 72 hours

Like acyclovir, BRL 39123 was not active against TK⁻ strains of herpes simplex virus, although it did inhibit the replication of an acyclovir-resistant strain of herpes simplex type 1 virus which carries a mutation in the DNA polymerase gene (CL101, P_2C_5). Furthermore, in virus yield reduction tests, which directly measure the quantity of virus produced and may relate more closely to efficacy *in vivo*, BRL 39123 proved to be 2-3 fold more active than acyclovir against type 1 strains of herpes simplex virus and equally active against type 2 strains (Table 3).[12,13]

Table 3

Antiviral Activity of BRL 39123 and Acyclovir in Virus
Yield Inhibition Tests in MRC-5 Cells

Concentration (µg/ml) for 99% Reduction
of Infectious Virus 24hr After Infection

	HSV-1 (3)	HSV-2 (4)
BRL 39123	0.4	0.7
ACV	1.0	0.6

( ) Number of clinical strains tested

Some additional properties of BRL 39123 which were
discovered subsequently are of major interest.

In Figure 3, the results are given of an experiment
in which MRC-5 cells infected with herpes simplex type 2
virus were exposed to 30µg/ml concentrations of BRL
39123 or acyclovir for various lengths of time,
beginning 2 hours after infection. Cell-free virus
titres produced in the cultures were measured 25 hours
after infection. Brief treatment with BRL 39123
resulted in considerably lower virus yield than a
similar period of treatment with acyclovir. Indeed, for
maximum reduction in virus yield, the presence of
acyclovir throughout the entire incubation period was
required.

Figure 3

Effect of Duration of Treatment with BRL 39123 and
Acyclovir on HSV-2 Replication in MRC-5 Cells

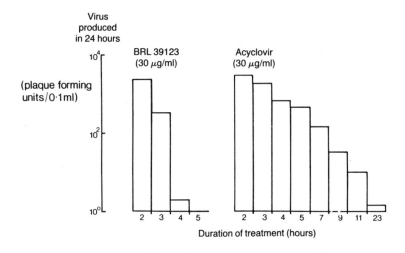

In another experiment (Figure 4) MRC-5 cells were treated with either BRL 39123 or acyclovir for 18 hours after infection with herpes simplex type 2 virus. The extracellular compound was then removed, incubation continued and infectivity titres of supernatants from the cultures determined 2, 4, and 6 days after infection. The amount of virus produced by infected cells after treatment with acyclovir at 30µg/ml rapidly increased upon removal of the compound to give levels similar to those released by untreated cells. However, after treatment with BRL 39123 even at 3µg/ml, infectivity titres declined.[13] This persistent antiviral activity with BRL 39123 has also been observed against herpes simplex virus type 1 and in other cell lines.

Figure 4

Persistence of Activity After 18 hour Treatment of HSV-2
Infected MRC-5 Cells with BRL 39123 and Acyclovir

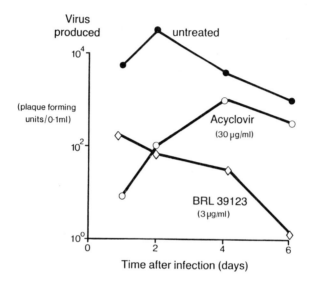

The mode of action of BRL 39123, like that of acyclovir, is believed to involve phosphorylation by a herpesvirus-specified thymidine kinase and subsequent conversion to a triphosphate, which is an inhibitor of the viral DNA polymerase. The explanation of its quicker onset of action and persistent effect lies in the more rapid formation of higher concentrations of its triphosphate in herpesvirus-infected cells (Figure 5) and the much slower rate of degradation of its triphosphate, as compared with the rate for acyclovir triphosphate, following removal of extracellular acyclonucleoside (Figure 6).

Figure 5

Formation of the Triphosphates of BRL 39123 and Acyclovir in MRC-5 Cells Infected with HSV-1

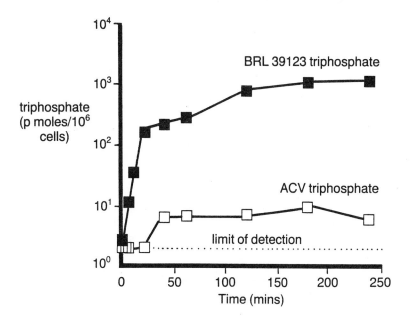

This persistence of activity in the absence of extracellular compound suggests that it may not be necessary to maintain high concentrations of BRL 39123 in the blood for a prolonged duration in order to achieve efficacy _in vivo_. Indeed, we have shown that, using less frequent dosing schedules, BRL 39123 is more effective than acyclovir in some animal infection models. In animal and uninfected human volunteer studies the toxicological profile of BRL 39123 has proved to be acceptable, and we hope that the advantageous properties that we have identified in cell culture and animal experiments will also become apparent in clinical trials which are now commencing.

Figure 6

Degradation of the Triphosphates of BRL 39123 and
Acyclovir in MRC-5 Cells Infected with HSV-1

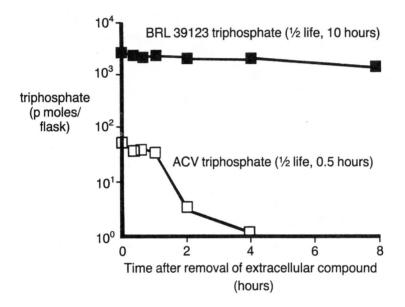

In recent years we have carried out extensive investigations on the synthesis and evaluation of analogues of BRL 39123 and other antiviral acyclonucleosides. As part of this programme we decided to prepare a series of acyclonucleosides in which the atom attached to N9 of guanine is oxygen.

Our initial target compounds were the analogues of acyclovir, ganciclovir and BRL 39123, buciclovir and i-NDG shown in Figure 7.

Figure 7
9-Alkoxyguanines

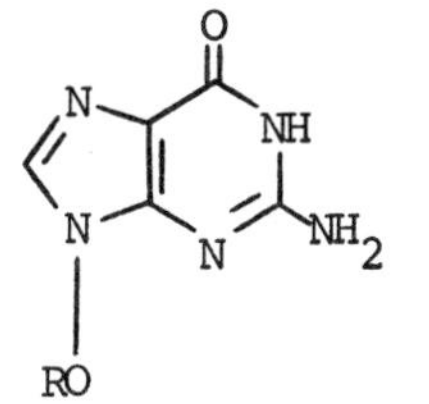

$$R$$
$$HO(CH_2)_3-$$
$$(HOCH_2)_2CHCH_2-$$
$$HOCH_2CHCH_2-$$
$$\qquad\quad OH$$
$$HOCH_2CHCH_2CH_2-$$
$$\qquad\quad OH$$
$$C_6H_5CH_2-$$

The only N9-alkoxyguanine to have been reported previously was 9-benzyloxyguanine, which had been prepared in 1974 as an intermediate in an unequivocal synthesis of 9-hydroxyguanine.[14] The synthesis reported is shown in Figures 8 and 9.

The guanine derivative was obtained via a 1-benzyl-oxyimidazole intermediate (Figure 8). This was prepared by reaction of benzyloxyamine with triethylorthoformate to give the N-benzyloxyformimidate, which was then condensed with aminocyanoacetamide. Unfortunately, the latter reaction proceeded in only 2% yield. A slightly improved variation involved condensation of benzyloxy-amine hydrochloride with the formimidate from amino-cyanoacetamide. However, this only increased the yield to 15% and the subsequent 5-stage conversion of the 5-amino-4-carbamoylimidazole to the required 9-hydroxy-guanine (Figure 9) was achieved in 30% overall yield.

Figure 8
Synthesis of 5-Amino-1-benzyloxy-4-carbamoylimidazole

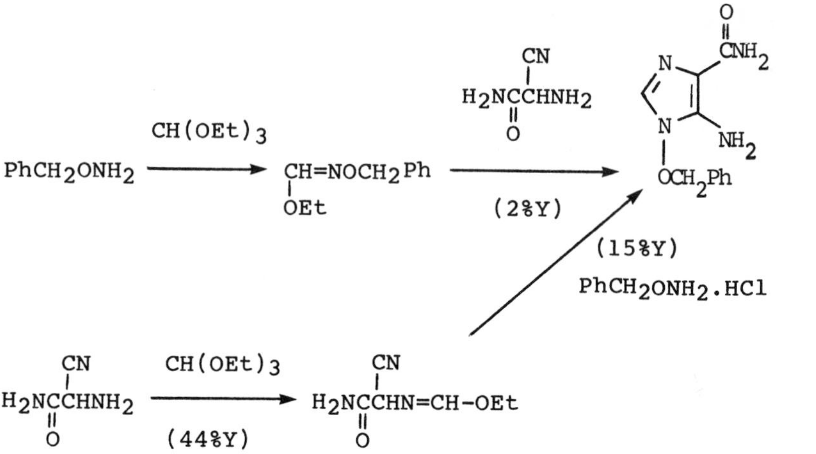

Figure 9
Synthesis of 9-Hydroxyguanine

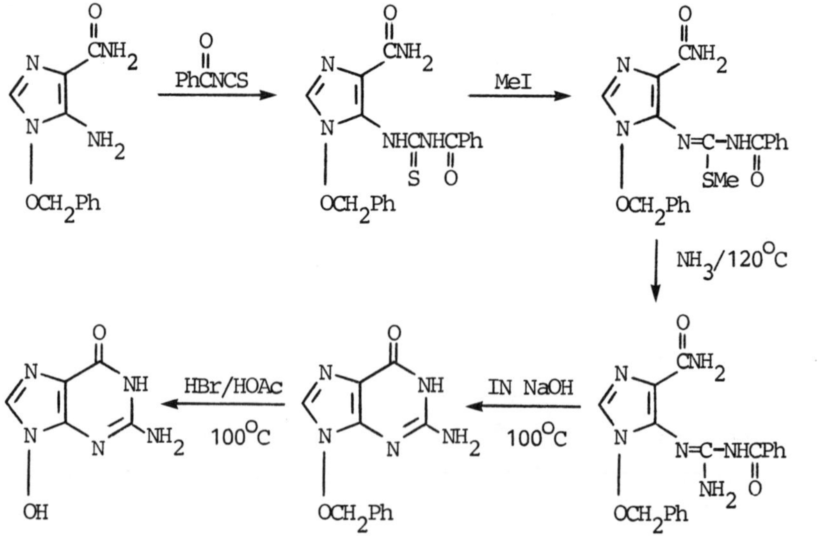

30% overall Y

Despite these low yields, we decided to commence work on the series using adaptations of the published procedures (Figure 10).

Figure 10
Synthesis of 9-(3-Hydroxypropoxy)guanine

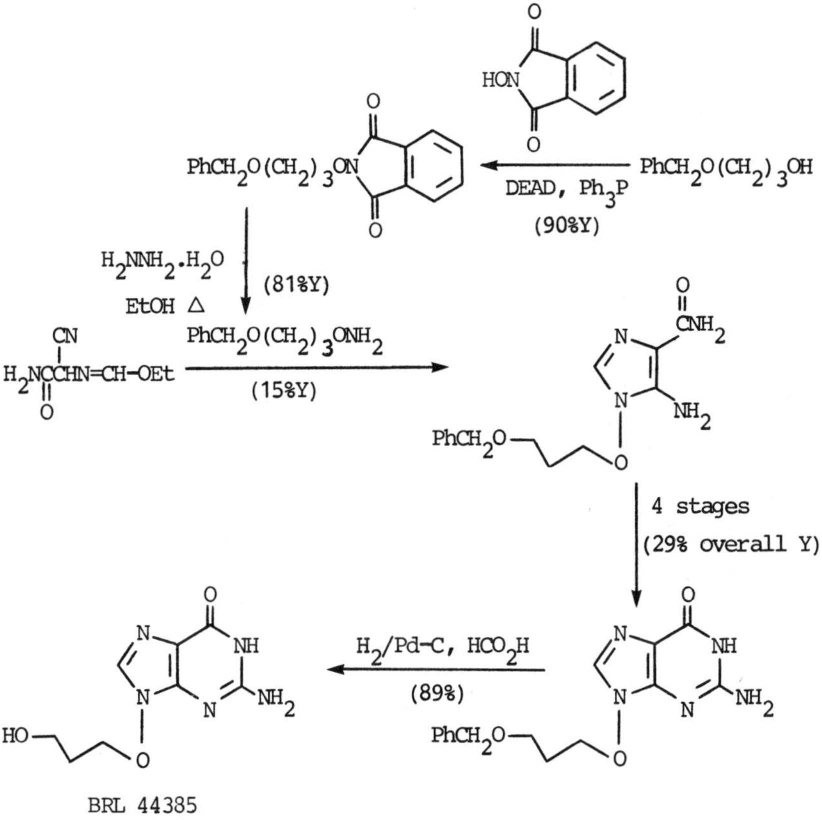

BRL 44385

We were able to prepare 3-benzyloxypropoxyamine in high overall yield by condensation of the monobenzyl ether of propane-1,3-diol with N-hydroxyphthalimide,[15] followed by reaction with hydrazine hydrate. The alkoxyamine was then reacted with the formimidate from aminocyanoacetamide to provide the intermediate imidazole in 15% yield. Using the same sequence of reactions as had been used to prepare 9-benzyloxy-guanine, the imidazole was converted to 9-(3-benzyloxy-propoxy)guanine in 29% overall yield. Catalytic hydrogenolysis then afforded the required 3-hydroxypropoxy derivative BRL 44385.

This novel acyclonucleoside showed a very interesting spectrum of anti-herpesvirus activity in cell culture (Table 4).

Against herpes simplex virus types 1 and 2, BRL 44385 is about 3 times more potent than acyclovir. Against varicella zoster virus it is about 5 times more potent and against Epstein-Barr virus about 30 times more potent. No activity was noted against cytomegalovirus, using BRL 44385 concentrations up to 100µg/ml. BRL 44385 showed no toxicity for confluent cell monolayers and concentrations of the compound that affected the growth of uninfected MRC-5 cells were more than one hundred times higher than the concentrations required for antiviral activity.

Encouraged by this data, we set about the task of improving the synthesis of BRL 44385 (Figure 11). Brief reaction of the protected alkoxyamine with N,N-dimethyl-formamide dimethyl acetal afforded the dimethyl-formamidine derivative in 92% yield. Reaction with aminocyanoacetamide hydrochloride in methanol at room temperature then gave the required formamidine. This was isolated as a mixture of geometric isomers in 70% yield and its cyclisation to the imidazole derivative was investigated. A 63% yield of the imidazole was achieved by treatment with boron trifluoride etherate in 1,2-dimethoxyethane at 60-80°C.

Table 4

Comparison of the Antiviral Activity of BRL 44385 and Acyclovir in Cell Culture

	X	Y
BRL 44385	O	CH_2
Acyclovir	CH_2	O

Virus[a]	Cell	IC_{50} (μg/ml) BRL 44385	Acyclovir
HSV-1 (HFEM)	Vero	0.3	0.9
HSV-1 (SC-16)	MRC-5	0.4	1.4
HSV-2 (MS)	Vero	0.2	0.5
HSV-2 (MS)	MRC-5	0.2	0.6
VZV (Ellen)	MRC-5	0.8	3.8
EBV	P$_3$HR-1	0.3	10
CMV (AD169)	MRC-5	>100	28
	MRC-5[b]	85	>100

[a] In Tables 4-9, IC_{50} values recorded against HSV-1 (HFEM) in Vero cells, against HSV-2 (MS) in Vero and MRC-5 cells and against VZV (Ellen) and CMV (AD169) in MRC-5 cells were obtained in plaque reduction assays. IC_{50} values recorded against HSV-1 (SC-16) in MRC-5 cells were for inhibition of virus-induced cytopathic effect. The IC_{50} values against EBV were obtained in a fluorescent antibody assay determining inhibition of capsid antigen expression.

[b] Inhibition of growth of uninfected cells over 72 hours.

Figure 11

Improved Synthesis of

5-Amino-1-(3-benzyloxypropoxy)-4-carbamoylimidazole

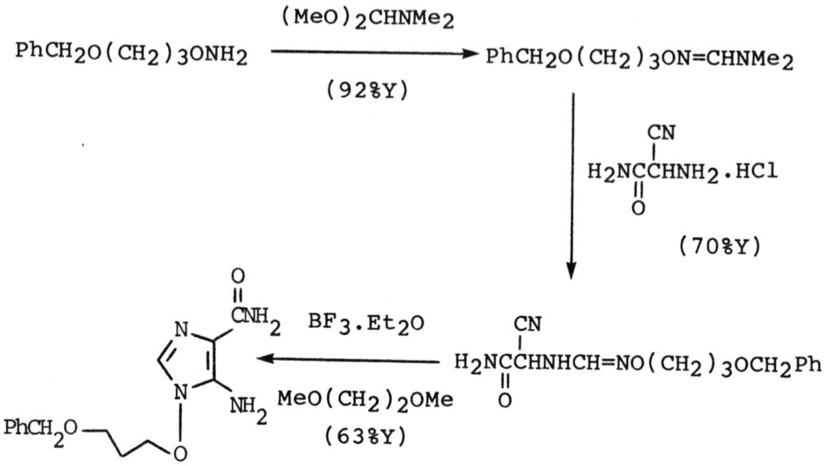

$$PhCH_2O(CH_2)_3ONH_2 \quad \xrightarrow[\text{(92\%Y)}]{(MeO)_2CHNMe_2} \quad PhCH_2O(CH_2)_3ON=CHNMe_2$$

$$\downarrow \quad \begin{array}{c} CN \\ | \\ H_2NCCHNH_2 \cdot HCl \\ \| \\ O \end{array}$$

(70%Y)

$$\xleftarrow[\text{MeO(CH}_2)_2OMe]{BF_3 \cdot Et_2O} \quad \begin{array}{c} CN \\ | \\ H_2NCCHNHCH=NO(CH_2)_3OCH_2Ph \\ \| \\ O \end{array}$$

(63%Y)

At this stage the imidazole route to our 9-alkoxy-guanines looked rather more attractive. However, there was still the problem of conversion of the imidazole to the guanine derivative which we had never achieved in better than 30% overall yield.

For this reason we had been concurrently investigating an alternative synthesis via pyrimidine intermediates (Figure 12).

Although displacement of chloride from 2,5-diamino-4,6-dichloropyrimidine with an alkoxyamine does not occur readily, after formylation of the amino groups this can be achieved in satisfactory yield. Closure of the imidazole ring can then be accomplished simply by fusion, but better yields are reliably obtained by heating with diethoxymethyl acetate.[16] Hydrolysis of the 6-chloro group under acidic conditions, followed by catalytic hydrogenolysis of the benzyl protecting group and removal of the N-formyl group with aqueous ammonia, then afforded BRL 44385 in 39% overall yield.

Figure 12
Alternative Synthesis of 9-(3-Hydroxypropoxy)guanine

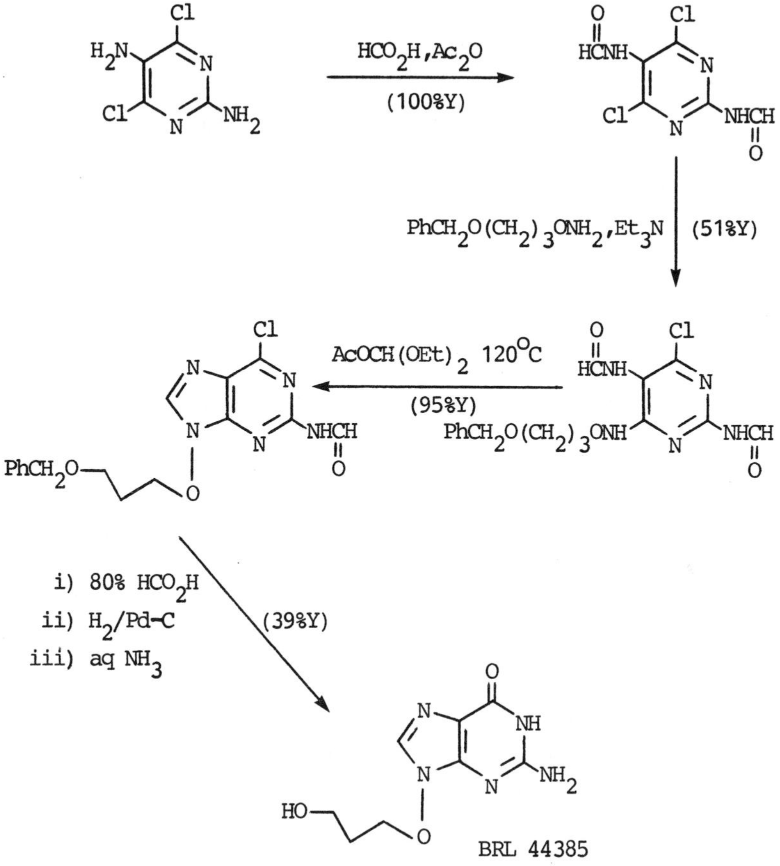

BRL 44385

This synthesis is much shorter and more efficient than the imidazole route and was used subsequently for the preparation of the other 9-alkoxyguanines which we had identified as target compounds.

The first of these was the bis-hydroxymethyl analogue (Figure 13).

Figure 13
Synthesis of 9-(3-Hydroxy-2-hydroxymethylpropoxy)guanine

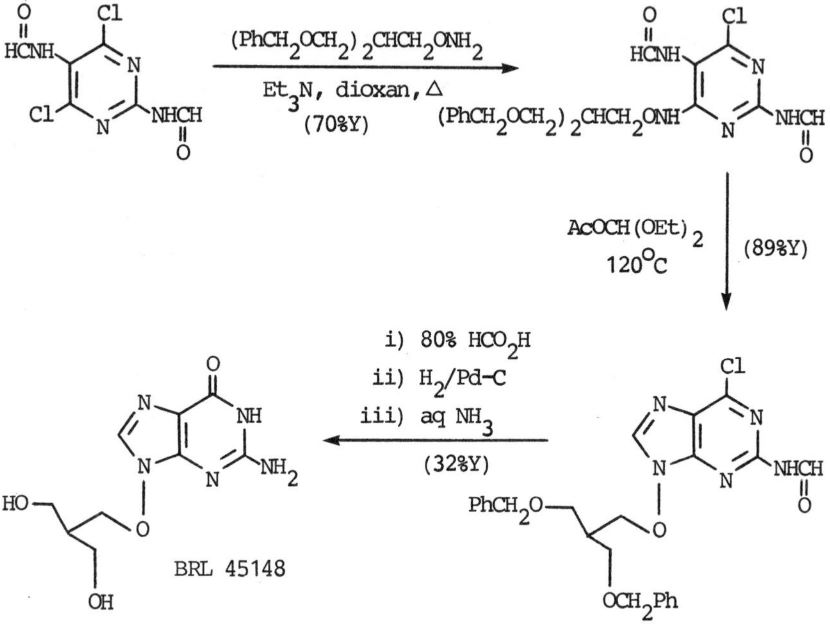

The benzyl-protected alkoxyamine was again obtained in high yield by condensation of the corresponding alcohol with N-hydroxyphthalimide and cleavage of the product with hydrazine hydrate. Displacement of a chlorine from the dichlorodiformamidopyrimidine and subsequent cyclisation to the purine occurred in high yields. The subsequent stages were also identical to those used previously in the synthesis of BRL 44385 and provided the required acyclonucleoside BRL 45148 in 32% overall yield from the chloroformamidopurine.

The anti-herpesvirus activity of BRL 45148 (Table 5) resembles that of BRL 39123 much more closely than that of ganciclovir.

Table 5
Comparison of the Antiviral Activity of BRL 45148, BRL 39123 and Ganciclovir in Cell Cultures

	X	Y
BRL 45148	O	CH_2
BRL 39123	CH_2	CH_2
Ganciclovir	CH_2	O

Virus	Cell	IC_{50} ($\mu g/ml$)		
		BRL 45148	BRL 39123	Ganciclovir
HSV-1 (HFEM)	Vero	1.5	1.6	0.4
HSV-1 (SC-16)	MRC-5	1.3	1.6	0.2
HSV-2 (MS)	Vero	1.5	1.7	0.4
HSV-2 (MS)	MRC-5	1.5	0.9	0.1
VZV (Ellen)	MRC-5	2.9	3.5	1.3
EBV	P3HR-1	13	10	–
CMV (AD169)	MRC-5	>100	57	1.3
	MRC-5*	>100	>100	100

* Inhibition of growth of uninfected cells over 72 hours

Its potency is virtually identical to that of BRL 39123 against herpes simplex virus types 1 and 2, varicella zoster virus and Epstein Barr virus. It is slightly less active than BRL 39123 against cytomegalo-virus, against which ganciclovir has potent activity.

In none of these tests were BRL 45148 concentrations up
to 100µg/ml toxic for the cell monolayers and, like BRL
39123, BRL 45148 had little effect upon the growth of
rapidly dividing MRC-5 cells.

The next compound that we set out to prepare was
9-(2,3-dihydroxypropoxy)guanine (Figure 14).

Figure 14

Synthesis of (RS)-9-(2,3-Dihydroxypropoxy)guanine

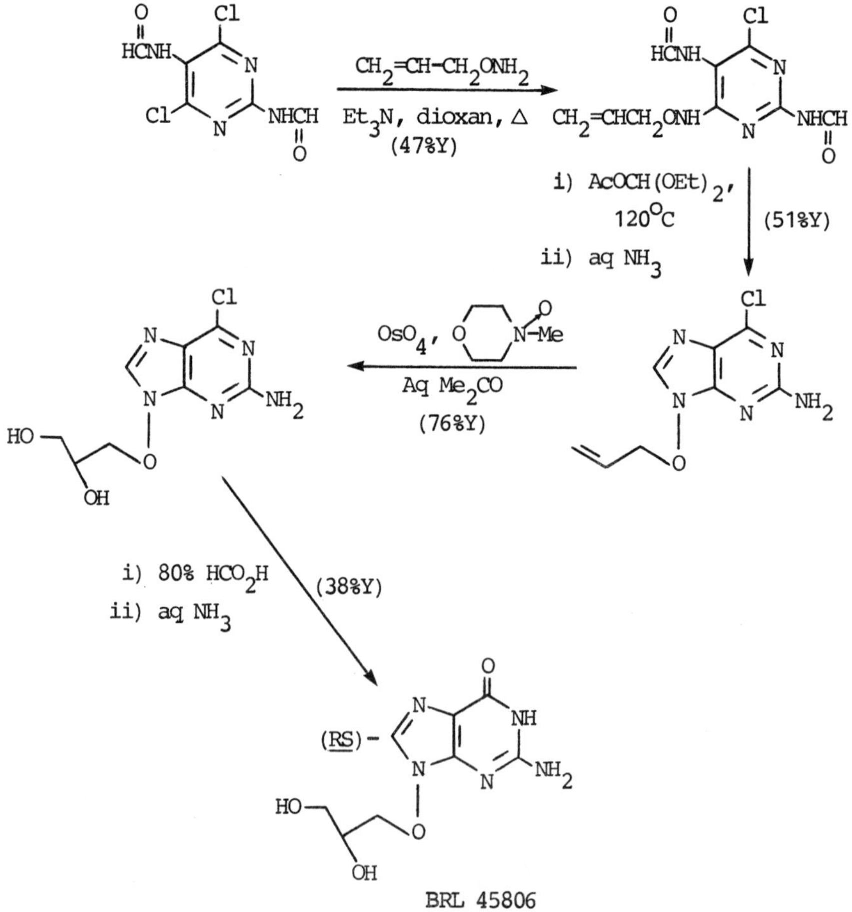

BRL 45806

For synthesis of the racemic acyclonucleoside we were able to use the commercially available O-allyl-hydroxylamine. Reaction with the dichlorodiformamido-pyrimidine and subsequent cyclisation to the purine derivative occurred as described for the other members of the series. In this case the N-formyl group was removed at this stage with aqueous ammonia. The 9-allyloxy-2-amino-6-chloropurine was then treated with osmium tetroxide and N-methylmorpholine oxide in acetone[17] to provide the diol in 76% yield. Upon acidic hydrolysis this was converted to the racemic acyclo-nucleoside BRL 45806.

Once again, BRL 45806 proved to be a highly selective anti-herpesvirus agent (Table 6).

Table 6

Comparison of the Antiviral Activity of BRL 45806 and (RS)-DHBG in Cell Culture

		X
BRL 45806		O
(RS)-DHBG		CH$_2$

		IC$_{50}$ (μg/ml)	
Virus	Cell	BRL 45806	(RS)-DHBG
HSV-1 (HFEM)	Vero	8.6	9.2
HSV-1 (SC-16)	MRC-5	1.9	7.0
HSV-2 (MS)	Vero	1.6	7.9
HSV-2 (MS)	MRC-5	0.7	1.4
VZV (Ellen)	MRC-5	2.3	13

In cell culture tests it inhibited the replication of herpes simplex virus types 1 and 2 and varicella zoster virus, with generally lower IC_{50}'s than were obtained with (RS)-DHBG.   At concentrations up to 100μg/ml, BRL 45806 was not toxic for the cell monolayers.

In the case of DHBG the (R)-enantiomer (buciclovir) is reported to be about 6-fold more potent than the (S)-enantiomer.   We were therefore interested in obtaining both enantiomers of 9-(2,3-hydroxypropoxy)guanine and determining their relative antiviral activity.

Figure 15
Synthesis of Enantiomers of 2,2-Dimethyl-1,3-dioxolan-4-ylmethoxyamine

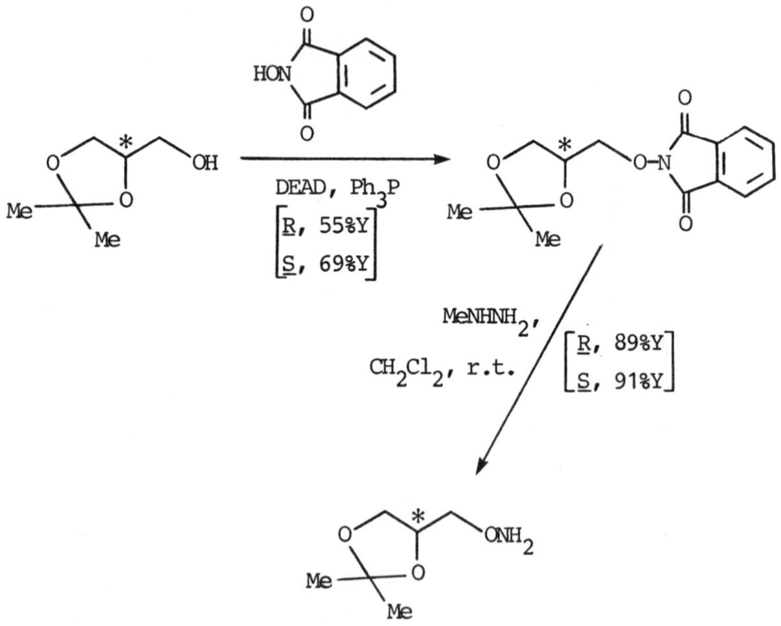

For this, we developed a synthesis starting from the commercially available enantiomers of solketal (Figure 15). These were converted to alkoxyamines using procedures that have been described. A minor variation was, however, used in cleavage of the phthalimide. It found that this occurred very efficiently using N-methylhydrazine in methylene chloride at room temperature.

Figure 16
Synthesis of Enantiomers of
9-(2,3-Dihydroxypropoxy)guanine

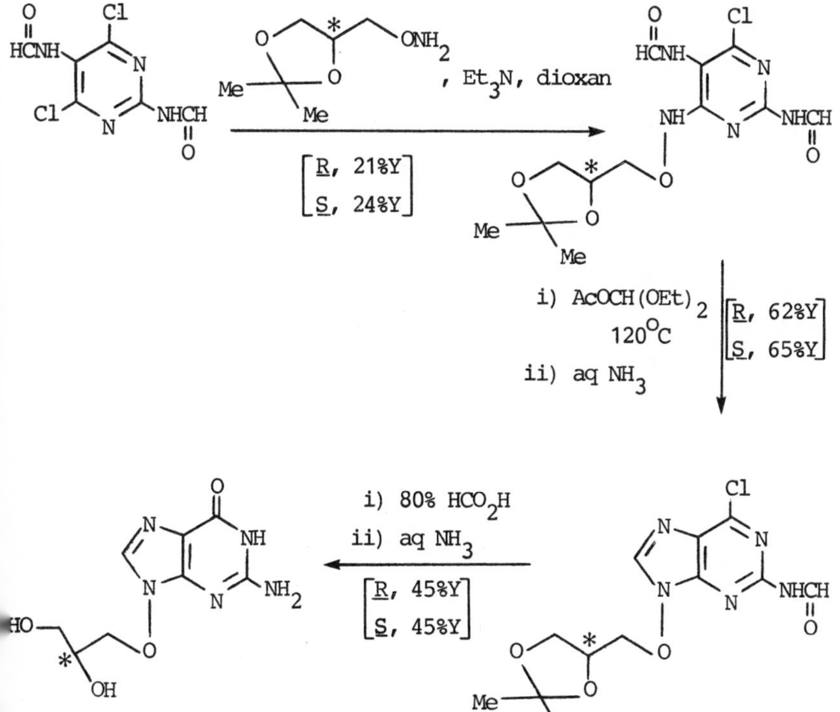

(R) BRL 46716
(S) BRL 46976

Displacement of chloride from the dichlorodiform-amidopyrimidine with these alkoxyamines was not as efficient as with other alkoxyamines that we had used, but the subsequent stages in the synthesis proceeded satisfactorily (Figure 16) and both enantiomers of the 9-alkoxyguanine were obtained.

In tests against herpes simplex virus types 1 and 2 in cell culture the (S)-enantiomer BRL 46976 proved to be very much more active than the (R)-enantiomer BRL 46716 (Table 7). It is interesting that the difference in activity between the two enantiomers is much greater than for the (R) and (S) enantiomers of DHBG.

Table 7

Antiviral Activity of BRL 46716 and BRL 46976 in Cell
Culture

Virus	Cell	IC_{50} ($\mu g/ml$)	
		BRL 46716	BRL 46976
HSV-1 (SC-16)	MRC-5	78	0.7
HSV-2 (MS)	MRC-5	>100	0.1

A synthetic approach similar to that used in the preparation of racemic 9-(2,3-dihydroxypropoxy)guanine was also employed for the synthesis of racemic 9-(3,4-dihydroxybutoxy)guanine (Figure 17).

Figure 17
Synthesis of (RS)-9-(3,4-Dihydroxybutoxy)guanine

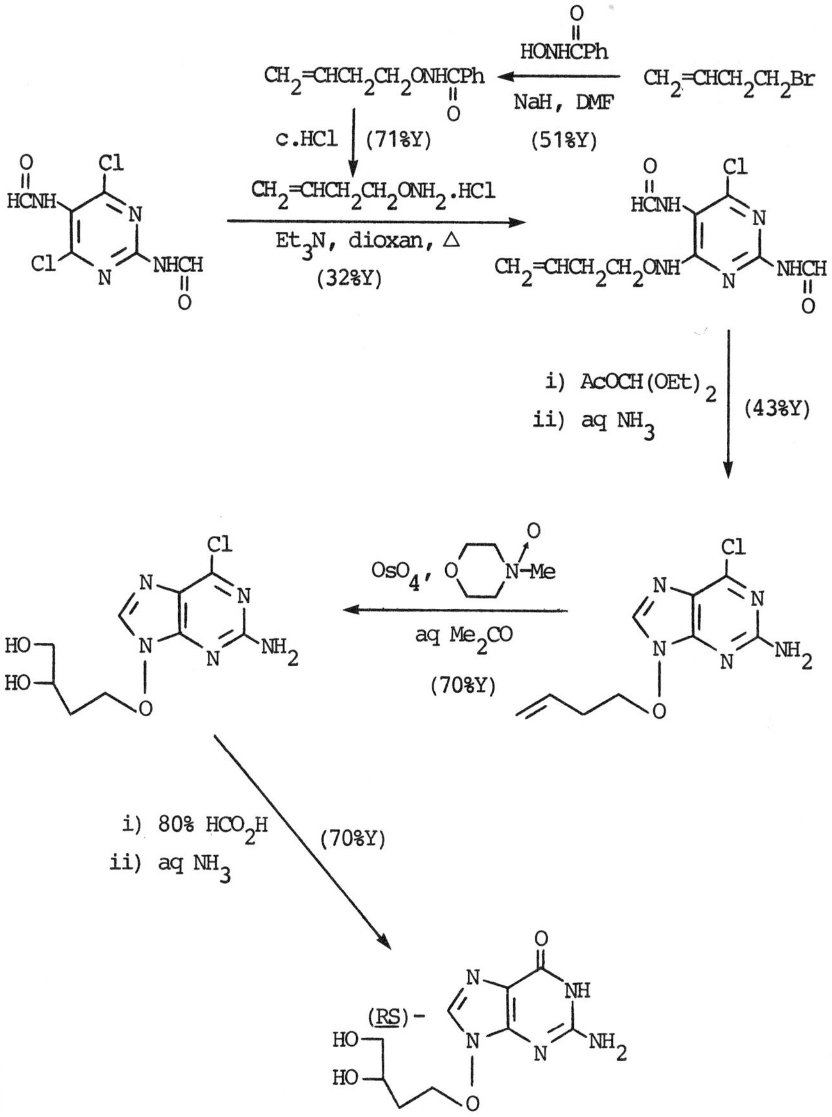

BRL 46717

In this case the unsaturated alkoxyamine was prepared from 4-bromo-1-butene by reaction with benzohydroxamic acid and hydrolysis with concentrated hydrochloric acid.[18] The subsequent stages leading to the 9-alkoxy-2-amino-6-chloropurine proceeded as in the synthesis of the propenoxy homologue. The diol was then obtained using osmium tetroxide and N-methylmorpholine oxide and efficiently converted to the racemic guanine derivative BRL 46717.

Table 8

Comparison of the Antiviral Activity of BRL 46717 and (S)-iNDG in Cell Culture

	X	Y
BRL 46717	O	CH_2
(S)-iNDG	CH_2	O

		IC_{50} (μg/ml)	
Virus	Cell	BRL 46717	(S)-iNDG[9]
HSV-1 (SC-16)	MRC-5	67	–
HSV-1 (Schooler)	MRC-5	–	1.0
HSV-2 (MS)	MRC-5	>100	–
HSV-2 (Curtis)	MRC-5	–	2.0

In cell culture tests against herpes simplex virus types 1 and 2, BRL 46717 had only low activity (Table 8). Although we did not have (RS)-iNDG available for comparison, previously reported data[9] for different virus strains in the same cells indicates that, in this case, the 9-alkoxymethylguanine is much more potent.

Another selective anti-herpesvirus agent that has been reported previously is 9-(4-hydroxy-2-hydroxy-methylbutyl)guanine, a positional isomer of BRL 39123. We and other workers[19] have independently found that this acyclonucleoside is a potent inhibitor of the replication of varicella zoster virus. We were therefore interested in preparing its 9-alkoxyguanine analogue also.

Figure 18
Synthesis of (RS)-1,4-Di-(4-methoxybenzyloxy)but-2-oxyamine

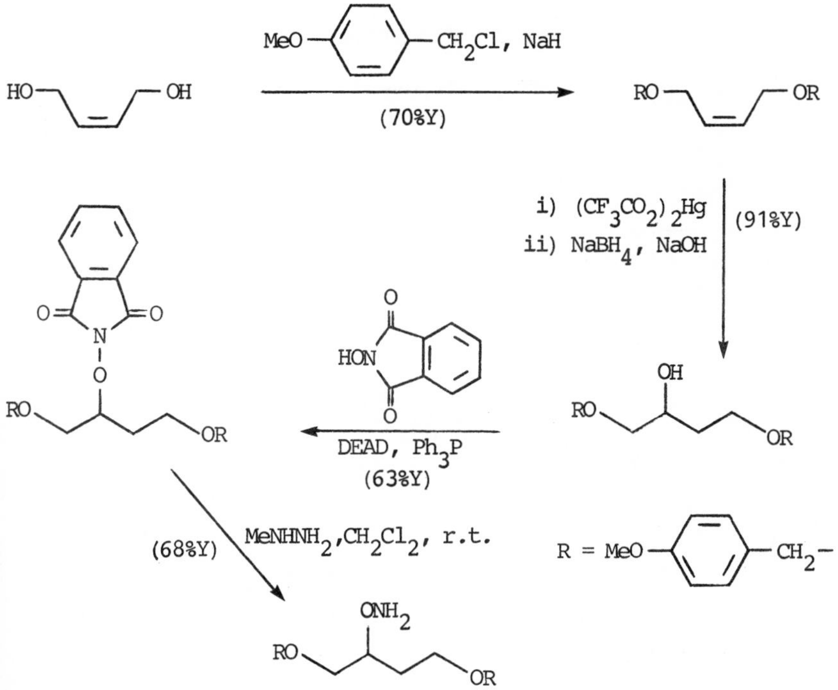

In this case p-methoxybenzyl groups were chosen for
hydroxyl protection of the intermediate alkoxyamine
(Figure 18). The di-p-methoxybenzyl ether of 2-butene-
1,4-diol was reacted with mercuric trifluoroacetate in
aqueous tetrahydrofuran followed by sodium borohydride
to afford the intermediate alcohol in high yield.   This
alcohol was then condensed with N-hydroxyphthalimide and
the product cleaved to afford the protected alkoxyamine
using procedures that have already been described for
preparation of other alkoxyamines.

Figure 19
Synthesis of (RS)-9-(3-Hydroxy-1-hydroxymethyl-
propoxy)guanine

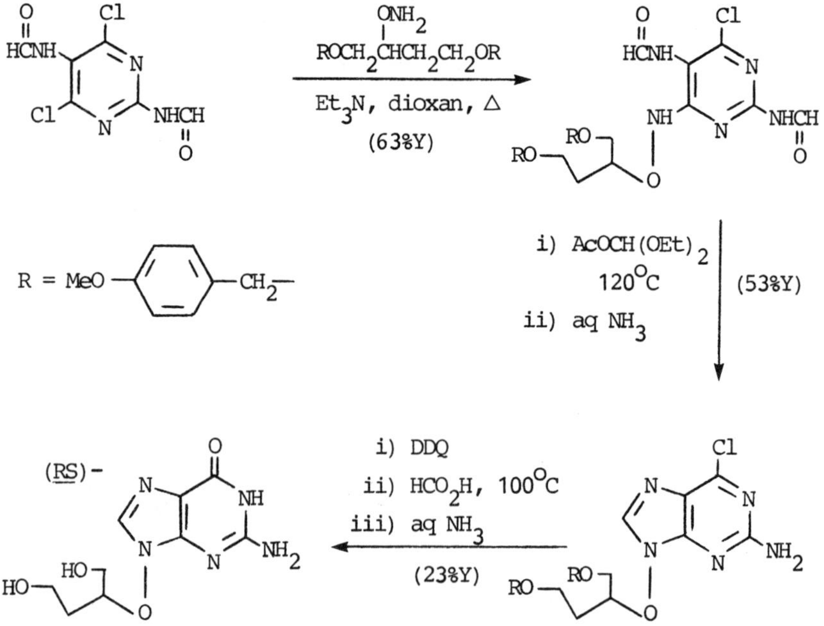

BRL 46011

Displacement of chloride from the dichlorodiformamidopyrimidine was readily accomplished with this alkoxyamine, and cyclisation to the aminochloropurine was also achieved without difficulty (Figure 19). Oxidative removal of the p-methoxybenzyl protecting groups with 2,3-dichloro-5,6-dicyano-1,4-benzoquinone and subsequent acid hydrolysis then afforded the 9-alkoxyguanine BRL 46011.

However, as with the i-NDG analogue, in cell culture BRL 46011 was found to possess considerably less activity against herpesviruses than the corresponding 9-hydroxyalkylguanine BRL 42377 (Table 9).

Table 9

Comparison of the Antiviral Activity of BRL 46011 and BRL 42377 in Cell Culture

	X
BRL 46011	O
BRL 42377	CH_2

		IC_{50} (μg/ml)	
Virus	Cell	BRL 46011	BRL 42377
HSV-1 (HFEM)	Vero	>100	8.7
HSV-1 (SC-16)	MRC-5	11	3.3
HSV-2 (MS)	MRC-5	>100	2.1
VZV (Ellen)	MRC-5	>100	0.2

It is, therefore, now apparent that structure/anti-herpesvirus activity relationships in the novel series of 9-hydroxyalkoxyguanines that have been discussed do not exactly parallel those for 9-hydroxyalkyl- and 9-hydroxyalkoxymethyl-guanines. Some compounds, for example the 3-hydroxypropoxy derivative BRL 44385 and the (S)-2,3-dihydroxypropoxy derivative BRL 46976, are substantially more active than their alkoxymethyl and alkyl analogues acyclovir and buciclovir, respectively. The bis-hydroxymethyl compound BRL 45148 has activity similar to that of BRL 39123, but is less potent than ganciclovir. The 3,4-dihydroxybutoxy derivative BRL 46717 and the 3-hydroxy-1-hydroxymethylpropoxy deriva-tive BRL 46011 have substantially reduced antiviral activity in comparison with i-NDG and the hydroxyalkyl derivative BRL 42377, respectively.

At present, no explanation of these differences is available. However, 9-alkoxy derivatives of guanine are undoubtedly a class of compounds that have not previous-ly been investigated and which have considerable potential for development as antiviral agents. Additionally, new compounds that may prove to be of use as drugs for the treatment of other human diseases may well be discovered by evaluation of the biological properties of other 9-alkoxypurines.

Acknowledgements

The authors wish to acknowledge the valuable contributions made to these studies by many Beecham colleagues. The useful improvements in procedures for synthesis of 1-alkoxy-5-amino-4-carbamoylimidazoles described are attributable to Dr. A. Parkin. Antiviral tests were carried out by Dr. T.H. Bacon, Mr. C. Patience and Mr. D. Sutton. Dr. R.A. Vere Hodge obtained the biochemical data reported for BRL 39123.

References

1. H.J. Schaeffer, L. Beauchamp, P. de Miranda, G.B. Elion, D.J. Bauer and P. Collins, *Nature*, 1978, *272*, 583.

2. G.B. Elion, P.A. Furman, J.A. Fyfe, P. de Miranda, L. Beauchamp and H.J. Schaeffer, *Proc. Natl. Acad. Sci. U.S.A.*, 1977, *74*, 5716.

3. P. Collins and D.J. Bauer, *J. Antimicrobial Chemother.*, 1979, *5*, 431.

4. C.K. Chu and S.J. Cutler, *J. Heterocyclic Chem.*, 1986, *23*. 289.

5. Collaborative DHPG Treatment Study Group, *N. Engl. J. Med.*, 1986, *314*, 801.

6. D.H. Shepp, P.S. Dandliker, P. de Miranda, T.C. Burnette, D.M. Cederberg, L.E. Kirk and J.D. Meyers, *Annals of Internal Med.*, 1985, *103*, 368.

7. O.L. Laskin, C.M. Stahl-Bayliss, C.M. Kalman and L.R. Rosecan, *J. Infectious Dis.*, 1987, *155*, 323.

8. B. Öberg and N.G. Johansson, *J. Antimicrobial Chemother.*, 1984, *14* (Suppl. A), 5.

9. W.T. Ashton, L.F. Canning, G.F. Reynolds, R.L. Tolman, J.D. Darkas, R. Liou, M.M. Davies, C.M. De Witt, H.C. Perry and A.K. Field, *J. Med. Chem.*, 1985, *28*, 926.

10. M.R. Harnden and R.L. Jarvest, European Published Patent Application 0141927A (to Beecham Group p.l.c.), 22nd May, 1985.

11. M.R. Harnden and R.L. Jarvest, *Tetrahedron Letters*, 1985, *26*, 4265.

12. M.R. Harnden, R.L. Jarvest, T.H. Bacon and M.R. Boyd, *J. Med. Chem.* in press.

13. M.R. Boyd, T.H. Bacon, D. Sutton and M. Cole, *Antimicrob. Agents Chemother.*, in press.

14. A.A. Watson, J. Org. Chem., 1974, 39, 2911.

15. E. Grochowski and J. Jurczak, Synthesis, 1976, 10, 682.

16. J.A. Montgomery and C. Temple Jr., J. Amer. Chem. Soc., 1957, 79, 5238.

17. V. Van Rheenen, D.Y. Cha and W.M. Hartley, Org. Synth., 1978, 58, 43.

18. P. Mamalis, J. Green and D. McHale, J. Chem. Soc., 1960, 229.

19. A.-C. Ericson, A. Larsson, F.Y. Aoki, W.-A. Yisak, N.G. Johansson, B.Öberg and R. Datema, Antimicrob. Agents Chemother., 1985, 27, 753.

Recent Developments in Avermectin Research

H. Mrozik

MERCK SHARP & DOHME RESEARCH LABORATORIES, PO BOX 2000,
RAHWAY, NEW JERSEY 07065, USA

The discovery of the avermectins in the mid seventies was the direct result of a screening effort for natural products with anthelmintic properties[1]. High activity against the rodent nematode *Nematospiroides dubius* accompanied by a good margin of safety was observed at Merck with a broth obtained from the Kitasato Institute. Isolation and structure determination of the new antiparasitic principles[2] showed a chemical relationship to the milbemycins[3], a group of macrocyclic lactones with insecticidal and acaricidal effects for the control of agricultural and horticultural pests. Avermectin B1a is the one of the eight closely related 16-membered lactones produced by *Streptomyces avermitilis* with the best overall antiparasitic and insecticidal activities. Minor structural variations at the 5, 22, 23, and 25 positions account for the analogs A1a through B2b. Avermectin B1 with generic name abamectin is under development as an agricultural insecticide and miticide[4]. For animal health, however, its semisynthetic 22,23-dihydro derivative ivermectin[5] was selected as an endo and ectoparasiticide of unprecedented potency. Ivermectin is superior to avermectin B1 with a more balanced anthelmintic spectrum and higher safety. Of particular interest are its systemic and topical activities against parasitic arthropods of cattle and sheep. No activities against flatworms, protozoa, fungi or bacteria are observed. Ivermectin is currently used in liquid or paste formulations by oral or subcutaneous routes for the control of gastrointestinal and systemic nematodes, grubs, lice, mites, bots, and certain ticks of cattle, sheep, swine, and horses at 0.2 to 0.3 mg/kg[6]. Prevention of dog heartworm infections is accomplished by

AVERMECTIN   STRUCTURES

AVERMECTIN   A : $R_5 = OCH_3$ B : $R_5 = OH$

1 : $X = -CH=CH-$ 2 : $X = -CH_2-\overset{\overset{OH}{|}}{C}H-$

a : $R_{25} =$ [structure]   b : $R_{25} =$ [structure]

IVERMECTIN : $R_5 = OH$ $X = -CH_2-CH_2-$ $R_{25} =$ [structure] AND [structure]

MILBEMYCIN STRUCTURES

MILBEMYCIN ALFA SERIES (SANKYO)[3a,3b]

R_5 = OH , OCH_3

R_{22} = H , OH

R_{23} = H , $OCOCH(CH_3)(CH_2)_3CH_3$

R_{25} = CH_3 , C_2H_5

ANTHELMINTIC F-28249 (AMERICAN CYANAMID)[3c,3e]

ANTIBIOTICS S 541 (GLAXO)[3d]

R_5 = OH , OCH_3

R_{22} = H R_{23} = OH

R_{25} =

			R_5	R_{22}	R_{23}	R_{25}	
MILBEMYCIN	ALFA$_1$	(A$_3$)	R_5 = OH	R_{22} =	R_{23} = H	R_{25} =	CH_3
MILBEMYCIN	ALFA$_3$	(A$_4$)	R_5 = OH	R_{22} =	R_{23} = H	R_{25} =	CH_2CH_3
MILBEMYCIN	D		R_5 = OH	R_{22} =	R_{23} = H	R_{25} =	$CH(CH_3)_2$
ANTHELMINTIC	F-28249-ALPHA		R_5 = OH	R_{22} = H	R_{23} = OH	R_{25} =	$C(CH_3)CHCH(CH_3)_2$

a program of monthly doses of 0.006 mg/kg due to the very high susceptibility of the infective larvae of *Dirofilaria immitis*, and even a bimonthly treatment with 0.05 mg/kg is fully effective[7]. In dogs, however, doses of 0.2 mg/kg as used regularly in the other animals and which would be effective against the intestinal parasites of dogs must be avoided because of severe adverse reactions in some dogs, especially of the collie breed[8]. Adult heartworms like other adult filarial worms with few exceptions such as certain horse *Onchocerca* species are not killed by any doses of ivermectin.

The efficacy of ivermectin against gastrointestinal nematode infections of man has just been reported[9]. So far it has mainly been the subject of a number of studies in human patients with *Onchocerca volvulus* infections. This parasite is responsible for river blindness, a disease endemic to Africa and central America. A single oral dose of 0.1 - 0.2 mg/kg results in almost complete disappearance of microfilariae from skin, and the effect persists for 6-12 months[10]. The slow death of the microfilariae presumably is responsible for the relativly mild side effects usually encountered due to immunological reactions to the decaying parasites in the tissue. The reproductive system of adult worms recovered from treated individuals showed serious degeneration, which explains the absence of microfilariae from skin for many months following treatment. Little is known yet about the effects against *Wuchereria*[11] and *Brugia*, some filarial parasites causing the dreaded elephantiasis disease.

Absorption of ivermectin occurs readily by oral and subcutaneous routes, though to greater degree by the latter. The drug is excreted mostly unchanged in the faeces, which may contain enough drug to prevent the development of certain fly larvae[12]. The highest drug residues are found in liver and fat, but are well below acceptable levels by the end of the withdrawal period of 28 to 35 days. Peak blood concentrations resulting from a 0.2 mg/kg dose depending on the route of administration and on the formulation is of the order of 0.01 to 0.05 μg/ml. Acute toxicities (LD_{50}) of ivermectin on oral administration are 40-50 mg/kg for male and female adult rats but only 2.3 mg/kg for infant rats.

The mode of action of avermectins is the subject of an increasing number of studies[13]. The use of a variety of biological specimens, experimental designs, and the recording of effects at greatly varying concentrations tends to complicate the picture. The avermectins act in a sense like GABA, opening the chloride ion channels of nerve membranes or keeping them open, which causes a flaccid paralysis in many invertebrates. These effects are observed at peripheral and central nerve endings containing mainly GABA-sensitive but also in some instances GABA insensitive chloride ion channels. Some of these effects can be reversed by picrotoxin, known to close chloride ion channels, and to a lesser degree by bicucullin, a GABA receptor antagonist. Neither avermectins, picrotoxin, or benzodiazepins bind directly to the GABA receptor, but they have their separate binding sites in close proximity probably on the same protein, presumably the chloride ion channel inside the membrane. The binding constants of avermectins to rat brain, the free living avermectin-sensitive nematode *Caenorhabditis elegans*[14], and to cockroach[15] brain have been determined and are in the order of 10-20 for rat brain and 0.1 to 1.0 nanomolar for invertebrate receptor preparations. These binding constants relate closely to the antiparasitic potency of avermectin derivatives and are readily obtained. Potent antiparasitic activities are shown only by compounds which bind very well. Interesting is the weaker binding of certain aglycone derivatives including some milbemycins to the mammalian brain receptor, pointing possibly to an increased differentiation of toxicity to mammalian host versus invertebrate parasite[14,15]. This, however, is only one aspect of differential toxicity. In mammals GABA receptors are mainly confined to the central nervous system, and the concentration of ivermectin in mammalian brain is very low. This is probably the most significant fact explaining the good tolerence of ivermectin in almost all vertebrates including humans.

The structural relationship of the avermectins and milbemycins prompted us to investigate the anthelmintic activities of avermectin monosaccharides, aglycones, and 13-deoxyaglycones as the closest analogs of the milbemycins. When ivermectin, monodesoleandrosyl-22,23-dihydroavermectin B1a (called monosaccharide for short), 22,23-dihydroavermectin B1a aglycone and 13-deoxy-22,23-dihydroavermectin B1a aglycone were tested against a mixed infection of gastrointestinal parasites in sheep it was found that the monosaccharide retains high potency, but the aglycone lost most of its anthelmintic activities[16]. It was therefore surprising to find good anthelmintic activities for 13-deoxy aglycones and certain milbemycins. Further tests against ticks in cattle, however, revealed a clear advantage for the compounds substituted by the oleandrosyl disaccharide at the 13 position of the macrocyclic lactone[17]. The ready availability of 22,23-dihydroavermectin B1a aglycone facilitated the investigation of the effect of various substituents of the 13-position on the biological activities of the derivatives[18]. The multisubstituted macrocyclic ring forces the C-13 allylic-homoallylic alcohol rigidly into a conformation where a 13-α leaving group cannot readily interact with the π-electron of the allylic 14,15-double bond. Therefore the reactivity of the aglycone at C-13 is not as expected for an allylic alcohol. Manganese dioxide oxidation of 22,23-dihydro-B1a aglycone gives readily the 5-oxo-13-alcohol. ⸱ The 13-α-O-(2-nitrophenylsulfonyl) derivative after protection of the 5-hydroxy and without isolation of the unstable sulfonylester reacts with chloride or iodide ions to give the 13-β-S$_N$2 substitution products.   Solvolyses of the more stable 13-α tosylate or mesylate give presumably under homoallylic participation the 13-α methoxy and chloro derivatives.   The 13-β-iodide is a versatile intermediate.   It has the π-bond orbitals of the allylic 14,15-double bond in an orientation which allows participation with the orbital of the leaving group.   Amines react with this exclusively under allylic rearrangement to give 15-amino-substituted 13,14-ene derivatives.   A base such as DBU causes the hydrodeiodination to the conjugated 8,9,10,11,12,13,14,15-tetraene, albeit with concomitant shift of the 3,4 double bond into the conjugated 2,3 position.   A weaker base such as collidine, however, allows isolation of the 12,13-dehydro aglycone with natural configuration retained at C-2, while heating with aqueous collidine gives mainly the 13-β-aglycone, epimeric with the natural product, accompanied by the 12,13-dehydro product and the 15-hydroxy-13,14-ene allylic substitution product.   Oxidation of the aglycone gives a 13-oxo derivative, which is reduced with NaBH$_4$ stereospecifically back to the aglycone with natural configuration at C-13 and reductively aminated almost exclusively to the 13-α-amine.   The 13-oxime and 13-α and β-methoxyethoxymethylene derivatives were prepared by conventional means.   Since it was not possible to determine the stereochemistry of the 15 substituents readily from their ^{1}H NMR spectra, we prepared some model compounds via a 3,3-shift of the 13-α- and 13-β-thionocarbonate derivatives. The structures of the rearranged 15-methylthio-13,14-enes were then determined by their NOESY NMR spectra[19].   The conversion of milbemycins to 13-epi-avermectin aglycones and their modifications have recently also been reported by other groups[20].

The new avermectin derivatives were tested for their antiparasitic and insecticidal activities in assays against the gastrointestinal nematode *Trichostrongylus colubriformis* in gerbils[21], the two-spotted spider mite (*Tetranychus urticae*) on bean plants[22], and neonate southern armyworm (*Spodoptera eridania*) larvae on drug-treated bean leaves[23].   In addition their binding constants to avermectin receptors isolated from the free living nematode *C. elegans* and from rat brain were determined[14].   Generally the highest potencies of the order of avermectin B1 or ivermectin or slightly less were observed for aglycones having 13-substituents such as H (milbemycins and 13-deoxyaglycones), methoxyethoxymethylene ethers (possibly mimicking an oleandrose), halogen (particularly fluorine), β-hydroxy and methoxime. On the other hand, compounds arising from allylic rearrangement having the double bond in 13,14-position are totally inactive. While the correlation of the activities in the four biological systems mentioned is reasonably good, there are some notable exceptions : 1) the low potency of avermectin B1 and ivermectin against the southern armyworm; 2) good receptor binding of 13-ketone and 13-α-alcohol combined with only fair *T. colubriformis* activity and very weak insecticidal activities against two-spotted spider mite and southern armyworm; 3) tight binding of the 12,13-dehydro derivative without any notable *in vivo* activities, possibly due to the instability of the compound.

It appears that nature provided a highly optimized structure, and that it is very easy to loose activity and almost impossible to improve on it. Avermectin B1 and ivermectin are highly potent with a good safety margin and a wide spectrum of activities against nematodes (roundworms) and arthropods (insects, ticks, mites). It would be interesting to extend the spectrum to include flukes and tapeworms, but this is unlikely to succeed, since not a hint of activity against these species has been observed so far. It is also believed that flatworms do not depend to the same extent on the GABA system for nerve transmission. From available data, however, it appeared that potency against southern armyworm and other lepidoptera larvae varied considerably and that increased activities against those economically important species could be achieved. While carrying out extensive modifications at the 4"-position, the only free hydroxy group of the disaccharide substituent, we observed that good activities could be retained with many variants. Eventually we found in 4"-amino-4"-deoxyavermectin B1 the desired derivative with more than 400 times increased potency against southern armyworm[24].

Several research groups recently were concerned with the synthesis of avermectin glycosides[25]. We prepared an avermectin containing a third oleandrose at the 4"-position which showed activities comparable but not superior to avermectin B1[26]. On the other hand, slight changes such as the loss of only one methyl from the oleandrose 3"-methoxy group, a compound obtained by blocking the biosynthetic methylation step, drastically reduced activity[27].

It has been observed that avermectins in thin films, as obtained by spraying agricultural crops, decompose rapidly particularly in strong sunlight[28]. This is not neccessarily bad, since it also assures the rapid removal of residues from the environment. Avermectins are also bound very tightly to soil so that runoff of residues toxic to certain aquatic life such as *Daphnia magna* or fish does not pose a problem when used as directed. Laboratory UV irradiation experiments with avermectins in dilute solution in a quartz tube showed rapid isomerisations of the 8,9- and 10,11-$\underline{E}$-double bonds to the $\underline{Z}$-configurations and subsequent decomposition under loss of the 245 nm diene UV absorption to a gross mixture of undefined products[29]. The diene function appears clearly as an undesirable structural feature as far as stability is concerned. On the other hand, the 3,4,8,9,10,11,22,23-octahydroavermectin B1 is completely inactive, suggesting strongly that at least part of the diene is needed for a biologically active derivative. Further light was shed on the structural requirement for bioactivity by preparation and testing of the 8,9-oxide[24]. Although this compound has reduced potency as an anthelmintic agent it is equipotent with avermectin B1 as a contact miticide, and superior in a residual miticidal assay, as could be expected since it is without the light-absorbing diene chromophor. This persistence was, however, not translated into a practical asset when tested in the field, possibly due to ready hydrolysis to an inactive 8,9-dihydroxy derivative. Subsequently it was investigated whether an 8,9-cyclopropyl group might be a useful, stable substitute for the oxide[30]. This compound, however, had only weak biological activities.

Despite the complete loss of biological activities of octahydroavermectin B1 we thought it interesting to explore the biological activities of derivatives with one of the diene double bonds reduced. The catalytic hydrogenation of avermectin B1 using Wilkinson's catalyst with selectivity for the only disubstituted 22,23-syn-double bond had of course given ivermectin almost exclusively[16]. Much less success was experienced in an attempt to prepare 10,11-dihydroavermectins through a hydrogenation procedure. We found, however, in N-bromacetamide a reagent with selectivity for the 10,11-double bond of avermectins, which gave in good yield the 11,10-bromohydrin[31]. Further chemical modifications afforded the desired 10,11-dihydro- and 10,11-dihydro-10-fluoroavermectins. Their good insecticidal activities warrant further investigations under actual field conditions.

REFERENCES

1. W.C.Campbell, R.W.Burg, M.H.Fisher, and R.A.Dybas in ACS Symposium Series, No. 255, Pesticide Syntheses Through Rational Approaches, P.S.Magee, G.K.Kohn, and J.J.Menn, Editors, American Chemical Society, Washington, D.C. 1984, pp 5-20.

2a. M.H.Fisher and H.Mrozik in Macrolide Antibiotics, S.Omura Editor, Academic Press, Inc. 1984, pp 553-606.

2b. H.G.Davies and R.H.Green, Natural Product Reports, 1986, 3, pp 87-121.

3a. Y.Takiguchi, H.Mishima, M.Okuda, M.Terao, A.Aoki, and R.Fukuda, J. Antibiot. 1980, 33, 1120.

3b. H.Mishima, J.Ide, S.Muramatsu, and M.Ono, J. Antibiot. 1983, 36, 980.

3c. I.B.Wood, J.A.Panksvich, G.T.Carter, M.J.Torrey, and M.Greenstein, 1986, Eur. Pat. Appl. EP 170006.

3d. J.B.Ward, H.M.Noble, N.Porter, R.A.Flatton, and D.Noble, 1986, UK Pat. Appl. GB 2166436A.

3e. G.T.Carter, J.A.Nietsche, and D.B.Borders, J. Chem. Soc., Chem. Commun., 1987, 402.

4. R.A.Dybas and A.St.J.Green, Proceedings 1984 British Crop Protection Conference, Brighton, England, 1984, 31, 947.

5a. W.C.Campbell, M.H.Fisher, E.O.Stapley, G.Albers-Schonberg, and T.A.Jacob, Science 1983, 221, 823.

5b. W.C.Campbell, Parasitology Today, 1985, 1, 1.

6. W.C.Campbell and G.W.Benz, J. Vet. Pharmacol. Ther. 1983, 7, 1.

7. W.C.Campbell, Seminars in Veterinary Medicine and Surgery (Small Animal), 1987, 2, 48.

8. J.D.Pulliam, R.L.Seward, R.T.Henry et al., Vet. Med. 1985, 33.

9. D.Nalin, M.Aziz, D.Neu, G.Ruiz-Palacios, and C.Naquira, Abstracts of the Annual Meeting of the American Society for Microbiology, 87th Annual Meeting Atlanta, Ga. 1987, A-101, 17.

10. K.Y.Dadzie, A.C.Bird, K.Awadzi, H.Schulz-Key, H.M.Gilles, and M.A.Aziz, Brit. J. Ophthalmology, 1987, 71, 78.

11. S.Diallo, M.Aziz, I.Diop-Mar, O.N'Dir, and P.Gaxotte, unpublished.

12a. A.Miller, S.E.Kunz, D.D.Oehler, and R.W.Miller, J. Econ. Entomol. 1981, 74, 608.

12b. C.D.Schmidt, Environ. Entomol. 1983, 12, 455.

13. D.J.Wright, in Neuropharmacology and Pesticide Action, M.G.Ford, G.G.Lunt, R.C.Reay, and P.N.R.Usherwood Editors, Ellis Horwood Ltd., Chichester, 1986, pp 174-202.

14. J.M.Schaeffer et al. unpublished.

15. L.Huang et al. unpublished.

16. J.C.Chabala, H.Mrozik, R.L.Tolman, P.Eskola, A.Lusi, L.H.Peterson, M.F.Woods, M.H.Fisher, W.C.Campbell, J.R.Egerton, and D.A.Ostlind, J. Med. Chem. 1980, 23, 1134.

17. J.R.Egerton et al. unpublished.

18a. J.C.Chabala, M.H.Fisher, and H.H.Mrozik, 1979, U.S. Pat. 4,171,314 and US 4,173,517.

18b. B.O.Linn and H.H.Mrozik, 1986, U.S. Pat. 4,587,247.

18c. B.O.Linn and H.H.Mrozik, 1986, U.S. Pat. 4,579,864.

19. H.Mrozik, P. Eskola, and R.A.Reamer, unpublished.

20a. J.C.Gehret, 1985, Eur. Pat. Appl. EP. 144285; C.A. 104: 68676w.

20b. B.Frei and A.C.O'Sullivan, 1986, U.K. Pat. Appl. GB 2167751.

20c. K.Sato, T.Yanai, T.Kinoto, and S Mio, 1986, Eur. Pat. Appl. EP. 184308.

21. D.A.Ostlind, unpublished.

22. F.A.Preiser, unpublished.

23. T.Anderson, unpublished.

24. H.Mrozik et al. unpublished.

25a. K.C.Nicolaou, R.E.Dolle, D.P.Papahatjis, and L.J.Randall, J. Am. Chem. Soc. 1984, 106, 4189.

25b. S.Hanessian, A.Ugolini, D.Dube, P.J.Hodges, and C.Andre, J. Am. Chem. Soc. 1986, 108, 2776.

25c. C.Bliard, F.C.Escribano, G.Lukacs, A Olesker, and P.Sarda, J. Chem. Soc., Chem. Commun., 1987, 368.

26. M.J.Wyvratt et al. unpublished.

27. M.D.Schulman, D.L.Valentino, O.D.Hensens, D.Zink, M.Nallin, L.Kaplan, and D.A.Ostlind, J. Antibiot. 1985, 38, 1494.

28. Y.Iwata, J.G.MacConnell, J.E.Flor, I.Putter, and T.M.Dinoff, J. Agric. Food Chem. 1985, 33, 467.

29. H.Mrozik, P.Eskola, G.Reynolds, B.Arison, G.Smith and M.H.Fisher, in preparation.

30. M.J.Wyvratt, Am.Chem.Soc. 189th National Meeting, Miami Beach, Fl., May 1985, Abstr. ORG-213.

31. T.L.Shih et al. in preparation.

Fluconazole: a Novel Systemically Active Antifungal Agent

K. Richardson,* K. Cooper, M. S. Marriot, M. H. Tarbit,
P. F. Troke, and P. J. Whittle

CENTRAL RESEARCH, PFIZER LIMITED, RAMSGATE ROAD,
SANDWICH, KENT CT13 9NJ, UK

In 1978 a programme was initiated at Pfizer Central Research in Sandwich to seek an agent suitable for the treatment of life-threatening systemic fungal infections.

Fungi are all around us. For example, we have Candida in our gastro-intestinal tract and we breathe in fungi such as Cryptococcus and Aspergillus. Despite this, these fungi don't normally cause infections because our body's defence systems usually protect us. However, it was realised that certain important medical advances were being accompanied by increasing numbers of patients being put at risk to these infections. For example, the medical profession was making major clinical advances and hence preventing patients dying from leukaemia and certain types of cancer, and there were increasing numbers of patients receiving organ transplants. All of these patients were immune-suppressed to some extent and were susceptible to both bacterial and fungal infections. Although there were many powerful and safe antibacterial agents, this was not the case with antifungals. The agents available were amphotericin B and 5-fluorocytosine. Amphotericin can be life-saving but it has to be given by slow intravenous infusion, and side-effects (fever, nausea, renal toxicity) tend to make it an agent of last resort - when it is frequently too late. 5-Fluorocytosine is orally absorbed but it is only active against a limited range of fungi and resistance can arise during chronic treatment. There was therefore a need for a safe, convenient agent to treat these infections and we anticipated that in the future there would be an even greater need, as immune-suppression became routine. (In 1978 of course, we could not predict the advent of

- **Oral and i.v.**
- **Safe**
- **Broad spectrum**
- **Therapy and prophylaxis**

Figure 1.   Profile of the ideal antifungal drug

AIDS and the accompanying increase in systemic fungal infections).

We felt that the ideal drug (Figure 1) for the treatment of systemic fungal infections should be effective both orally and intravenously. The usual treatment would be orally but some very ill patients have difficulty with oral dosage forms and hence an i.v. dosage form would be needed. Clearly, it should also be safe and be active against a wide range of fungi. A well tolerated drug would also allow prophylaxis of patients at risk, to prevent systemic

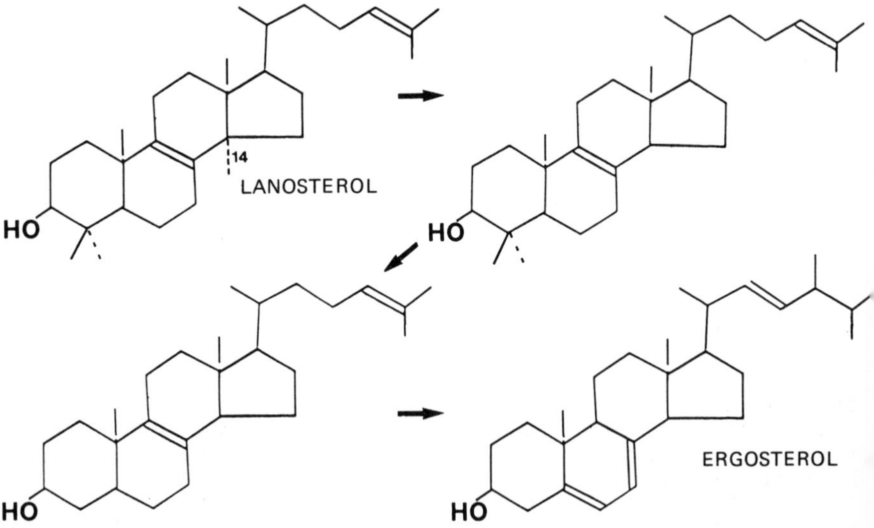

Figure 2.   C-14 Demethylation, a key step in the synthesis of ergosterol

infections occurring (e.g. during periods of immune-suppression for organ transplants or during periods of intensive chemotherapy in cancer patients). In addition, a very safe agent would offer the possibility of treating less-serious, but much more common, infections such as vaginal candidosis and dermatomycoses.

We considered the known classes of antifungal agent and decided that there were advantages in working in azoles.

(1) We had a substantial structure-activity knowledge from our work, leading to the topically active tioconazole.

(2) We knew that the azoles inhibited the C-14 demethylation step that is crucial for the synthesis of ergosterol (Figure 2), the essential sterol in the fungal membrane, and high selectivity for the fungal system could readily be achieved.

(3) Our work with tioconazole suggested that the imidazoles were only poorly effective against fungal infections when dosed orally because they were rapidly and extensively metabolised. For example, only 30% of an oral dose reached the systemic circulation in an intact form. We therefore anticipated that if we could overcome the metabolism then we could obtain an orally effective agent.

In mid-1978, these ideas received support from the announcement by Janssen that they had prepared an imidazole derivative (ketoconazole - Figure 3) that was active orally in several animal models of fungal infection. Although still susceptible to metabolism, it was less readily metabolised than earlier imidazole derivatives and it was important since it proved the concept that an azole derivative could have good oral absorption and be effective after oral dosing.

Our initial approach was to work in imidazole derivatives and we investigated a wide range of structural types including the dioxolanes, dithiolanes, tetrahydrofurans and tertiary alcohols shown in Figure 4. It proved relatively easy to obtain derivatives showing good in vitro potency but it was far more difficult to achieve good activity in animal models of fungal infections. As is well known, this difficulty arises because, in order to show in vivo efficacy, the agent

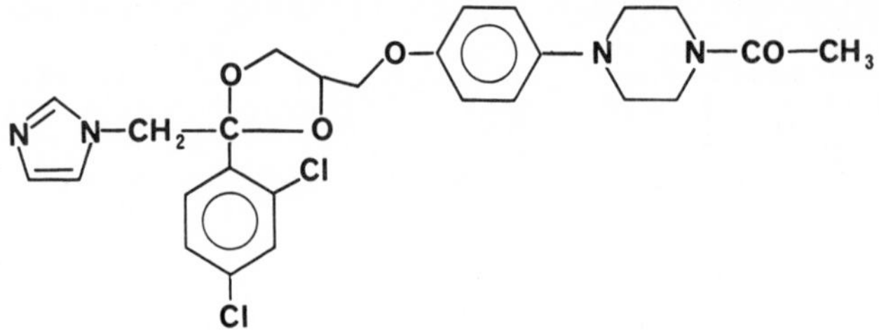

Figure 3. Ketoconazole (Janssen)

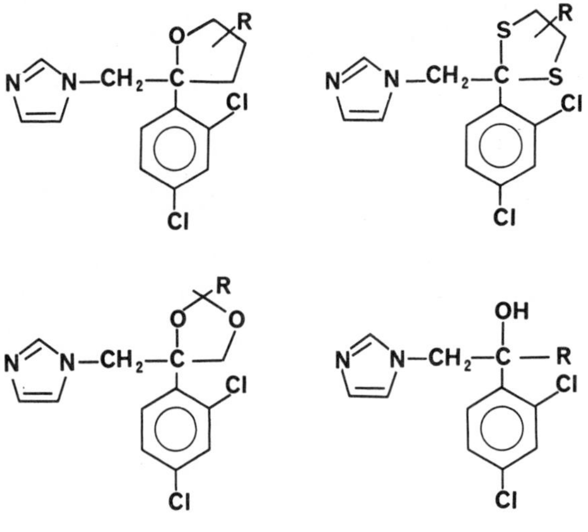

Figure 4. Structural types of imidazole derivatives synthesised

must be absorbed from the gastro-intestinal tract, pass unscathed through the liver and be delivered to the site of action. With imidazole antifungal derivatives a major portion of many compounds is lost by first-pass metabolism in the liver, before entering the systemic circulation. In addition, most of these derivatives are very lipophilic, leading to high protein binding (often >99%), and hence only very low levels of unbound drug (its active form) reach the site of infection. We concentrated our efforts on the tertiary alcohol series because these often gave activity in the animal infection models. A wide range of derivatives were prepared but we had great difficulty in achieving superior activity to ketoconazole. Pharmacokinetic studies indicated that these compounds were still susceptible to metabolism and we reasoned that we required a metabolically stable replacement for the imidazole unit. A series of derivatives were prepared where the imidazole had been replaced by a range of groups, including those shown in Figure 5. The only derivative offering any encouragement was the 1,2,4-triazole derivative, which was more active than the corresponding imidazole compound despite being 4 x less potent in vitro. This suggested to us that we had blocked one site of metabolism, possibly due to triazole being less nucleophilic and therefore more stable to attack.

The next step was to prepare a series of derivatives (Figure 6) where we varied the group R. We believed that, in order to achieve good in vivo activity, the R group should not only be resistant to metabolism but should be chosen such that the resulting derivatives would have low lipophilicity. A low lipophilicity was

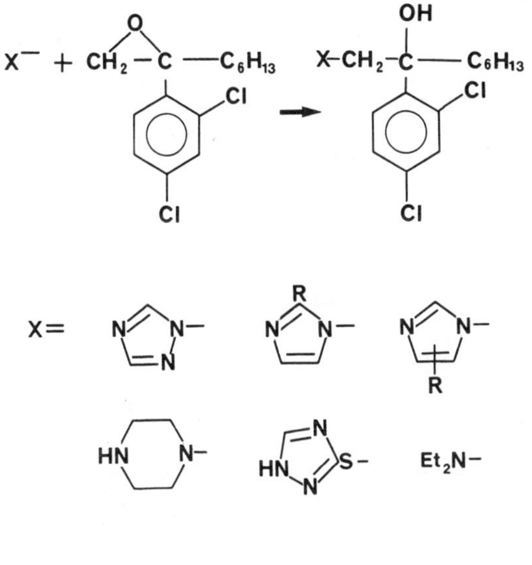

Figure 5.   Preparation of non-imidazole t-alcohols

predicted to be consistent with high blood levels and, since it should reduce protein binding, there should be high levels of unbound drug. The <u>first</u> polar, metabolically stable group chosen was a 1,2,4-triazole, resulting in the bis-triazole shown. This derivative, UK-47,265, showed outstanding efficacy in our systemic candidosis model. In this model, mice are given a lethal systemic <u>Candida</u> infection, and are then treated with test compounds or saline. The dose of compound required to cause 50% of treated animals to survive when all of the saline-treated animals are dead (2 days) is the ED_{50}. In this model, UK-47,265 was almost 100 times more potent than the standard agent, ketoconazole. This outstanding activity encouraged us to prepare a series of bis-triazoles where we replaced the dichlorophenyl unit by a range of groups.

These compounds were examined in the mouse systemic candidosis test and the results are shown in Table 1. As can be seen, the dichlorophenyl moiety was replaced by a range of substituted aryl groups, in addition to alkyl, cycloalkyl and heterocyclyl groups. The best activity (low ED_{50} values) was seen with halo-substituted phenyl derivatives, particularly 2,4-disubstituted analogues. In contrast, several 3- and/or 5-substituted derivatives had poor activity, and this was shown in a cell-free system, with the 3-chloro and 3,5-difluoro analogues, to be due to poor inhibition of the fungal C-14 demethylase enzyme.

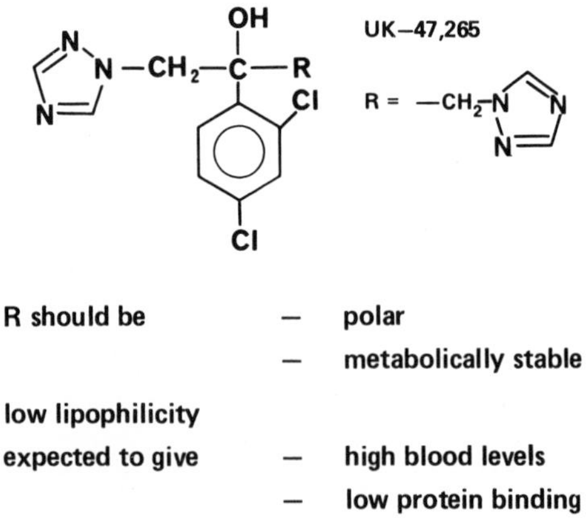

Figure 6. Triazole tertiary alcohol derivatives

Table 1. Evaluation vs. systemic candidosis in mice

$$\underset{N}{\overset{N}{\diagup}}N-CH_2-\underset{\underset{R}{|}}{\overset{\overset{OH}{|}}{C}}-CH_2-N\underset{N}{\overset{N}{\diagdown}}$$

R	(ring, F,F)	(ring, F,F)	(ring, F,F)	(ring, F,F)	(ring, F,F)	(ring, Cl,Cl)	(ring, Br)
ED50*	0.1	>10	>10	0.1	>10	>10	0.1	0.1

R	(ring, Cl)	(ring, Cl)	(ring, Cl)	(ring, Br)	(ring, F)	(ring, Cl,Cl)	(ring, CF₃)	(ring, I)
ED50	>10	0.1	0.2	>10	0.3	2.0	0.1	0.1

R	(ring, N, Cl)	(ring, F,F,F)	(ring, Me, Cl)	(ring, Cl)	(ring, Cl)	(ring, F,F,F)	(ring, Me, Me)	(ring, OMe)
ED50	0.4	0.6	>10	>10	0.1	0.8	>10	>10

R	(phenyl)	Et	(ring, CH₂Cl)	(cyclohexyl, H)	(thiophene, Cl)	(thiophene, Br)	(pyridine, N)	(pyridine, N)
ED50	5.7	>10	>10	>10	>10	1.9	>10	>10

ED_{50} = effective dose in mg/kg to protect 50% of the mice.

Table 2. Evaluation vs. dermatophytosis and vaginal candidosis in mice

R	(ring, F,F)	(ring, F,F)	(ring, Cl,Cl)	(ring, Br)	(ring, Cl)	(ring, Cl)	(ring, F)
Dermatophytosis	79% (5)	20% (20)	80% (10)	49% (20)	76% (20)	35% (10)	30% (10)
Vaginal Candidosis	80% (10)	N.D.	95% (10)	90% (10)	80% (10)	64% (10)	14% (10)

R	(ring, CF₃)	(ring, I)	(ring, N, Cl)	(ring, F,F)	(ring, Cl, F)	(ring, F,F)
Dermatophytosis	0% (20)	70% (10)	24% (20)	59% (20)	85% (20)	0% (20)
Vaginal Candidosis	34% (5)	30% (5)	N.D.	N.D.	84% (10)	N.D.

Values are % cures (at the dose in mg/kg). N.D. = not done.

Table 3. Pharmacokinetic evaluation in mice

R	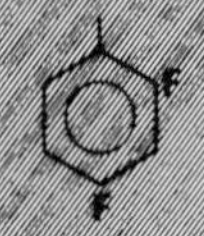	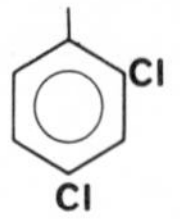		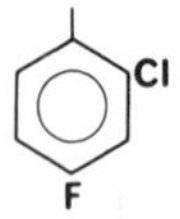
$T\frac{1}{2}$(h)	5.1	3.9	4.0	3.6
% Excreted in urine	75	29	12.5	29.4

Thirteen derivatives had outstanding activity and were progressed to animal models of vaginal candidosis[1] and dermatophytosis[2]. The results in Table 2 show the % clinical cures at the given dose (in mg/kg - p.o.).

The best four derivatives, which were all chloro- or fluoro-substituted phenyl derivatives, were progressed into pharmacokinetic studies (Table 3), from which the 2,4-difluorophenyl compound emerged as the only compound that combined water-solubility, long half-life and high urinary recovery (indicating clearance without metabolism).   This derivative, fluconazole, was examined in a wide range of fungal infection models, in both immune-normal and immune-suppressed animals, then

Fluconazole   150 mg single-dose
Clotrimazole   200 mg a day (3 days)

	Cures	
	1 week	6 weeks
Fluconazole	99	93
Clotrimazole	97	84

Figure 7. Clinical efficacy against vaginal candidosis

progressed through safety studies and into pharmacokinetic evaluation in man.

The results were highly encouraging[3]. Fluconazole was readily absorbed orally, had a plasma half-life of 25 hours, high blood levels, low protein binding and high urinary recovery. Fluconazole was then progressed into clinical evaluation against a range of fungal infections. Against vaginal candidosis (Figure 7) a single dose of 150 mg is highly effective when evaluated at both 1 week and 6 weeks post-treatment. Thus fluconazole is the first antifungal agent effective, for this indication, as a single oral dose.

Examination against a range of dermatomycoses showed that 50 mg/day given orally is highly effective against skin infections, regardless of the site, being at least equivalent to the best available agents (Figure 8).

Following these highly encouraging results, fluconazole was evaluated against Candida infections in leukemia, cancer and AIDS patients and, as can be seen from Figure 9, a 50 mg/day oral dose is highly effective against infections at a variety of sites.

One of the most exciting results with fluconazole has been its efficacy against cryptococcal meningitis in AIDS patients. This infection is usually fatal but a 200 mg oral dose of fluconazole, given once daily, has given a high cure rate (Figure 10). After cure, these patients are receiving a maintenance dose of 100 mg/day to prevent re-infection. A contributory factor to

50 mg/day orally

Tinea pedis
Tinea cruris
Tinea capitis
Pityriasis

} **Over 90% cures**

At least equivalent to best available agents — ketoconazole/griseofulvin/topicals.

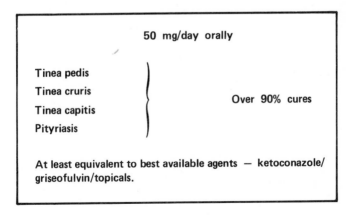

Figure 8. Clinical efficacy against dermatomycoses

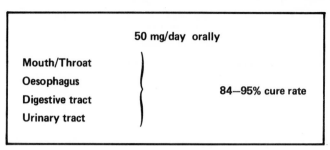

<u>Figure 9.</u>   Candida   infections   in   AIDS   and   cancer
<u>patients</u>

these excellent results is undoubtedly the fact that
the levels of fluconazole in cerebral spinal fluid
(CSF) are almost as high as its blood levels.

        Early results are also promising, against systemic
infections   due   to   <u>Coccidioides</u>,   <u>Blastomyces</u>   and
<u>Histoplasma</u> spp., which are endemic in parts of the
Americas, and therefore clinical studies are being
expanded.   Evaluation   against   infections   due   to
<u>Aspergillus</u> spp. is just beginning.

        <u>In summary</u>, after our initial studies learning
about the SAR of azole antifungals, we can see <u>three</u>
key stages leading to the identification of fluconazole
as a highly effective antifungal.

1.   <u>Replacement of imidazole by triazole</u> – this change
      not only blocks a principal metabolic site, but we
      subsequently   found   that   it   leads   to   improved
      selectivity   (e.g.   no   effect   on   testosterone
      synthesis).

2.   <u>Addition of the second triazole</u> – this achieved
      exactly what we had hoped for – it blocked
      metabolic   breakdown   of   this   portion   of   the
      molecule, raised blood levels and reduced protein
      binding.   This led to high, sustained levels of
      unbound drug.

3.   <u>Substitution of difluorophenyl</u> for dichlorophenyl
      caused a further reduction in lipophilicity (log P
      0.5 <u>vs.</u> 1.0), leading to reduced serum protein
      binding and increased water solubility (10 mg/ml,
      facilitating an i.v. formulation).

> **200 mg/day orally**
>
> **CSF levels = 80% of blood levels**
>
> **Cure rate = over 80%**
>
> **[100 mg/day to prevent re-infection]**

Figure 10.   Cryptococcal meningitis in AIDS

Thus, we appear to have achieved our objective from 1978. Fluconazole can be dosed once daily, either orally or intravenously, and is highly effective against all fungal infections studied to date, although efficacy against some fungi remains to be examined. It is effective in both immune-normal and immune-suppressed patients, and studies in over 4,000 patients suggest that it has a very good safety profile.

References

1.   K. Richardson, K.W. Brammer, M.S. Marriott and P.F. Troke, <u>Antimicrob. Agents Chemother</u>., 1985, <u>27</u>, 832.

2.   M.S. Marriott, R.J. Andrews, K. Richardson and P.F. Troke in Recent Advances in Chemotherapy, J. Ishigami, ed., U. of Tokyo Press, 1985, p.1936.

3.   M.J. Humphrey, S. Jevons and M.H. Tarbit, <u>Antimicrob. Agents Chemother</u>., 1985, <u>28</u>, 648.

Antimalarial Hydroxynaphthoquinones

A. T. Hudson

THE WELLCOME RESEARCH LABORATORIES, LANGLEY COURT,
BECKENHAM, KENT BR3 3BS, UK

Despite optimism in the 1950's and 1960's that malaria could be eradicated by vector control in combination with chemotherapy, the disease remains a major threat to health. Currently 45% of the world's population is at risk[1] and in tropical Africa alone more than 1 million malaria-related deaths occur annually.[2]

Malaria is caused by mosquito-transmitted infection with four species of protozoa - *Plasmodium falciparum, P. vivax, P. ovale* and *P. malariae.* The former parasite is the more prevalent accounting for some 85% of disease incidence. Unfortunately, this also causes the most virulent form of disease which rapidly leads to death in non-immune victims unless drug treatment is promptly initiated. Ironically, the widespread use and attendant misuse, of drugs to combat malaria has led to the emergence of drug-resistant strains of parasite. This has seriously compromised the effectiveness of the front line drugs chloroquine and pyrimethamine. There is now an urgent need for novel anti-malarial agents suitable for mass treatment and prophylaxis of *P. falciparum* infections. Unfortunately the likely commercial returns from such drugs are not sufficient to tempt most pharmaceutical companies into the area. However, significant advances in malaria research have been made in the last decade to give rise to optimism that new drugs can be discovered. In particular, screens have been developed which allow putative anti-malarials to be evaluated against *P. falciparum*, both *in vitro*[3] and in Aotus monkeys.[4] These assays have been used at the Wellcome Research Laboratories to investigate structure-activity relationships in a series of novel hydroxynaphthoquinones related to compounds first examined for anti-malarial activity some 40 years ago.

Interest in the anti-malarial properties of 2-hydroxy-3-alkyl-1,4-naphthoquinones originated in America in the early 1940's as a result of a massive collaborative programme between industry and academia to find a replacement for quinine.[5] In the course of screening large numbers of compounds with diverse structures against *P. lophurae* in ducks, the primary anti-malarial assay of the day, the hydroxyquinone hydrolapachol was found to have significant activity. This led Fieser *et al.* to synthesise further analogues numbering some 300 in all from which the following structure-activity conclusions were drawn[6]:

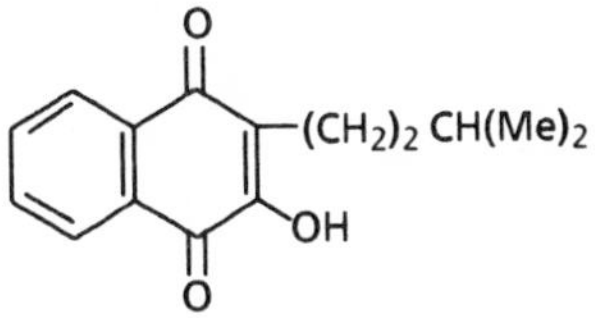

HYDROLAPACHOL

I. The 2-OH group is essential for activity; replacement by OMe, H, Me, NH_2, $NHCOCH_3$, Cl or SH results in complete or extensive loss of activity.

2. Introduction of substituents into, or reduction of, the aromatic nucleus diminishes activity.

3. The presence of heteroatoms, halogens or double bonds in the alkyl moiety is detrimental to activity.

4. Direct attachment of an alicyclic moiety to the 3-position is especially favourable for potent activity.

This last point is exemplified by the bicyclohexyl analogue (1) which was exceptionally potent in the *P. lophurae* assay - 25 times more active than quinine

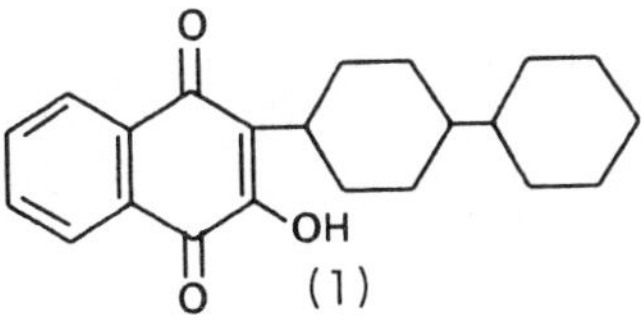

ED_{95}* = 0.67 mg/kg   cf. quinine ED_{95} = 17.35 mg/kg

* ED_{95} = the effective dose given orally 3 times per day for 4 days which produces a 95% reduction of P. lophurae parasitaemia in ducks relative to controls

As a result of this work a number of analogues were evaluated in man and all were found to be rapidly metabolised, primarily by oxidation of the alkyl side chain.[7]   Thus quinones such as (2) were converted to carboxylic acids whilst cycloalkyl analogues *e.g.* (3) were metabolised to alcohols - see Scheme 1.   The latter had weak but definite anti-malarial activity and long plasma half-lives.

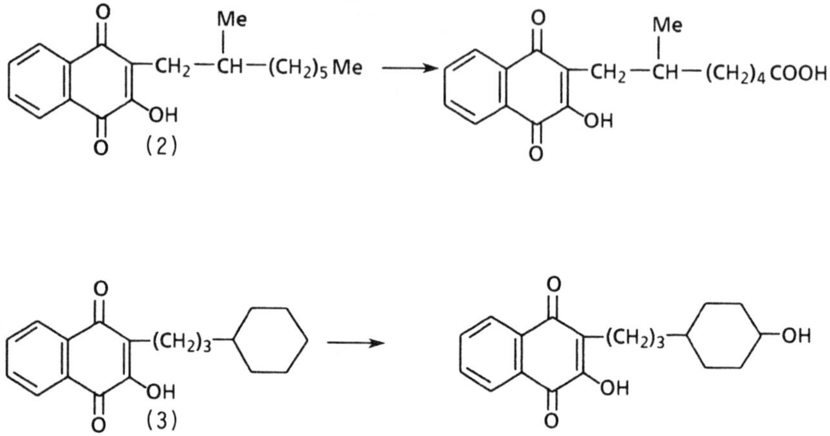

Scheme 1

The latter observation led Fieser *et al.* to synthesise lapinone.[8]   This quinone contained a hydroxylated side chain in order to impart metabolic stability.   The length of the alkyl residue was carefully chosen to offset the hydrophilic properties of the alcohol moiety.   In the *P. lophurae* assay this compound was found to have good activity but only when administered parenterally.   In a subsequent clinical trial lapinone was given intravenously to 9 patients with primary *P. vivax* infections and found to exert a marked therapeutic effect relieving all disease symptoms.[9] However because the drug had to be given intravenously at a fairly high dosage (2 g/day for 4 days) it was not developed further and with the advent of chloroquine and primaquine interest in the hydroxynaphthoquinones waned.

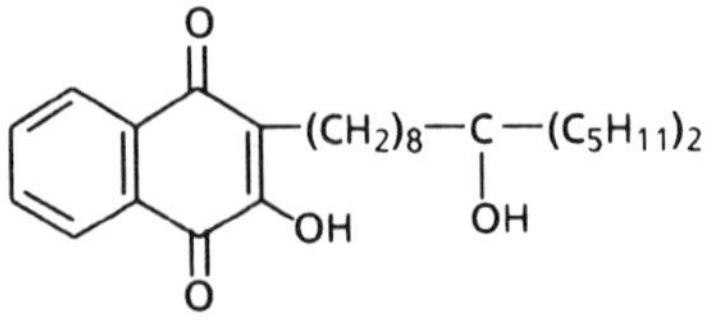

lapinone

As a result of the work of Fieser and his associates the pharmacokinetic behaviour of various hydroxynaphthoquinones was known in a wide range of avian and mammalian species.   Only mice were shown to metabolise the quinones similar to man.[10]   In the 1960's by the time the need for new anti-malarial drugs had become apparent the primary *P. lophurae* assay in ducks had been replaced with one using *P. berghei* in mice.   This was seen as the ideal system for evaluating hydroxynaphthoquinones and Fieser in association with scientists at Sterling-Winthrop again commenced work in the area.[11]   This led to the synthesis of menoctone a highly lipophilic quinone which was predicted to undergo oxidation in both humans and mice to a cyclocarbinol (4) with the correct physical properties for potent anti-malarial activity.

MENOCTONE, R = H, *ED_{50} = 12mg/kg   cf chloroquine ED_{50} = 5mg/kg
   (4)   R = OH

*ED_{50} = the effective dose given twice daily for 4 days intragastrically which produces a 50% reduction of P. berghei parasitaemia in mice relative to controls.

This compound was sufficiently active in the *P. berghei* mouse assay[12] to warrant evaluation in man.   However oral administration to patients with *P. falciparum* infections failed to show any therapeutic effects[13] and all interest in menoctone evaporated.

A novel approach to the design of quinone anti-malarials was then taken when Folkers *et al.* obtained circumstantial evidence that these compounds exerted their anti-parasitic effects by acting as ubiquinone antagonists.[14]   A variety of heterocyclic analogues of the hydroxynaphthoquinones were synthesised, *e.g.* (5), but insufficient activity was obtained in animal models to warrant testing in man.

Ubiquinone n = 1-10   (5)

Interest in the anti-parasitic properties of hydroxynaphthoquinones at the Wellcome Research Laboratories stemmed from the finding that

menoctone was highly active against *Theileria parva.*[15]   This parasite is
related to Plasmodia species and is responsible for the tick-borne cattle
disease endemic in East Africa, East Coast Fever.   Because menoctone
would be too expensive to produce for treatment of a Third World animal
disease structure-activity studies[16] were carried out in the hope of finding
a more cost-effective analogue.   As a result of this work the 2-cyclohexyl
derivative, parvaquone, was found to be highly active *in vitro* and was
shown in field trials to be effective in curing cattle with East Coast Fever.
This compound, under the trade name Clexon, was subsequently
marketed in 1983 as the first effective treatment for the disease.

parvaquone, Clexon

   The key factor to discovering parvaquone was the ability to carry out
structure-activity studies against the target parasite *in vitro*.   It was
therefore logical to extend this work to malaria as for the first time with this
class of compound activity could be optimised *in vitro* against the most
virulent human parasite, *P. falciparum*.   In embarking upon this work
consideration had to be given to solving such likely problems as poor oral
absorbtion and/or rapid metabolism which had been so evident from the
work of Fieser *et al.*   Given that a class of compound suffers these
disadvantages there are several approaches possible:

1.   Modify lipophilic/hydrophilic properties to facilitate absorbtion.

2.   Pro-drugs.

3.   Block major point of metabolic degradation.

4.   Increase inherent activity.

   The first 2 strategies were dependent upon having animal models
capable of predicting the behaviour of the hydroxyquinones in man and
these were clearly lacking.   The other 2 possibilities seemed far more
realistic targets.   Parasite activity could be measured and optimised
against *P. falciparum in vitro* whilst some idea of likely metabolic stability
could be gained from studies in mice as this species had previously been
claimed to metabolise the compounds similar to man.[10]
   Hydroxynaphthoquinones containing cyclohexyl groups were shown by
Fieser to be metabolised at the 4-position of the cyclohexyl ring to alcohols
with greatly reduced anti-plasmodial properties.[7]   We reasoned that it
should be possible to block such metabolism by introducing suitable

substituents into this position. Furthermore, we anticipated that since a hydroxyl moiety in the 4-position adversely affected activity, conversely a lipophilic 4-substituent might be beneficial. To test these hypotheses we decided to synthesise analogues of parvaquone with a variety of substituents in the 4-position.

Parvaquone itself has activity comparable to the standard anti-malarial drugs in the *P. falciparum* assay - see Table 1. Introduction of a hydroxyl group into the cyclohexyl unit as anticipated destroys activity - see (6)[17]. Replacement of this group with a fluoro or trifluoromethyl moiety results in compounds (7) and (8), respectively, with activity comparable to the parent drug. However when a *t*-butyl group is substituted into the 4-position of the cyclohexyl ring activity is spectacularly increased. The resulting compound BW58C is far more active than any of the anti-malarials currently in use. This activity reverts to that of parvaquone if the *t*-butyl moiety is moved to the 3-position - see (9). Similarly, positioning a methylene group between the quinone and cyclohexyl rings has a detrimental effect on efficacy, see (10) as does replacing the *t*-butylcyclohexyl unit with *t*-butylphenyl - (11). It is possible however to replace the *t*-butyl group in BW58C with a *t*-amyl or *t*-butoxy group without significantly affecting potency, see (12) and (13). Surprisingly the *n*-butoxy derivative (14) is far less active than (13). Several dialkyl analogues as exemplified by (15) were also synthesised but no improvement over the activity of parvaquone was seen.

All of the cyclohexylquinones shown in Table 1 were prepared by reaction of the appropriate cyclohexane carboxlic acid with 2-chloro-1,4-naphthoquinone in the presence of ammonium persulphate and silver nitrate according to the procedure of Jacobsen and Torssell.[18] The resulting chloroquinones were then hydrolysed to the desired hydroxynaphthoquinones as shown for BW58C in Scheme 2.

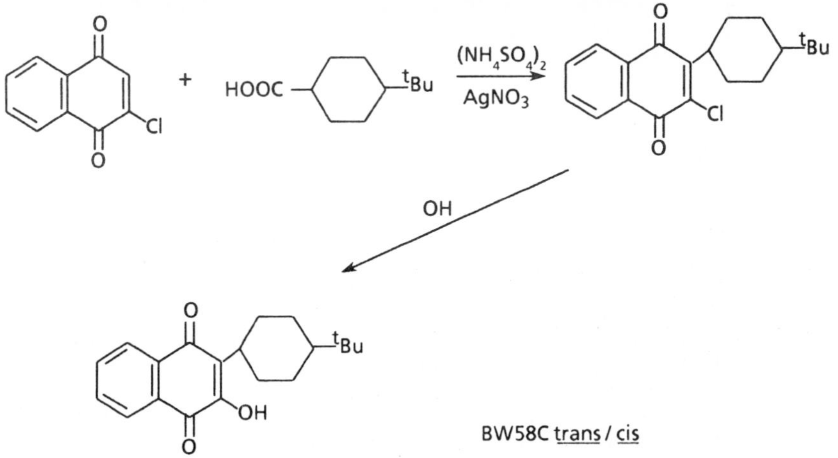

Scheme 2

Table 1 Activity of hydroxynaphthoquinones against P.falciparum[d] In vitro[17]

Compound	R	EC_{50} (M)[a]
Parvaquone	(cyclohexyl)	7.8×10^{-8}
(6)	(cyclohexyl)–OH	$> 1 \times 10^{-4}$
(7)	(cyclohexyl)–F	2.4×10^{-7}
(8)	(cyclohexyl)–CF_3	3.0×10^{-8}
BW58C	(cyclohexyl)–^tBu	5.7×10^{-11}
Cis Isomer		1.0×10^{-9}
(9) [b]	(cyclohexyl)^tBu	3.0×10^{-7}
(10)	–CH_2–(cyclohexyl)–^tBu	5.5×10^{-8}
(11)	(cyclohexenyl)–^tBu	2.2×10^{-7}
(12) [c]	(cyclohexyl)–^tAm	6.0×10^{-11}
(13)	(cyclohexyl)–O^tBu	9.5×10^{-11}
(14)	(cyclohexyl)–O^nBu	3.0×10^{-8}
(15)	(cyclohexyl)$<^{Et}_{Et}$	1.4×10^{-8}
Chloroquine		3.7×10^{-8}

a EC_{50} = concentration of drug required to reduce the proportion of parasite-
    infected cells to 50% of that of untreated controls
b   cis / trans isomer ratio = 1 / 4
c   cis / trans isomer ratio = 2 / 3
d   Strain FC - W - 1 / Nigeria

The drawback to this is that it always generates a mixture of *cis* and *trans* isomers when the 4 (or 3)-position is monosubstituted. These can be separated, usually by fractional crystallisation, and except for compounds (12) and (9) all results in Table 1 were obtained, where applicable, with *trans* isomers.[17]

Since the *trans* isomers should be favoured thermodynamically the conversion of *cis* to *trans* forms should be feasible *via* the mechanism shown in Scheme 3.

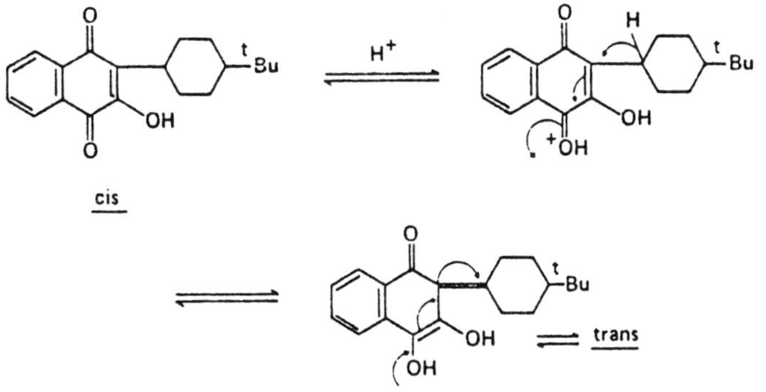

Scheme 3

When a 1/1 *cis/trans* mixture of BW58C was heated in concentrated sulphuric acid at 55⁰ for 24 hours the *cis* isomer disappeared and the pure *trans* isomer was obtained in 46% yield. Despite considerable efforts it was never possible to improve on this yield. This suggested the possibility that the *cis* isomer was being selectively destroyed rather than epimerised. This was confirmed when it was shown[17] that the *trans* isomer was completely unaffected by sulphuric acid treatment whereas the *cis* form decomposed to a mixture principally composed of 2 furanoquinones (16) and (17) in which the latter predominated. The formation of these compounds is rationalised in Scheme 4. Presumably the remarkable differences in stability between the 2 isomers result from steric strain in the *cis* form. This would be relieved by oxidation to the cyclohexenylquinone (18) which could then react further to form the observed products.

Although the sulphuric acid method was useful for producing substantial quantities of *trans* BW58C rapidly it was wasteful of material. A better method of obtaining the individual isomers of BW58C and other cyclohexylquinones of interest was found to be a stereospecific route based upon the so-called Hooker oxidation. This is a reaction discovered in the 1930's by Samuel Hooker for oxidising alkylquinones to the next lower homologue.[19] The reaction course is such that the stereochemistry of the alkyl chain is unaffected[20] and proceeds for BW58C as shown in Scheme 5.

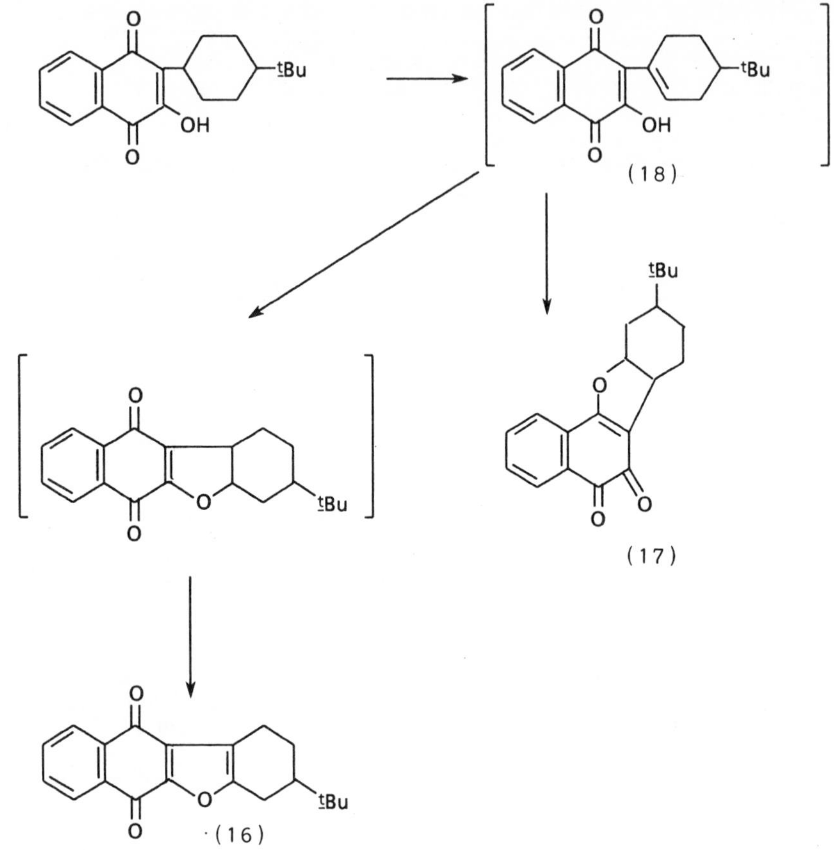

Scheme 4

The starting material for this route is the *trans* quinone (10). This was obtained by treating *trans* t-butylcyclohexylacetic acid with 2-chloro-1,4-naphthoquinone in the presence of ammonium persulphate and silver nitrate to give the chloroquinone which was then hydrolysed to (10). This compound was oxidised under alkaline condiditions, initially with hydrogen peroxide and then with copper sulphate, to give *trans* 58C in 70% yield with no trace of the corresponding *cis* isomer.[21]

The Hooker oxidation was used to synthesise the *cis* and *trans* isomers of various cyclohexylquinones. In every case the *trans* isomers were found to be more active than the *cis* against *P. falciparum* - e.g. cf. value of *trans* 58C (5.7×10^{-11}M) with *cis* 58C (1.0×10^{-9}M) in Table I.

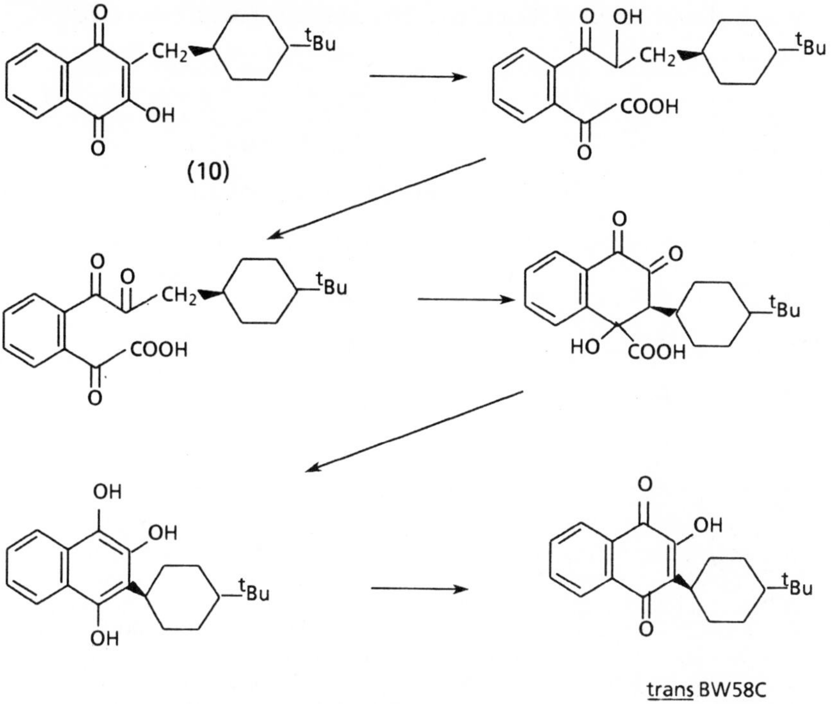

(10)

Scheme 5

trans BW58C

The promising activity of BW58C against *P. falciparum* justified testing this compound against other *Plasmodia* in various species. Unfortunately it was not possible to compare directly *in vitro* these parasites with *P. falciparum* as their life cycles and culture conditions differ.

Table 2. Activity of BW58C and chloroquine against rodent and chick
          Plasmodia in vivo

Host	Parasite	ED$_{50}$ mg/kg x 7 p.o.	
		BW58C	Chloroquine
Mouse	*P. Yoelii*	0.41	1.19
Mouse	*P. berghei*	0.46	1.33
Mouse	*P. chabaudi*	1.19	1.22
Rat	*P. berghei*	1.19	1.10
Chick	*P. gallinaceum*	1.90	1.11

It is evident from the *in vivo* studies with BW58C that it compares
favourably with chloroquine and is selectively toxic for *Plasmodia* in a
variety of hosts - see Table 2. Efforts were made to establish the
biochemical basis of this selectivity. In previous studies a correlation was
noted between the anti-plasmodial properties of hydroxynaphthoquinones
and their ability to inhibit mammalian electron transport processes by
acting as ubiquinone antagonists.[14] Electron transport systems have
been extensively studied in mammals where a central role has been
established for ubiquinone.[22] This quinone acts as an electron carrier in
a cytochrome-linked chain mediating the oxidation of NADH and succinate
- see Scheme 6 for a simplified schematic representation.

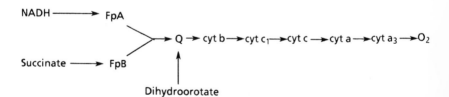

Q = ubiquinone, Fp = flavoprotein, cyt = cytochrome

Scheme 6

Pyrimidine Biosynthesis

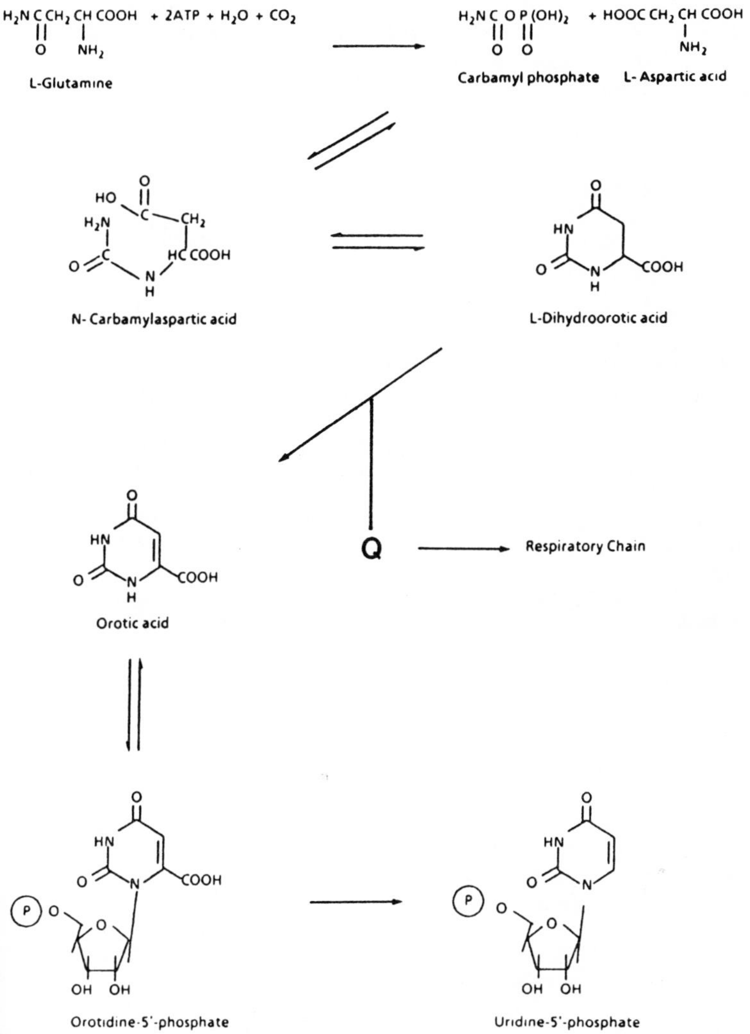

$H_2N\,C\,CH_2\,CH\,COOH$ + 2ATP + H_2O + CO_2

L-Glutamine

$H_2N\,C\,O\,P(OH)_2$ + HOOC CH_2 CH COOH

Carbamyl phosphate   L-Aspartic acid

N-Carbamylaspartic acid

L-Dihydroorotic acid

Orotic acid

Q ⟶ Respiratory Chain

Orotidine-5'-phosphate

Uridine-5'-phosphate

Scheme 7

In mammals one of the major functions of this process is to generate energy which is tapped off and stored in the form of ATP. This role is unlikely to be of any importance to *Plasmodia* as these parasites are homolactate fermentors and generate their energy by the anaerobic catabolism of glucose.[23] However another major function of electron transport systems which is common to both mammals and *Plasmodia* is the involvement with pyrimidine biosynthesis.[24-27] In the *de novo* synthesis of pyrimidines dihydroorotate is oxidised to orotate in a reaction catalysed by dihydroorotate dehydrogenase and linked to the respiratory chain by ubiquinone - Scheme 7.

These systems were isolated from rat liver and *P. yoelii* and used to assay BW58C. The results obtained[28] show clear differences between the mammalian and plasmodial systems. The EC_{50} against the parasite preparation was 7.0×10^{-10}M whilst against the corresponding rat liver system a value of 1.5×10^{-7}M was obtained. This significant differential in activity explains the selective toxicity of the drug *in vivo*.

In order to study further the mode of action of BW58C the effects on pyrimidine biosynthesis in *P. falciparum* were measured. It was shown[29] that 8×10^{-4}M of BW58C had no effect on the energy status of the parasite as shown by unchanged ATP levels. In contrast 1×10^{-10}M of drug caused a dramatic decrease in the concentrations of pyrimidine nucleotides, specifically uridine triphosphate, over a 4 hour incubation period. The concentration of this nucleotide was reduced to one quarter that of untreated parasites. At the same time the levels of the pyrimidine intermediate carbamylaspartate rose by over 700% from a concentration of 0.25 mM to 1.8mM. An increase in the concentration of dihydroorotate was not observed as might be predicted from BW58C inhibiting dihydroorate dehydrogenase. This is because the equilibrium constant of dihydroorotase greatly favours the formation of carbamylaspartate.[30]

The above results provide convincing evidence that the anti-malarial properties of BW58C arise by inhibition of the parasite respiratory chain - linked enzyme dihydroorotate dehydrogenase. This produces a blockade of pyrimidine biosynthesis which, unlike mammals, the parasites are unable to relieve by the up-take of pre-formed pyrimidines.[31]

As a consequence of the promising biological and biochemical data on BW58C it was selected for evaluation in humans in a Phase I clinical trial. Toxicology studies in 2 animal species (rat, marmoset) established that the drug was sufficiently safe to allow single dose oral administration of up to 8 mg/kg to healthy volunteers. At this level it was found to be well tolerated and no drug-related changes in haematology or clinical chemistry were observed. Peak plasma levels of drug were seen after 1.5 hours reaching 6.0×10^{-7}M. Concentrations greater than 10^{-7}M were maintained for approximately 8 hours. Unfortunately extensive metabolism was observed to the alcohol (19) and carboxylic acid (20) and other unidentified derivatives - see Scheme 8. These compounds were present in significant quantities in plasma and urine with elimination half lives (T0.5) of up to 17 hours - compare BW58C T O.5 ll hours. In addition these compounds were also present in a conjugated form, most likely as glucuronides. For example the mass spectra of the conjugate of (19) was consistent with (21).

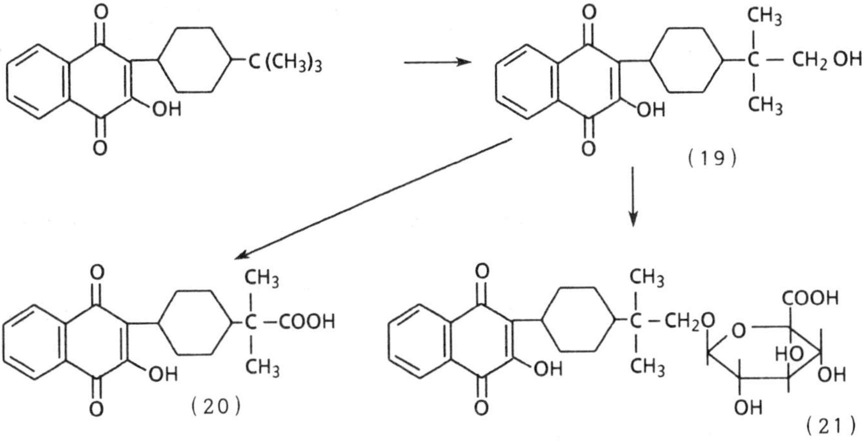

Scheme 8. Metabolism of BW58C

These metabolites including their conjugates were also seen in rats but here the degree of metabolism and extent of metabolite conjugation were far less. Such quantitative differences were readily apparent - in humans 8 mg/kg of BW58C gave rise to red urine whereas in rats and other species no coloration was observed at doses many times higher. This was attributed to the fact that the formation of water soluble hydroxynaphthoquinones such as the conjugated metabolite (21) occurred to a far greater extent in humans than any other species studied. Consequently the hydroxynaphthoquinone content of human urine was proportionally higher resulting in the observed colour differences.

In order to evaluate BW58C in patients with malaria it was judged necessary to maintain anti-plasmodial levels over a 5 day period. Since the drug has a half life of only 11 hours this would require twice daily dosing. It was feared that this would lead to a build up of metabolites and their conjugates whose toxicity in man could not be predicted from animal studies. As a result of this further clinical studies with this compound were terminated and a search was initiated for an analogue which would be far more resistant to metabolism in man whilst retaining the potent anti-plasmodial properties of BW58C. Because of the difficulty in extrapolating studies in animals to humans with this series of compounds it was decided to investigate the metabolism of selected analogues using hepatic microsomes from various species including man. This technique has found increasing use in recent years and is limited only by the availability of human liver.

Initially BW58C was incubated with microsomes at 37^0 in the presence of NADPH. After 30 minutes the reaction was terminated and the mixtures extracted with ethyl acetate which was then analysed by thin-layer chromatography. This showed microsome preparations from a variety of species to metabolise BW58C to the alcohol (19).[32] However, the extent to which this reaction occurred varied considerably as *in vivo* man was the most efficient promoter of this oxidation and rat the least as indicated in Table 3.

Table 3.

Comparative Metabolism of BW58C, BW59C and BW566C with Liver Microsomes

Species	Degree of Metabolism		
	BW58C	BW59C	BW566C
Man	+ + +	+ + +	-
Monkey (aotus)	+ +	+	-
Monkey (cynomolgus)	+	-	-
Marmoset	+	NT	-
Dog	+	NT	-
Mouse	+	-	-
Rat	+	NT	-
IC_{50} P.falciparum* Invitro	7.0×10^{-10}	6.0×10^{-10}	$7.0 \times 10^{-1\cdot}$

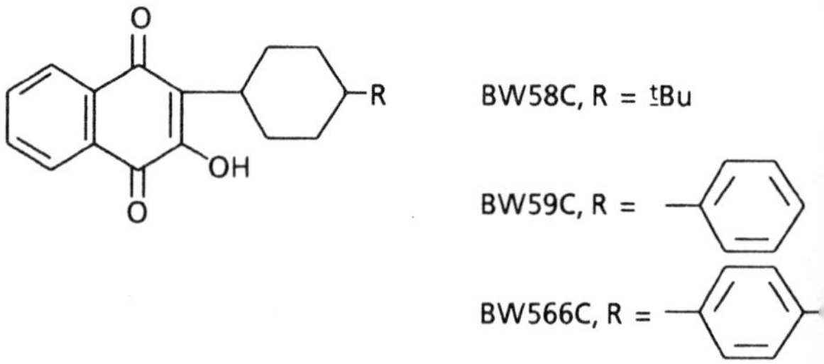

BW58C, R = $\underline{t}$Bu

BW59C, R =

BW566C, R =

* Cloned strain T9/96

In view of these results other hydroxynaphthoquinones which had shown promising activity in the *P. falciparum* assay were treated with microsomes and their susceptibility to degradation was measured. This revealed the phenylcyclohexylquinone BW59C to be resistant to metabolism by mouse and monkey preparations. However in the presence of human microsomes it was extensively degraded to an unknown metabolite. In contrast the chlorophenyl analogue BW566C was particularly striking in that it was inert in every preparation tested. This compound was subsequently evaluated against rodent and chick *Plasmodia in vivo* and found to be more active than BW58C - see Table 4. Significantly in view of its stability to Aotus microsomes BW566C was highly effective against *P. falciparum* in *Aotus trivirgatus* monkeys, 7 daily doses of I mg/kg effecting cures.

It is predicted that quinones such as BW566C should be better candidates for clinical trials than BW58C as the toxicology and efficacy studies conducted in animals should be extrapolated to humans with greater confidence. Hopefully the strategy of using *in vitro* techniques to optimise activity against the target parasite and minimise metabolism in the target host will be successful and after 4 decades of research a hydroxynaphthoquinone will finally find use in the fight against malaria.

Table 4. Activity of BW566C against rodent and chick
Plasmodia in vivo

Host	Parasite	ED_{50} mg/kg x 7 p.o.		
		BW566C	BW58C	Chloroquine
MOUSE	*P. yoelii*	0.03	0.41	1.19
MOUSE	*P. chabaudi*	0.06	1.19	1.22
CHICK	*P. gallinaceum*	0.22	1.90	1.11

Acknowledgements

· The contribution is gratefully acknowledged of all staff at the Wellcome Research Laboratories, Beckenham who have participated in the anti-malarial programme. In particular thanks are due to M. J. Pether, A. W. Randall (Chemistry), M. Fry, D. J. Hammond, B. Hill (Biochemistry), M. Dickens, C. R. Jones, G.Land,B. C. Weatherley (Metabolism and Pharmacokinetics), V. S. Latter, M. Pudney, M. A. Smith (Biology) and D. B. A. Hutchinson (Clinical).

References

1.   WHO Report of the Steering Committees of the Scientific Working Groups on Malaria June 1980 -June 1983.   TDR/MAL/SC-SWG(80-83)/83.3

2.   *Lancet* 1975, 2:15-6.

3.   W. Trager and J. B. Jensen, *Science* 1976, 193, 673.

4.   L. H. Schmidt, *Am. J. Trop. Med. Hyg.*, 1978, 27, 718.

5.   For review see A. T. Hudson, in Handbook of Experimental Pharmacology Vol. 68/11, Antimalarial Drugs, Eds. W. Peters and W. H. G. Richards, Springer-Verlag, Berlin, 1984, p. 343.

6.   L. F. Fieser and A. P. Richardson, *J. Am. Chem. Soc.*, 1948, 70, 3156.

7.   L. F. Fieser, F. C. Chang, W. G. Caulsen, C. Heidelberger, A. M. Seligman, *J. Pharmacol. Exp. Ther.*, 1948, 94, 85.

8.   L. F. Fieser, E. Berliner, F. J. Bondhus, F. C. Chang, W. G. Dauben, M. G. Ettlinger, G. Fawaz, M. Fields, M. Fieser, C. Heidelberger, H. Heymann, A. M. Seligman, W. R. Vaughan, A. G. Wilson, E. Wilson, M. I. Wu, M. T. Leffler, K. E. Hamlin, R. J. Hathaway, E. J. Matson, E. E. Moore, M. B. Moore, R. T. Rapala and H. E. Zaugg, 1948, *J. Am. Chem. Soc.*, 70, 3151.

9.   G. Fawaz and F. S. Haddad, *Am. J. Trop. Med. Hyg.*, 1951, 31, 569.

10.   L. F. Fieser, H. Heymann and A. M. Seligman, *J. Pharmacol. Exp. Ther.*, 1948. 94, 112.

11.   L. F. Fieser, J. P. Schirmer S. Archer, R. R. Lorenz and P. I. Pfaffenbach, *J. Med. Chem.*, 1967, 10, 513.

12.   D. A. Berberian and R. G. Slighter, *J. Parasitol.* 1968, 54, 999.

13. WHO Tech. Rep. Ser. 1973, 529, 70.

14. For review see T. H. Porter and K. Folkers, *Angew. Chem. (Engl.)*, 1974, 13, 559.

15. N. McHardy, A. J. B. Haigh and T. T. Dolan, *Nature (London)*, 1976, 261, 698.

16. P. Boehm, K. Cooper, A. T. Hudson, J. P. Elphick and N. McHardy, *J. Med. Chem.*, 1981, 24, 295.

17. A. T. Hudson, M. J. Pether, A. W. Randall, M. Fry, V. S. Latter and N. McHardy, *Eur. J. Med. Chem.*, 1986, 2l(4), 271.

18. N. Jacobsen and K. Torssell, *Liebigs Ann. Chem.* 1972, 763, 135.

19. S. C. Hooker and A. Steyermark, *J. Am. Chem. Soc.*, 1936, 58, 1202.

20. See Ref. 8, p. 3186.

21. A. T. Hudson and A. W. Randall, Eur. Pat. Application, 1983, EP 77,551; *Chem. Abstr.*, 1983, 99, 87848y.

22. F. L. Crane, *Arch. Biochem. Biophys.*, 1960, 87, 198.

23. For review see I. W. Sherman, *Microbiol. Rev.*, 1979, 43 (4), 453.

24. M. E. Jones, *Ann. Rev. Biochem.*, 1980, 49, 253.

25. J. Kennedy, *Arch. Biochem. Biophys.*, 1973, 157, 369.

26. W. E. Gutteridge, D. Dave and W. H. G. Richards, *Biochim. Biophys. Acta*, 1979, 582, 390.

27. L. L. Bennett, D. Smithers, L. M. Rose, D. J. Adamson and H. J. Thomas, *Cancer Res.*, 1979, 39, 4868.

28. A. T. Hudson, A. W. Randall, M. Fry, C. D. Ginger, B. Hill, V. S. Latter, *Parasitology*, 1985, 90, 45.

29. D. J. Hammond, J. R. Burchell and M. Pudney, Mol. *Biochem. Parasitol.*, 1985, 14, 97.

30. D. Keppler and A. Hostege in Metabolic Compartmentation, Ed. H. Sies, Academic Press, London, 1982, p. 147.

31. J. J. Jaffee and W. E. Gutteridge, *Congr. Int. Protozool.*, 1973, I, 23.

32. M. Dickens - Unpublished results.

Approaches to the Synthesis of Antibacterial Compounds

E. J. Thomas

THE DYSON PERRINS LABORATORY, UNIVERSITY OF OXFORD,
SOUTH PARKS ROAD, OXFORD OX1 3QY, UK

Many compounds of pharmaceutical interest are chiral.
The development of asymmetric syntheses of such
compounds is an area of considerable importance at
present because frequently only one enantiomer has the
required biological activity, the other either being
inactive or associated with unwanted side-effects. We
are interested in developing asymmetric syntheses of
chiral natural products, and now wish to describe our
results to date in two areas of possible pharmaceutical interest, namely approaches to the asymmetric
synthesis of thiotetronic acids and milbemycins-
avermectins.

1 THIOTETRONIC ACIDS

The chemistry of thiotetronic acids has been studied
intermittently since Benary first synthesized the
parent system (1) in 1913.[1] However interest in this
area has recently increased because of the isolation
and characterization of several thiotetronic acids
which have antibacterial activity, including
thiolactomycin (2),[2] thiotetromycin (3),[3] and U-68,204
(4).[4] A synthesis of <u>racemic</u> thiolactomycin has been
described,[5] but prior to our work no asymmetric
synthesis in this area had been reported.[6]

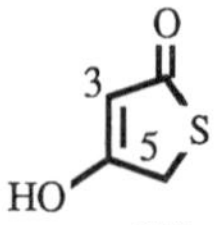

(1)

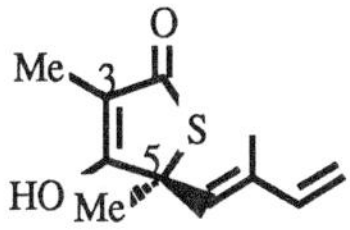

(2) 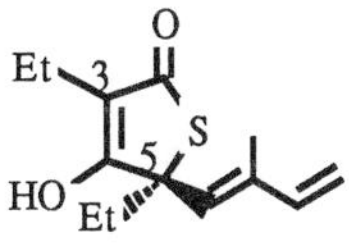   (3) 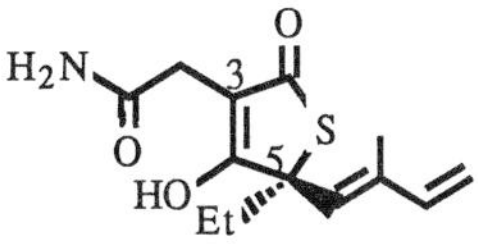   (4)

The antibiotics (2) - (4) possess a chiral centre
at C(5) which in the natural products has the
(S)-configuration shown, and which corresponds to a
tertiary thiol. Our problem was therefore to develop
an asymmetric synthesis of a suitable tertiary thiol
for subsequent conversion into the thiotetronic acid
antibiotics (2) - (4). Ideally the synthesis would
need to provide intermediates of high optical purity,
and should be flexible to permit access to a wide
range of analogues. It occurred to us that one
approach to this problem would be to use an allylic
rearrangement of a chiral alcohol derivative as shown
in Scheme 1. In particular hydroxy-ester (5) should
be readily accessible from (S)-lactic acid, and on
rearrangement could provide access to tertiary thiol
derivatives (6). Such rearrangements are known for
several classes of allylic alcohol derivatives
including sulphenates, (7)⇌(8),[7] and xanthates, (9)⇌
(10).[8] We chose first to investigate rearrangements
of allylic xanthates.

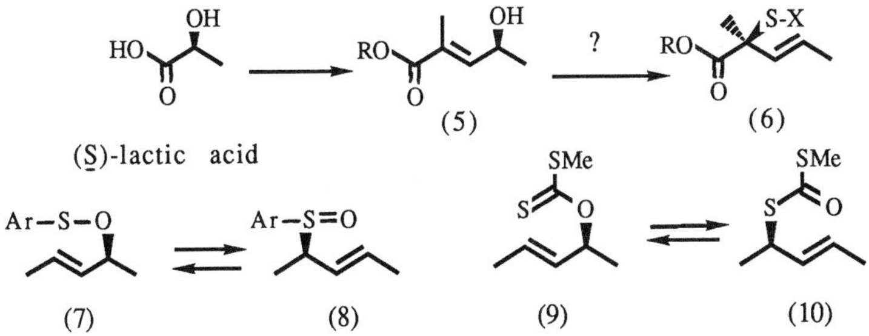

(S)-lactic acid   (5)   (6)

(7)   (8)   (9)   (10)

Scheme 1

Thus ethyl (S̲)-lactate was converted into the
hydroxy ester (13) as shown in Scheme 2 via alcohol
protection and reduction using diisobutylaluminium
hydride (DIBAL), followed by stereoselective Wittig
condensation and deprotection (overall yield ca.
50%). Alcohol (13) was then converted into its
xanthate (14) by sequential treatment with NaH, CS₂,
and methyl iodide. This xanthate was stable at room
temperature, and could be purified by column
chromatography. However, on distillation in a
Kugelrohr at 145°C, 0.4 mm, it rearranged cleanly to
provide dithiocarbonate (15).

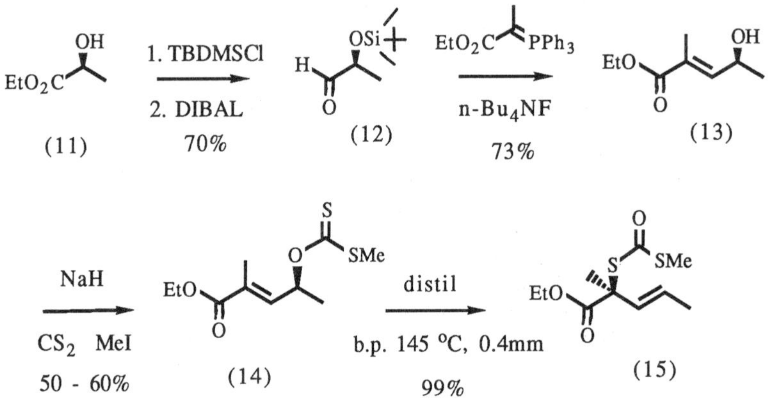

Scheme 2

The rearrangement of xanthate (14) was extremely
efficient (99%) and stereoselective. Only the
dithiocarbonate shown, with the (E̲)-double-bond, could
be detected by high field (300 MHz) ¹H n.m.r. of the
distilled product. To establish the optical purity
of the rearranged dithiocarbonate (15), its
enantiomer (18) was prepared as shown in Scheme 3.
Mitsunobu inversion of alcohol (13) gave (17) which
was converted to (18) via xanthate rearrangement. The
enantiomeric dithiocarbonates (15) and (18)
were then converted into the corresponding S̲-benzyl
alcohols (16) and (19) by transprotection (K̲OH,
PhCH₂Cl) followed by DIBAL reduction, and the optical
purities of the alcohols estimated using their
Mosher's derivatives. These were clearly
distinguishable by ¹H and ¹⁹F n.m.r., and confirmed
that both alcohols were at least 98% optically pure,
showing that the allylic xanthate to dithiocarbonate

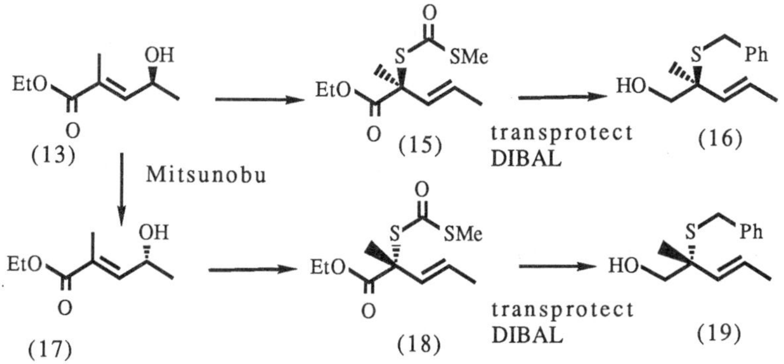

Scheme 3

rearrangements had taken place with essentially complete transfer of asymmetry. The absolute configurations of the rearranged dithiocarbonates were established by X-ray crystallography (see later) and were found to correspond to those shown in Schemes 2 and 3. The allylic xanthate to dithiocarbonate rearrangement may therefore be envisaged as taking place _via_ a six-membered, chair-like, cyclic transition state with the allylic methyl group adopting an equatorial position to avoid 1,3-diaxial interactions with the vinylic methyl substituent, as shown in Figure 1.

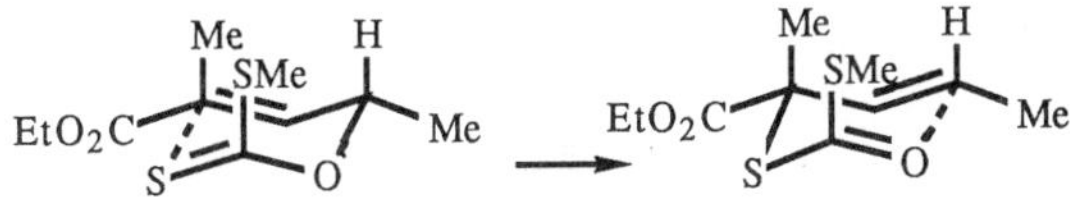

(14)   (15)

Figure 1. Possible transition state geometry for the allylic xanthate to dithiocarbonate rearrangement.

Having developed an asymmetric synthesis of the tertiary dithiocarbonates (15) and (18), it remained to demonstrate their use for thiotetronic acid synthesis. Because of the base sensitivity of the dithiocarbonate group, this was first converted into a p-methoxybenzyl thioether by treatment with KOH in

the presence of p-methoxybenzyl chloride. The
thioether-ester (20) from dithiocarbonate (15) was
then hydrolysed and treated with carbonyl diimidazole
to provide acyl imidazolide (21) which was used to
acylate methyl propanoate giving ketoester (22) as a
mixture of diastereoisomers. S-Deprotection and
cyclization was achieved in a single step by treatment
with trifluoroacetic acid (reflux, ca. 1 h) to provide
the thiolactomycin analogue (24) (40%). In CDCl$_3$ this
was found to be mainly the enol tautomer (24) in
equilibrium with a small amount (≅ 10%) of the keto
tautomers (23). Previously only enol tautomers have
been reported for thiotetronic acids.

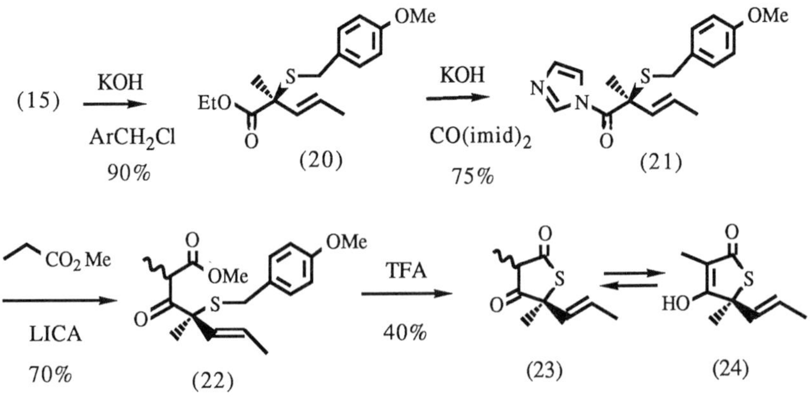

Scheme 4

The acyl imidazolide (21) was also condensed with
methyl phenylacetate and ethyl acetate to provide the
thiotetronic acids (25) and (27). The 3-phenyl
compound (25) was found by ^{1}H n.m.r. to be completely
enolic in both CDCl$_3$ and DMSO-d$_6$, whereas the
3-unsubstituted compound was predominantly enolic in
DMSO-d$_6$ and ketonic in CDCl$_3$. The structure and
absolute configuration of the 3-phenylthiotetronic
acid (25) was confirmed by X-ray crystallography.

These asymmetric syntheses of thiotetronic acids
(24), (25), and (27), show that the optically active
dithiocarbonate (15) is a useful intermediate for the
synthesis of a range of thiotetronic acids with
different substituents at C(3). The preparation of
analogues with varying substituents at C(5) now

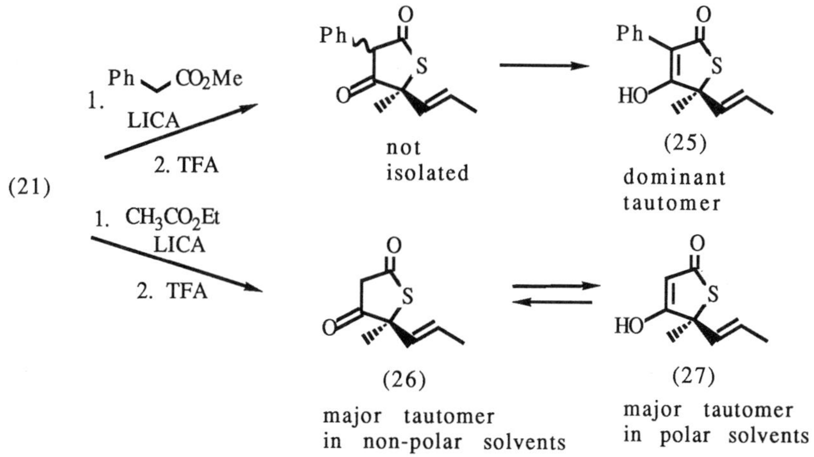

(21)

1. Ph CO₂Me
   LICA
2. TFA

1. CH₃CO₂Et
   LICA
2. TFA

not
isolated

(25)
dominant
tautomer

(26)
major tautomer
in non-polar solvents

(27)
major tautomer
in polar solvents

Scheme 5

requires either the rearrangement of different allylic xanthates or modification of the dithiocarbonate (15) or p-methoxybenzyl thioether (20). To date we have investigated the latter alternative.

Ozonolysis of the allylic thioether (20) gave aldehyde (28) quite efficiently (65%) after treatment of the intermediate ozonide with an excess of dimethyl sulphide. This aldehyde could be reduced to the hydroxyester (30) using NaBH₄, and was treated with isopropenylmagnesium bromide to provide adducts (31). These underwent stereoselective Claisen rearrangement with triethyl orthoacetate - acetic acid to provide the bis-ester (33). Alternatively Peterson and Wittig reactions were used to prepare the αβ-unsaturated aldehyde (29) and ester (32).

In an attempt to prepare (S)-thiolactomycin (2), diene (34) was prepared from aldehyde (29), and was converted into keto-ester (35) using the chemistry discussed above. However, on treatment with trifluoroacetic acid, polymerization occurred, and no thiolactomycin could be isolated. In contrast, alcohol (36) obtained by reduction of aldehyde (29) or ester (32) (via selective t-butyl ester hydrolysis and reduction of the corresponding acid chloride) was successfully converted into the thiotetronic acid

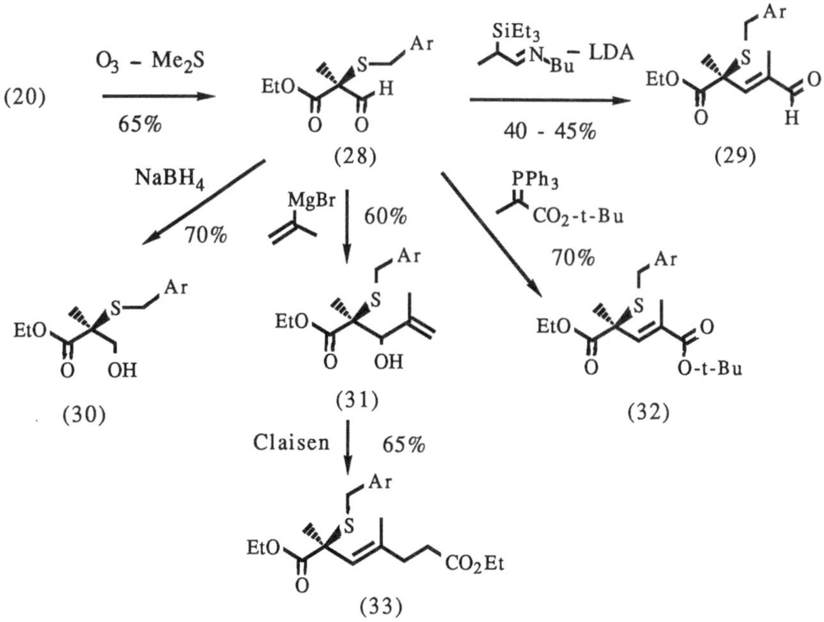

Scheme 6

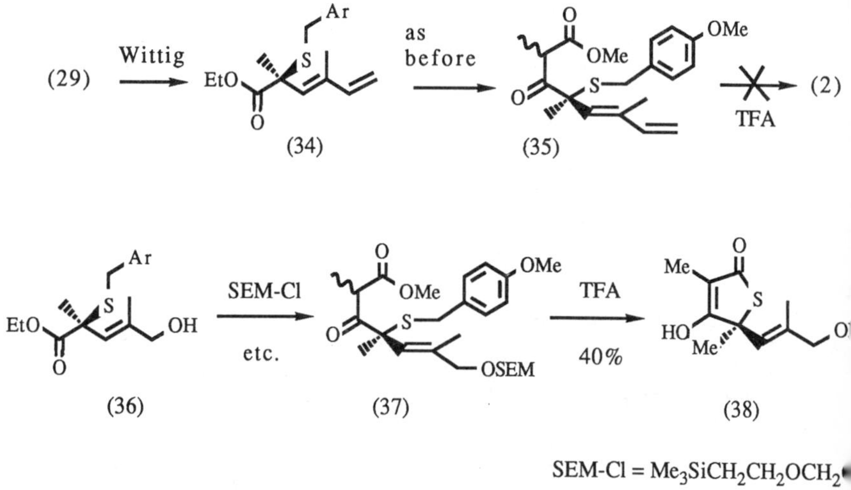

SEM-Cl = Me₃SiCH₂CH₂OCH₂

Scheme 7

(38). Attempts are now being made to convert (38) into (S)-thiolactomycin (2).

2 MILBEMYCIN-AVERMECTIN SYNTHESIS

The milbemycins and avermectins comprise an important group of macrocyclic natural products with useful biological activity which have already been introduced and discussed during the course of this meeting.[9,10] Representative structures are shown in Scheme 8. Because of their commercial importance, the synthesis of these compounds is being widely studied, and several syntheses of the aromatic milbemycin β_3 (40),[11] and one synthesis of an avermectin, have already been reported.[12] We have been interested in developing a synthetic approach to these compounds. Our first objective was to develop a strategy for the

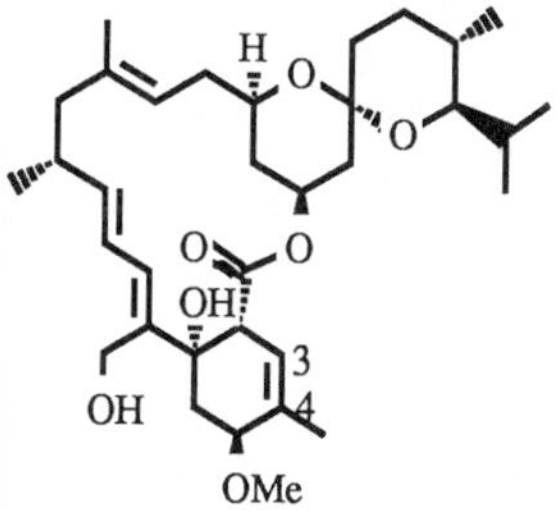

milbemycin E
(39)

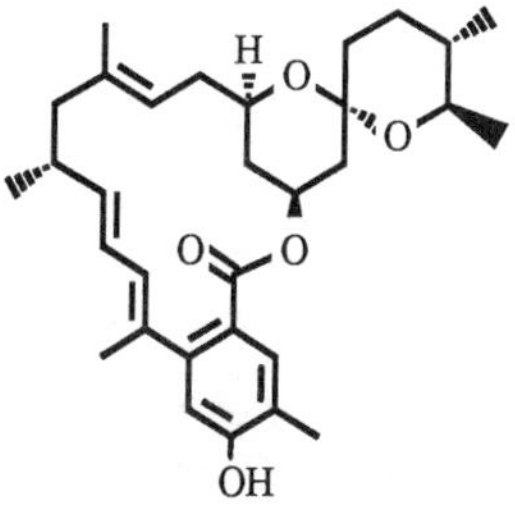

milbemycin β_3
(40)

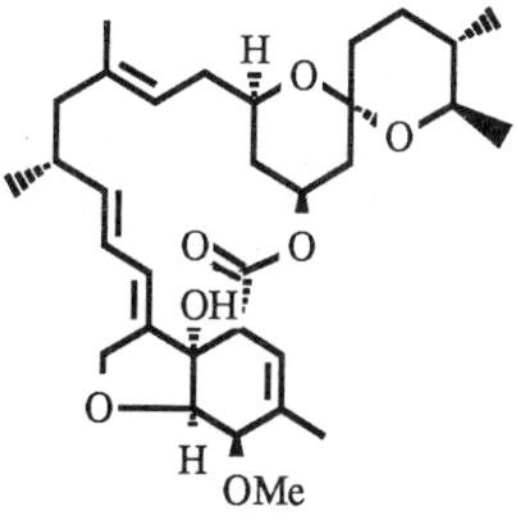

milbemycin α_2
(41)

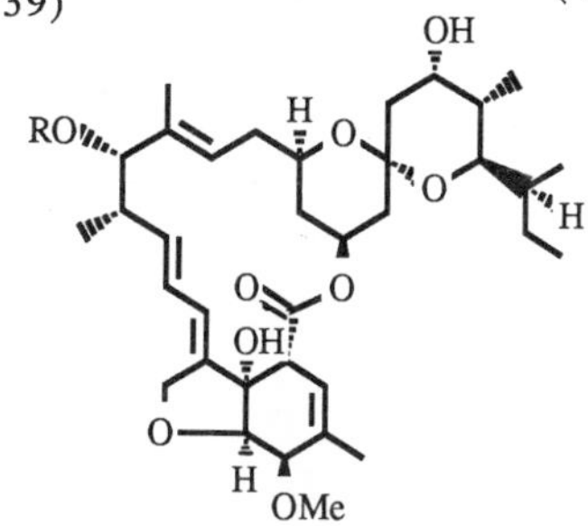

avermectin A_{2a}
R = α-L-oleandrosyl-
α-L-oleandrosyl
(42)

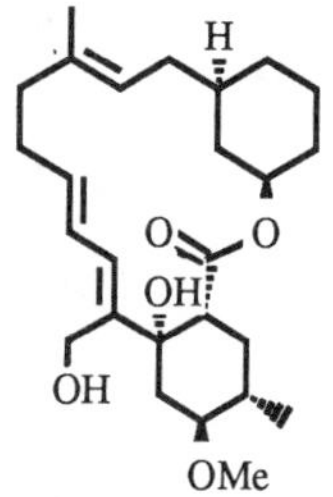

(43)

Scheme 8

synthesis of the non-aromatic β-milbemycins, and to
this end we have synthesized the macrocyclic analogue
(43).[13]   More recently we have developed procedures
for the regioselective introduction of the C(3)-C(4)
double-bond,[14] and for the asymmetric synthesis of the
spiroacetal fragment of these compounds.[15]

Development of a Strategy for β-Milbemycin Synthesis

The macrocyclic milbemycin analogue (43) was our
first target in this area. Its synthesis required us
to synthesize a lower hemisphere fragment with the
correct stereochemistry around the cyclohexane ring;
we also had to develop an approach to the dienyl
alcohol fragment, C(8) - C(11), and for
macrocyclization. Our approach to analogue (43) was
based upon work being carried out by Dr. M.D. Turnbull
at ICI Plant Protection Division. It had been
reported that the keto ester (44) condensed under
basic conditions with methyl vinyl ketone to give a
single adduct (45) isolated by crystallization from
the reaction mixture.[16] Making the assumption that
the reaction was thermodynamically controlled, Dr.
Turnbull surmized that the product had the same
relative stereochemistry at C(2) and C(7) (milbemycin
numbering) as the milbemycins, and postulated that it
preferred the conformation shown in Scheme 9. He
then used this reaction for lower hemisphere analogue
synthesis.[17]

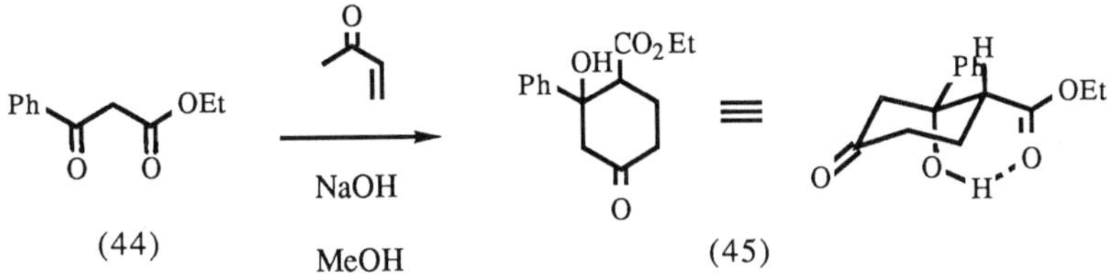

Scheme 9

In order to have access to the C(8)-C(11) dienyl
alcohol fragment we modified this approach by the use
of 3-furanyl keto esters. Thus methyl isopropenyl
ketone and the furanyl keto ester (46) were condensed
in the presence of base to give the hydroxy cyclo-
hexanone (47) isolated as a single diastereoisomer.
The structure and stereochemistry of this adduct were

initially assigned on the basis of spectroscopic data
and were subsequently confirmed by X-ray crystallography (see later). Stereoselective ketone reduction
to diol (48) was effected using sodium triacetoxyborohydride,[17,18] presumably <u>via</u> intramolecular hydride
donation as shown in Scheme 10, and the secondary
hydroxyl group was selectively methylated using
Meerwein's salt.

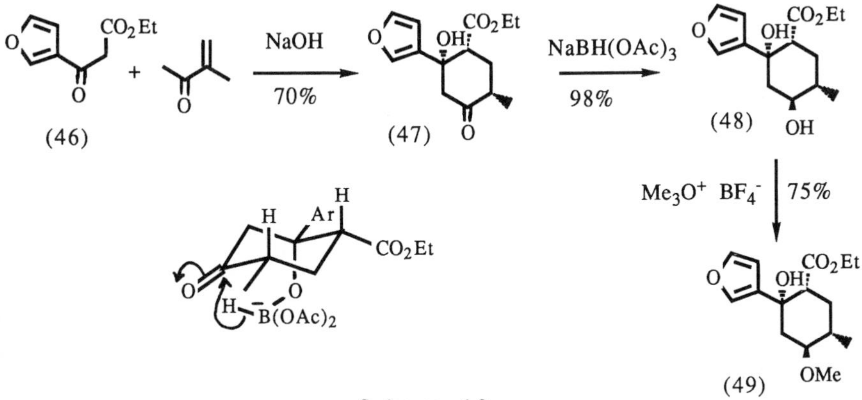

Scheme 10

Regioselective modification of the furan ring was
accomplished by oxidation using bromine in methanol,

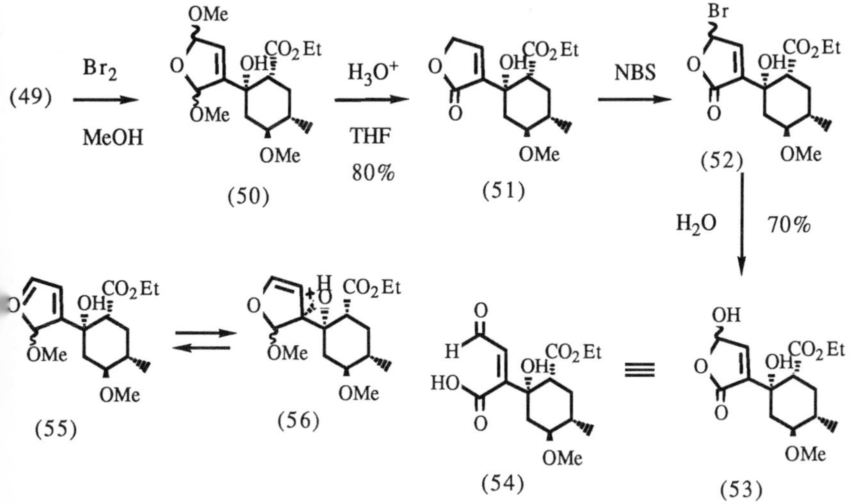

Scheme 11

and acidic hydrolysis of the intermediate dimethoxy-
dihydrofurans (50). This hydrolysis gave the
3-alkylbutenolide (51) which was isolated in good
yield as a single regioisomer. The regioselectivity
of this hydrolysis step was a pleasant surprise and
may have involved the selective participation of
oxonium ion (55) and/or the protonated epoxide (56).
Oxidation of the butenolide was then carried out using
N-bromosuccinimide (NBS), followed by hydrolysis, to
provide the hydroxybutenolides (53), which are
synthetic equivalents of aldehyde-acid (54).

The final stages of the synthesis of milbemycin
analogue (43) are outlined in Scheme 12. Wittig
coupling of the racemic hydroxybutenolides (53) with
the racemic phosphorane (57), which had been obtained
from the bicyclic lactone (60),[13] was efficient, but
gave a mixture of racemic diastereoisomers (58) and
(59). The relative stereochemistry of diastereoisomer
(58) corresponds to that in the milbemycins, but
diastereoisomer (59) has the incorrect combination of
upper and lower hemispheres. The formation of this
mixture was expected since little molecular
recognition was thought to be likely for the Wittig
condensation step. The newly formed double-bonds in
(58) and (59) were also obtained as mixtures of
(E)-and (Z)-isomers in which the (E)-isomers were the
minor components, but treatment of this mixture with a
catalytic quantity of iodine initiated double-bond
isomerization to provide the desired (E)-isomers (61)
and (62) exclusively. Before cyclization could be
attempted, suitable differentiation between the two
carboxyl substituents had to be achieved. Thus the
mixture of (61) and (62) was treated with trimethyl-
silylethanol under basic conditions to effect ester
exchange, followed by esterification of the free
carboxylic acid group with diazomethane. Removal of
the silyl protecting groups (F⁻, THF-H$_2$O then 1 M HCl)
gave the hydroxy-acids (63) and (64) which were
treated with N-methyl-2-chloropyridinium iodide-
triethylamine. Only one hydroxy acid cyclized, no
identifiable products being isolated from the other
diastereoisomer. X-Ray crystallography established
the structure of the cyclized product as that shown in
formula (65), and selective methyl ester reduction
gave the milbemycin analogue (43). The selective
cyclization of the hydroxy acid whose stereochemistry
corresponds to that in the milbemycins was not

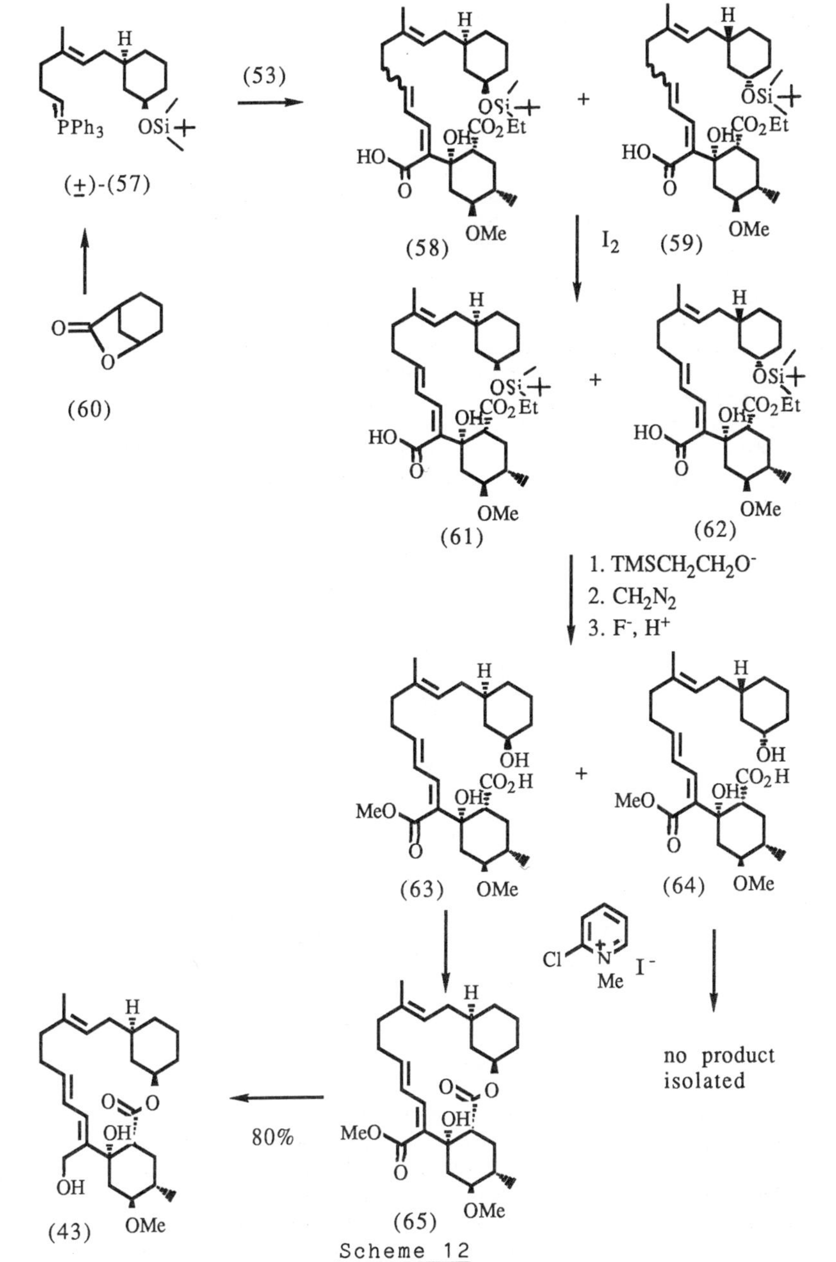

(53)
PPh3
OSi
(±)-(57)
(60)
(58)
(59)
HO
OH
CO2Et
OMe
I2
(61)
(62)
1. TMSCH2CH2O-
2. CH2N2
3. F-, H+
MeO
OH
CO2H
(63)
(64)
Cl
N
Me
I-
no product isolated
(65)
80%
(43)

Scheme 12

unexpected. Molecular models had shown that the
"unnatural" diastereoisomer (64) cannot easily adopt a
conformation suitable for cyclization.

Introduction of the C(3)-C(4) Double Bond

     This synthesis of macrolide (43) has established
a strategy for an approach to the non-aromatic
β-milbemycins. Our present work is concerned with
extending this work to complete natural product
syntheses in this area. One of our targets is
milbemycin E(39). Any synthesis of milbemycin E will
require us to develop a stereoselective spiroacetal
synthesis and a synthesis of the hydroxycyclohexenone
present in the lower hemisphere. Because of our
experience in using Robinson annelation-furan
oxidation chemistry to prepare analogue (43), we
decided to attempt to introduce the crucial C(3)-C(4)
double-bond into the Robinson annelation product (47)
(Scheme 13).

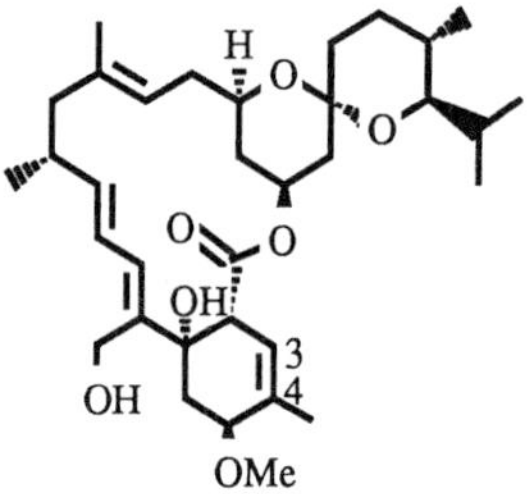

milbemycin E

(39)

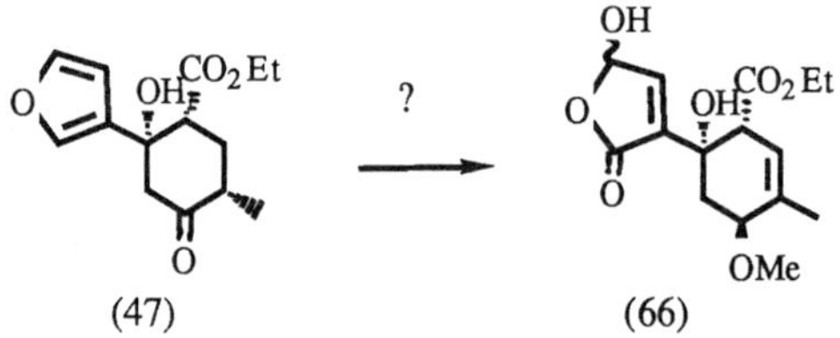

(47)   (66)

Scheme 13

     Treatment of hydroxycyclohexanone (47) with
trimethylsilyltriflate-triethylamine gave the more
substituted enol ether (67). This reacted with
benzene selenenyl chloride in the presence of
anhydrous tetra-n-butylammonium fluoride to give the
tertiary selenide (68). Stereoselective reduction
using sodium triacetoxyborohydride gave the 5β-alcohol

(69), but this on oxidative elimination gave
predominantly the exocyclic allylic alcohol (70)
containing only about 10% of the desired endocyclic
isomer (72). However, if the oxidative elimination was
carried out before reduction, endocyclic elimination
predominated. Thus the phenylselenyl ketone was
converted to the dihydroxycyclohexenol (72) in 70-75%
overall yield (Scheme 14). By using silylated

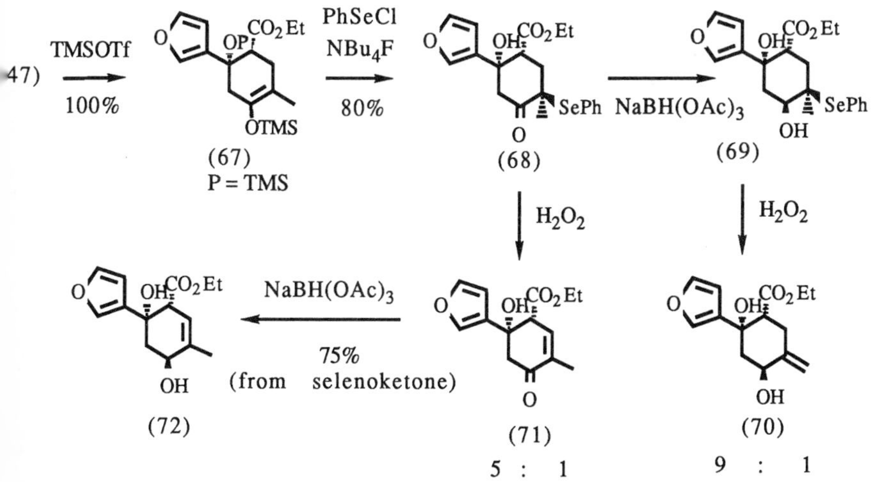

Scheme 14

furans,[19] hydroxybutenolides related to (66)
possessing the C(3)-C(4) double-bond, have been
obtained using this phenylselenation-oxidative
elimination approach.[14]

Asymmetric Synthesis of Milbemycin and Avermectin Spiroacetals

The spiroacetal fragments of the milbemycins and
avermectins have been subject to many synthetic
studies.[10] In our laboratory we have developed an
asymmetric synthesis of the milbemycin E spiroacetal
(73) from α-methyl-D-glucoside.[15] An alternative
approach to spiroacetal (73) is outlined in Scheme 15.
The spiroacetal should be formed on hydrolysis of the
trihydroxydithiane (74) which in turn should be
accessible from the suitably protected hydroxyepoxide
(75) and iodide (76) using dithiane anion chemistry.
This approach is attractive because it is convergent,

and should provide access to a wide range of
milbemycins and their analogues. However the
difficulty lies in developing an efficient
stereoselective route to epoxide (75). The obvious
intermediate for a synthesis of this is acetal-
aldehyde (77), which is readily available from (cheap)
(S)-malic acid,[20] but in its reactions with
nucleophiles, aldehyde (77) is known to show rather
poor diastereoface selectivity.

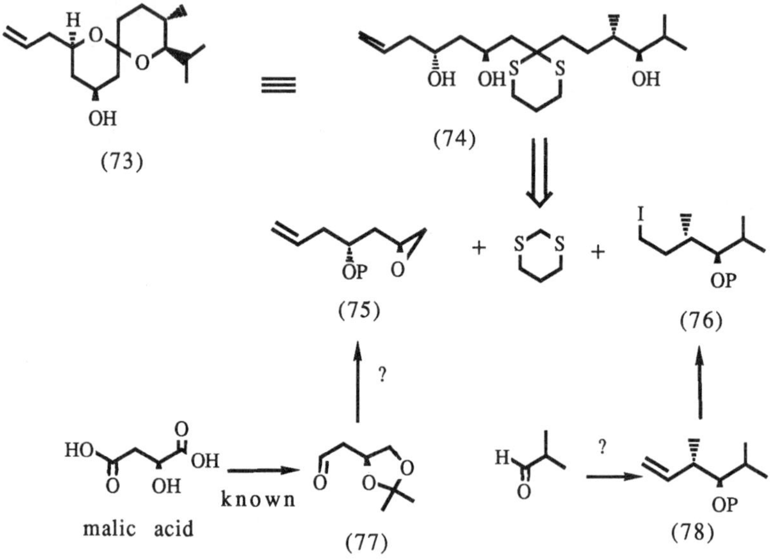

Scheme 15

    H.C. Brown has recently reported the preparation
and use of a series of diisopinocampheylborane
reagents which transfer allyl and crotyl groups to
aldehydes with high stereoselectivity. For example
methoxydi-(+)-isopinocampheylborane (79) and metal
salts of (E)-but-2-ene provide the crotyldiisopinocampheylborane (80) which reacts with aldehydes to
provide anti-adducts (81) with good enantio- and
diastereo-selectivity, probably via the cyclic
transition state shown in Scheme 16.[21] The analogous
allyldiisopinocampheylborane is also known to form
homoallylic alcohols from simple aldehydes with high
enantioselectivity.[22] We hoped to use these boron
reagents to develop efficient syntheses of epoxide
(75) and iodide (76).

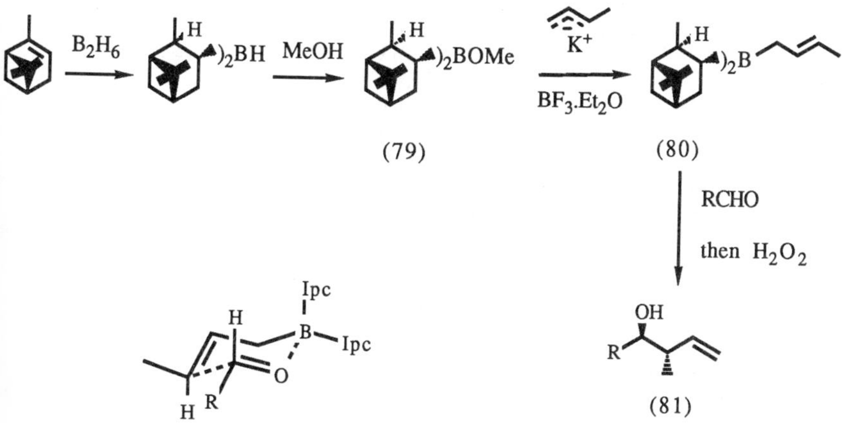

(79)   (80)

RCHO

then H$_2$O$_2$

(81)

Scheme 16

It was found that treatment of aldehyde (77), which had been obtained in four steps from (S)-dimethyl malate,[20] with the allyldiisopinocampheylborane prepared from (-)-α-pinene[22] gave a good yield of the homoallylic alcohol (82) containing only about 10% of its unwanted diastereoisomer. Alcohol protection, followed by acetal hydrolysis and epoxide formation, then gave the protected hydroxy epoxide (83).

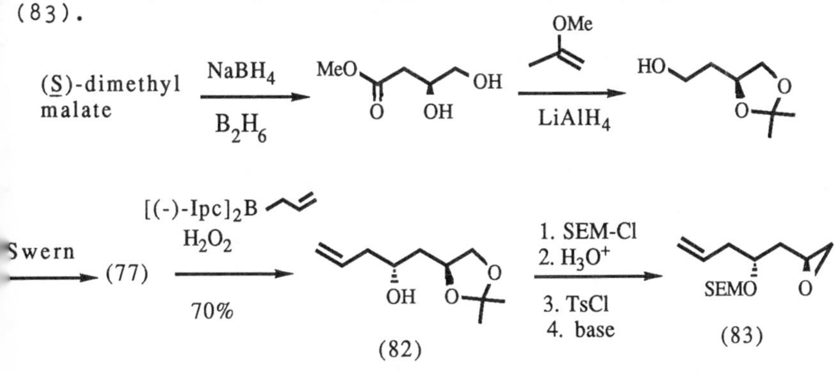

(S)-dimethyl malate $\xrightarrow[\text{B}_2\text{H}_6]{\text{NaBH}_4}$ ... $\xrightarrow{\text{LiAlH}_4}$ ...

Swern $\xrightarrow{}$ (77) $\xrightarrow[\text{70\%}]{[(-)-\text{Ipc}]_2\text{B} \diagup\diagdown \quad \text{H}_2\text{O}_2}$ (82) $\xrightarrow[\begin{array}{l}\text{1. SEM-Cl}\\\text{2. H}_3\text{O}^+\\\text{3. TsCl}\\\text{4. base}\end{array}]{}$ (83)

Scheme 17

The protected hydroxy iodide (88) was prepared in
four steps from 2-methylpropanal using the boron
reagent (80) as shown in Scheme 18. The reaction of
the crotyldiisopinocampheylborane (80) derived from
(+)-α-pinene[21] with 2-methylpropanal gave an 85:15
mixture of the <u>anti</u> and <u>syn</u> adducts (84) and (85) which
were conveniently separated by preparative g.l.c. The
major adduct (84) was found to have an enantiomeric
excess of greater than 90% (Mosher's derivative).
Protection of the secondary hydroxyl group and regioselective hydroboration-oxidation then gave the mono-
protected diol (87) which was converted into the iodide
(88) using iodine, triphenylphosphine and imidazole.

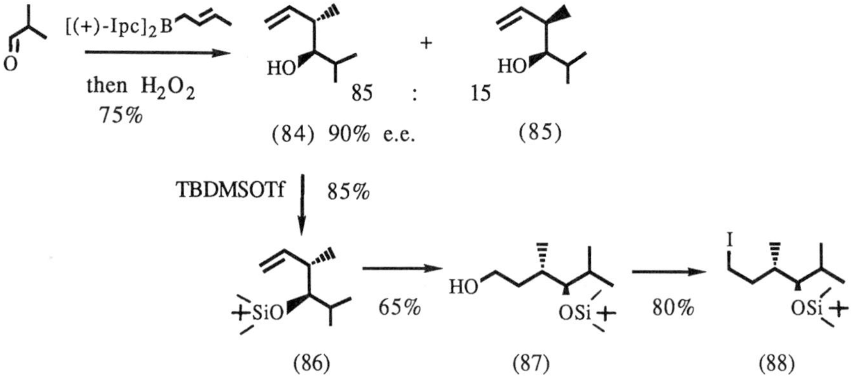

Scheme 18

Having prepared the epoxide (83) and iodide (88)
it remained to assemble the spiroacetal. Alkylation
of 1,3-dithiane with iodide (88) was achieved using
n-butyllithium as base to provide the monoalkylated
dithiane (89) in over 70% isolated yield. The second
dithiane alkylation using epoxide (83) required
t-butyllithium, and gave the bisalkylated dithiane
(90), also in good yield. Removal of the hydroxyl
protecting groups, dithiane hydrolysis, and spiroacetal formation were then achieved in a single step
using dilute aqueous hydrogen fluoride in
acetonitrile, to give the spiroacetal (73) identical
with the sample prepared earlier from α-methyl
D-glucoside.[15] This asymmetric spiroacetal synthesis
has the advantages of being stereoselective,
convergent, and short [seven linear steps from
aldehyde (77) and 2-methylpropanal].

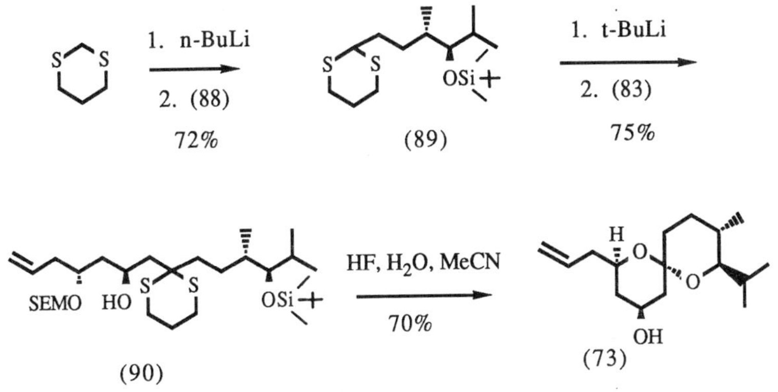

Scheme 19

The avermectin spiroacetals, e.g. the avermectin
A_{2a} spiroacetal (91), are more complex than the
milbemycin E spiroacetal (73) since they have
additional chiral centres at C(23) and in the C(25)
side-chain. Scheme (20) outlines an approach to the
avermectin spiroacetal (91) which parallels the
synthesis of spiroacetal (73) discussed above. Thus
spiroacetal (91) should be formed on hydrolysis of the
bisalkylated dithiane (92) which in turn should be
accessible from dithiane and epoxides (83) and (93).

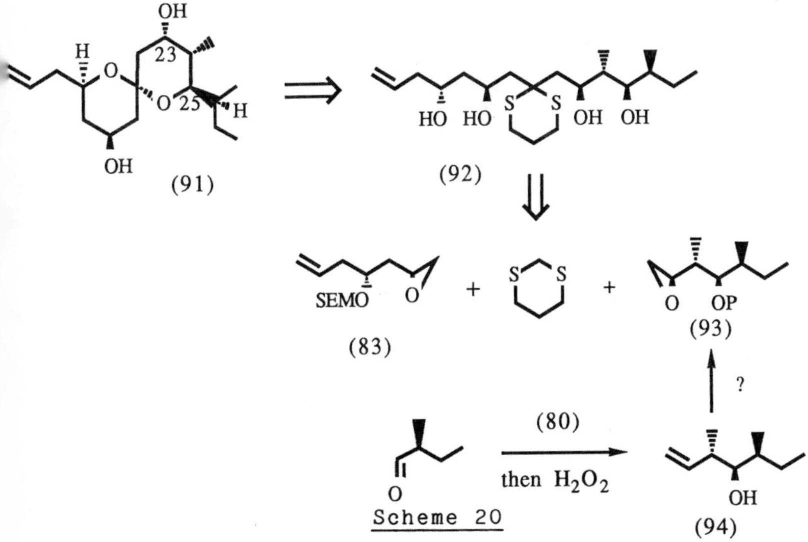

Scheme 20

The new aspect of this route would therefore be the
development of a synthesis of epoxide (93), one
possibility from (S̲)-2-methylbutanal involving the use
of the crotyldiisopinocampheylborane (80) followed by
stereoselective epoxidation.

   (S̲)-2-Methylbutanal is readily available by
chromic acid oxidation of (S̲)-2-methylbutanol, but in
our hands some racemization accompanied this
oxidation, and the enantiomeric excess of the aldehyde
was only ca̲. 70-80%. Treatment of this aldehyde with
the crotyldiisopinocampheylborane (80) derived from
(+)-α-pinene gave a mixture of adducts which contained
≅ 75% of the desired adduct (94) together with minor
diastereoisomers including (95). Interestingly the
major adduct (94) now had an enantiomeric excess of
>90% (Mosher's derivative), the increase in optical
purity being due to the preferential formation of
adduct (95), a diastereoisomer of (94) from the
(R̲)-aldehyde. [The enantiomer of adduct (94) could
only be formed if the minor (R̲)-aldehyde reacted with
the boron reagent (80) via̲ the reagent's less
favourable mode]. Adduct (94) was obtained pure, free
of the other diastereoisomers, by a combination of
flash chromatography and preparative g.l.c. (isolated
yield 45%).

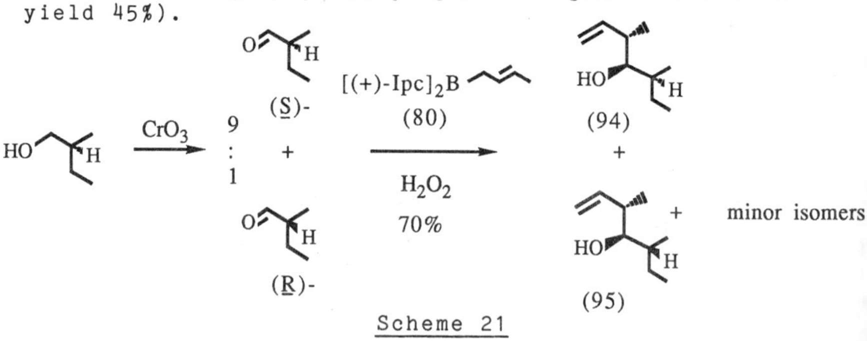

Scheme 21

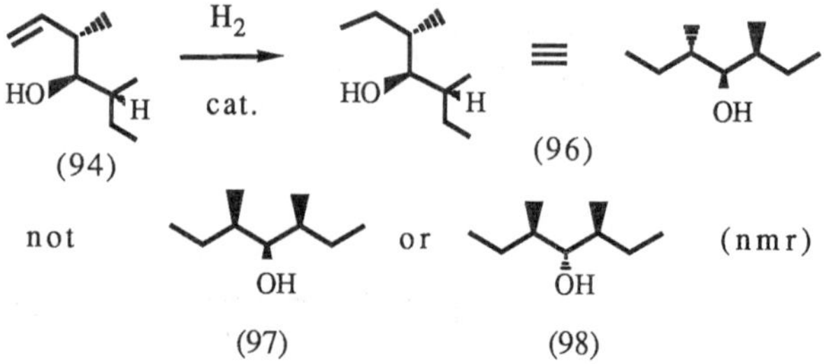

Scheme 22

The structure of the major adduct was checked by hydrogenation. This gave a 3,5-dimethylheptan-4-ol which was identified as isomer (96) since its two methyl groups were not equivalent by ^{1}H n.m.r. The alternative diastereoisomers (97) and (98) have planes of symmetry which make their two methyl groups equivalent by n.m.r. in each case. *

The stereoselective epoxidation of adduct (94) was then investigated. Preliminary studies of iodolactonization procedures were not promising, however direct epoxidation using t-butyl hydroperoxide catalysed by VO(acac)$_2$ was found to be very efficient and gave a 5:1 mixture of the easily separable epoxides (99) and (100) (> 90%). The stereochemistry of this epoxidation was assigned by analogy with the literature;[24] peracid epoxidation was not stereoselective.

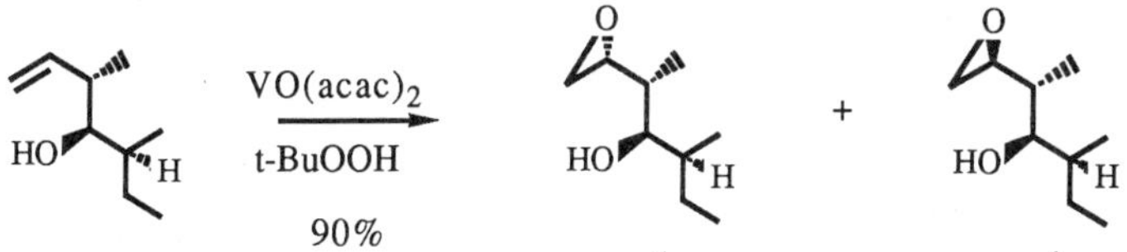

Scheme 23

The potential of epoxide (99) for spiroacetal synthesis was checked by treatment with the lithium salt of dithiane (101) generated using t-butyllithium. This gave a good yield of the bisalkylated dithiane (102) which was deprotected and hydrolysed under basic conditions (HgCl$_2$, CaCO$_3$) to provide the hydroxyspiroacetal (103). [Five steps from (S)-2-methylbutanal as summarized in Scheme 24]

---

* The absolute configuration of the hydrogenation product (96) follows from the fact that it was derived originally from (S)-2-methylbutanal. The syn-adduct (i) is the only other (S)-2-methylbutanal derived 3,5-dimethylhept-1-en-4-ol that could give rise to hydrogenation product (96) and was dismissed as the crotylborane adduct since crotylboranes derived from trans-but-2-ene are known to give anti-adducts.

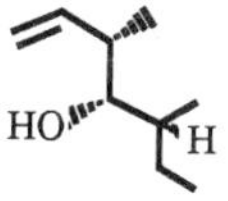

(i)

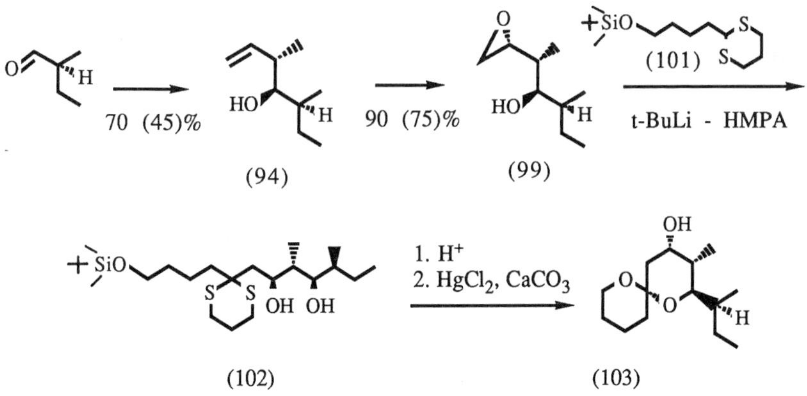

(94)   (99)

(102)   (103)

Scheme 24

Finally the hydroxyepoxide (99) was incorporated into a synthesis of the avermectin A_{2a} spiroacetal (91) as outlined in Scheme 25. Thus treatment of (99) with lithiated dithiane gave the hydroxydithiane (104) protected as its acetonide (105). Deprotonation using t-butyllithium and alkylation by epoxide (83) gave the bisalkylated dithiane (106) which was deprotected and cyclized in a single step, using a solution of HF-pyridine in dichloromethane, to provide spiroacetal (91) as a single diastereoisomer (73%).

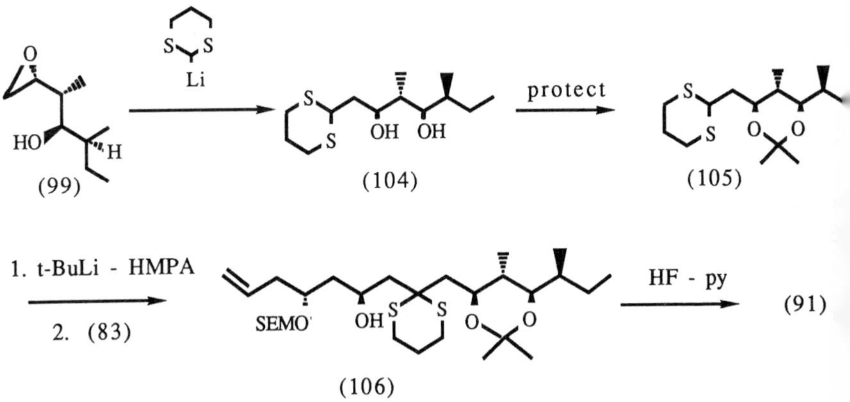

(99)   (104)   (105)

(106)

Scheme 25

This work has established a flexible and convergent approach to the milbemycin and avermectin spiroacetals. Our present work is concerned with incorporating these spiroacetal fragments into complete natural product syntheses.

Acknowledgements

I should like to thank my collaborators who have carried out this work, in particular M.S. Chambers (thiotetronic acids), M.J. Hughes (milbemycin macrolide model), N.A. Stacey (introduction of the milbemycin C(3)-C(4) double-bond), P.D. Steel (Spiroacetal synthesis), and E. Merifield (spiroacetal synthesis) whose work I have discussed today, together with other members of my group who have contributed to these and other research programmes. I should also like to thank Dr. M.D. Turnbull for many helpful discussions, and ICI Plant Protection Division and the SERC for financial support. Finally I should like to thank Dr. D.J. Williams (Imperial College, London) for X-ray crystal structures.

References

1. E. Benary, Berichte, 1913, 46, 2103.

2. H. Sasaki, H. Oishi, T. Hayashi, I. Matsuura, K. Ando, and M. Sawada, J.Antibiotics, 1982, 35, 396.

3. S. Omura, A. Nakagawa, R. Iwata, and A. Hatano, J.Antibiotics, 1983, 36, 1781.

4. L.A. Dolak, T.M. Castle, S.E. Truesdell, and O.K. Sebek, J.Antibiotics, 1986, 39, 26.

5. C.-L. Wang and J.M. Salvino, Tet.Let., 1984, 25, 5243; K. Tsuzuki and S. Omura, J.Antibiotics, 1983, 36, 1589.

6. M.S. Chambers, E.J. Thomas, and D.J. Williams, J.C.S.Chem.Comm., 1987, 1228.

7. R.K. Hill, in Asymmetric Synthesis Volume 3, ed. J.D. Morrison, Academic Press, London, 1984, p. 503.

8.   T. Taguchi, Y. Kawazoe, K. Yoshihira, H.
      Kanayama, M. Mori, K. Tabata, and K. Harano,
      Tet.Let., 1965, 2717; S.G. Smith,
      J.Am.Chem.Soc., 1961, 83, 4285; R.J. Ferrier and
      N. Vethaviyasar, J.C.S.Chem.Comm., 1970, 1385;
      K. Harano and T. Taguchi, Chem.Pharm.Bull., 1972,
      20, 2357; T. Nakai and A. Ari-izumi, Tet.Let.,
      1976, 2335; Y. Ueno, H. Sano, and M. Okawara,
      Tet.Let., 1980, 231, 1767.

9.   H. Mrozik, this volume.

10.   H.G. Davies and R.H. Green, Nat.Prod.Reports,
      1986, 3, 87.

11.   A.G.M. Barrett, R.A.E. Carr, S.V. Attwood, G.
      Richardson, and N.D.A. Walshe, J.Org.Chem., 1986,
      51, 4840; R.A. Baker, R.H.O. Boyes, D.M.P. Broom,
      M.J. O'Mahoney, and C.J. Swain, J.C.S. Perkin I,
      1987, 1613; R.A. Baker, M.J. O'Mahoney, and C.J.
      Swain, ibid, p. 1623; S.R. Schow, J.D. Bloom,
      A.S. Thompson, K.N. Winzenberg, and A.B. Smith,
      J.Am.Chem.Soc., 1986, 108, 2662; S.D.A. Street,
      C. Yeates, R. Kocienski, and S.F. Campbell,
      J.C.S.Chem.Comm., 1985, 1386; C. Yeates, S.D.A.
      Street, P. Kocienski, and S.F. Campbell, ibid, p.
      1388; D.R. Williams, B.A. Barner, K. Nishitani,
      and J.G. Phillips, J.Am.Chem.Soc., 1982, 104,
      4708.

12.   S. Hanessian, A. Ugolini, D. Dube, P.J. Hodges,
      and C. Andre, J.Am.Chem.Soc., 1986, 108, 2776.

13.   M.J. Hughes, E.J. Thomas, M.D. Turnbull, R.H.
      Jones, and R.E. Warner, J.C.S.Chem.Comm., 1985,
      755.

14.   S.V. Mortlock, N.A. Stacey, and E.J. Thomas,
      J.C.S.Chem.Comm., 1987, 880.

15.   G. Khandekar, G.C. Robinson, N.A. Stacey, P.G.
      Steel, E.J. Thomas, and S. Vather,
      J.C.S.Chem.Comm., 1987, 877.

16.   C.D. DeBoer, J.Org.Chem., 1974, 39, 2426.

17.   M.D. Turnbull, G. Hatter, and D.E. Ledgewood,
      Tet.Let., 1984, 25, 5449.

18. A.K. Saksena and P. Mangiaracina, <u>Tet.Let.</u>, 1983, <u>24</u>, 273.

19. S. Katsumura, K. Hori, S. Fugiwara, and S. Isoe, <u>Tet.Let.</u>, 1985, <u>26</u>, 4625.

20. S. Saito, T. Hasegawa, M. Inaba, R. Nishida, T. Fujii, S. Nomizu, and T. Moriwake, <u>Chem.Lett.</u>, 1984, 1389.

21. H.C. Brown and K.S. Bhat, <u>J.Am.Chem.Soc.</u>, 1986, <u>108</u>, 5919.

22. H.C. Brown and P.K. Jadhav, <u>J.Am.Chem.Soc.</u>, 1983, <u>105</u>, 2092.

23. E.J. Badin and E. Pascu, <u>J.Am.Chem.Soc.</u>, 1945, <u>67</u>, 1352; W. Kirmse and H. Arold, <u>Berichte</u>, 1971, <u>104</u>, 1800.

24. E.D. Mihelich, K. Daniels, and D.J. Eickhoff, <u>J.Am.Chem.Soc.</u>, 1981, <u>103</u>, 7690.

A Rational Approach to the Design of Antihypertensives: *X*-Ray Studies of Complexes between Aspartic Proteinases and Aminoalcohol Renin Inhibitors

J. B. Cooper, S. I. Foundling, and T. L. Blundell*

LABORATORY OF MOLECULAR BIOLOGY, DEPARTMENT OF
CRYSTALLOGRAPHY, BIRKBECK COLLEGE, LONDON WC1E 7HX, UK

R. J. Arrowsmith, C. J. Harris, and J. N. Champness

THE WELLCOME RESEARCH LABORATORIES, LANGLEY COURT,
BECKENHAM, KENT BR3 3BS, UK

1   The Renin-Angiotensinogen System

The octapeptide angiotensin II is a potent vasoconstrictor and
mediator of aldosterone secretion. The processing of its precursor,
angiotensinogen, is accomplished by two peptidases. The highly
specific aspartic proteinase, renin, cleaves the precursor yielding
the N-terminal decapeptide (AI) from which the C-terminal dipeptide i
cleaved by angiotensin converting enzyme (ACE) giving AII as shown in
Figure 1. Inhibitors of ACE, for example, captopril, are effective i
lowering blood pressure [1] and hence inhibition of the preceding
step, catalysed by renin, has been the subject of much investigation.

The renin-angiotensin system

1	2	3	4	5	6	7	8	9	10	11	12	13
Asp	Arg	Val	Tyr	Ile	His	Pro	Phe	His	Leu	Val	Ile	His . . .

(Angiotensinogen)

↓ Renin

Asp Arg Val Tyr Ile His Pro Phe His Leu   Val Ile His . . .

(Angiotensin I)

↓ Angiotensin converting enzyme

Asp Arg Val Tyr Ile His Pro Phe   His Leu

(Angiotensin II)

a) Vasoconstriction

b) Retention of salt by kidneys

Figure 1   The cleavage specificities of renin and ACE

2   Design of Transition-State Isosteres

So far, all the known potent renin inhibitors have been obtained by modification of substrate sequences in which the scissile Leu-Val bond of angiotensinogen is replaced by various non-hydrolysable transition-state analogues [2]. For example, the reduced bond analogue ($-CH_2-NH-$) has been used to give inhibitor H-142 [3] which inhibits human renin with an IC_{50} value of 10 nM (Figure 2). The natural transition-state analogue statine ($-CHOH-CH_2-CO-NH-$) has also been incorporated giving inhibitor L-363,564 (Figure 2) [4]. The potency of keto analogues ($-CO-CH_2-$), e.g. H272 (Figure 2) [3], can be improved substantially by electron-withdrawing substituents, e.g. $-CO-CF_2-$ [5], which stabilise the hydrate (gem-diol). The hydroxyethylene analogue ($-CHOH-CH_2-$) of H-261 (Figure 2) produces binding more than one order of magnitude tighter than the reduced analogue [6]. These results suggest an important role for a hydroxyl group in the analogue. However, the ideal aminal analogue $-CHOH-NH-$ is unstable. Interposing a methylene group between the $-CHOH-$ and $-NH-$ groups could allow the resulting aminoalcohol analogue ($C\alpha-CHOH-CH_2-NH-C\alpha$) to retain favourable interactions with the enzyme [7].

In general, the observed degree of inhibition is heavily dependent on the length of the peptide. Successively shortening an inhibitor from the N-terminal side causes a stepwise loss of potency [7]. However, the large size and peptidic nature of the most potent inhibitors explains their poor transport across the gut and their susceptibility to the digestive enzymes of the intestinal tract. Hence, smaller molecules with fewer peptide links are essential for drug development. This led us to synthesise a number of tetrapeptide inhibitors based on the sequence at the C-terminal side of the peptide bond, e.g. BW624 and BW625 which differ only in the chirality at the aminoalcohol hydroxyl position (Table 1).

Table 1

	P_1	P_1'	P_2'	P_3'	P_4'	IC_{50} (renin)
Human Angiotensinogen	...Leu -CO-NH-	Val	Ile	His...		
BW624	Leu -CHOH-CH$_2$-NH-Val (S)	Ile	Phe	OMe		100µM
BW625	Leu -CHOH-CH$_2$-NH-Val (R)	Ile	Phe	OMe		6µM

The greater selectivity of the enzyme for the R-form is unusual since the S-form is preferred with statine or hydroxyethylene inhibitors [8]. This difference may be due to the absence of an N-terminal polypeptide which allows greater freedom at the catalytic centre.

Sequences of the renin substrate and inhibitors

Position	P6	P5	P4	P3	P2	P1		P1'	P2'	P3'	IC50 (µM) for human renin
	(5)	(6)	(7)	(8)	(9)	(10)		(11)	(12)	(13)	
Human angiotensinogen	...Ile	His	Pro	Phe	His	Leu	$-$ CO $-$ NH $-$	Val	Ile	His Asn ...	
H142	Pro	His	Pro	Phe	His	Leu	$-$ CH$_2$$-$ NH $-$	Val	Ile	His Lys	0.01
L-363,564	Boc	His	Pro	Phe	His	Leu	$-$ C $-$ CH$_2$$-$ CO $-$ NH $-$ (H above, OH below C)	Leu	Phe	NH$_2$	0.01
H272		His	Pro	Phe	His	Leu	$-$ CO $-$ CH$_2$$-$	Val	Ile	His	0.5
H261	Boc	His	Pro	Phe	His	Leu	$-$ C $-$ CH$_2$$-$ (H above, OH below C)	Val	Ile	His	0.0007

Figure 2   Inhibitors containing various transition state analogues

3   X-Ray Studies
We have recently achieved the co-crystallization of both
diastereoisomers, BW624 and BW625, with endothiapepsin, an aspartic
proteinase homologous to renin [9]. In this paper we report
preliminary structural data on the BW624-endothiapepsin complex for
which X-ray data have been collected to 2.0A using an Enraf Nonius
F.A.S.T. area detector. As the co-crystals were isomorphous with the
native enzyme, we were able to solve the structure using the
difference Fourier method and least squares refinement (R = 0.21).
Crystals of the BW625 complex were not isomorphous but we have used
molecular replacement methods to obtain a promising difference map at
2.2A. Data collection to higher resolution is in progress and we will
report the structures of both inhibitors in full later.

4   Conformation of the BW624 complex with endothiapepsin
The inhibitor, BW624, binds in one half of the extended active-
site cleft between the two lobes of the enzyme, occupying the $S_1 - S_3'$
binding pockets. The hydroxyl group of the aminoalcohol analogue lies
within hydrogen bonding distance of, and approximately coplanar with,
both essential carboxyl groups of aspartates 32 and 215. This strong
interaction, which has been seen in several other renin inhibitor
complexes [10,11,12], would appear to account for the high potency of
analogues bearing a hydroxyl group.

The P_1, P_1' and P_2' sidechains of the inhibitor are shielded from
the solvent by a loop - the 'flap' - that lies over the catalytic
centre (Trp 71 - Gly 82). The mainchain of the flap forms two
hydrogen bonds with the inhibitor, as shown in Figure 3. The amide
nitrogen of Gly 76 donates a hydrogen bond to the carbonyl oxygen of
P_1', and carbonyl of Ser 74 accepts a hydrogen bond from the peptide
nitrogen of P_3'. The amide of P_2' also donates a weak hydrogen bond
to the carbonyl of Gly 34, which is one of the absolutely conserved
glycines. This residue, together with its equivalent in the other
lobe, Gly 217, allows the chain following each aspartate to fold
sharply so that the hydroxylic sidechains at positions 35 and 218 can
interact with the essential carboxyl groups via hydrogen bonds [13].
The carbonyl of Gly 217 also participates in a weak, but possibly
charge-assisted, hydrogen bond with the free N-terminal amino group of
the inhibitor.

The leucine sidechain of P_1 lies in a hydrophobic pocket formed
by Tyr 75, Phe 111 and Leu 120, as well as two presumably protonated
aspartic acid residues, 30 and 77. The S_1' pocket, as defined by
studies with other inhibitors [10], is filled by the valine sidechain.
This indicates an important difference between aminoalcohol and
statine inhibitors. Statine, which is thought to be a dipeptide
analogue, possesses two extra mainchain atoms relative to an amino
acid, and hence causes a frameshift in the C-terminal side of the
inhibitor, forcing the P_1' sidechain to occupy the S_2' pocket.
Although the aminoalcohol has one extra atom relative to a peptide
link the frameshift is not so large as with statine, allowing the P_1'

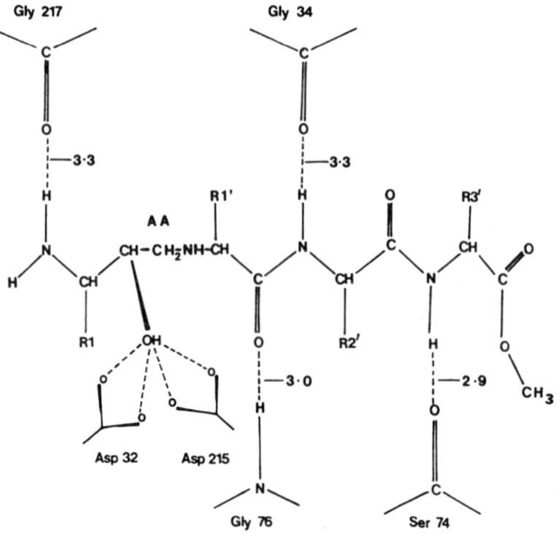

<u>Figure 3</u>   A schematic diagram of the hydrogen bonds between BW624 an
endothiapepsin (AA, aminoalcohol). Donor-acceptor distances are show
in A units

sidechain to lie in the S_1' pocket which consists of a cluster of
isoleucines (213, 297, 299 and 301). Similarly, the P_1' valine
carbonyl forms a hydrogen bond with the peptide nitrogen of Gly 76,
which is also observed with inhibitors which do not cause a
frameshift, e.g. reduced bond. Hence, despite the insertion of one
methylene group in the mainchain, the aminoalcohol moiety would appea
to replace a single amino acid rather than a dipeptide.

The P_2' Ile lies sandwiched between the active site flap (at Il
73) and the sidechain of Leu 128. The sidechains at P_1' and P_2'
appear to lie against the aromatic ring of Phe 189 which is
conservatively replaced by tyrosine in human renin. The P_3'
phenylalanine is rather exposed, but makes van der Waals contact wit
Gly 76 and Ile 297. The aromatic ring is tilted towards the
hydrophobic S_1' pocket far more than the polar sidechains at P_3' in
other inhibitor complexes [10]. We have reported that a number of
pockets in endothiapepsin, S_2 and S_5, are large enough to allow
different sidechains to optimise interactions with the enzyme. The
methoxy-blocking group, although exposed, is clearly defined by the
electron density. It forms van der Waals interactions with Ile P_2'
and Ser 74.

All of the hydrogen bonds to the inhibitor involve the maincha
of the enzyme and hence are probably conserved in renin. Of the
sidechains that interact with the inhibitor at S_1, Phe 111 and Tyr
are conserved in both enzymes. Other conservative changes include
120 at S_1 to valine (in renin), Phe 189 to Tyr, Ile 213 to Leu (bo
S_1') and Ile 73 to Leu (S_2'). Some larger differences between

endothiapepsin and renin may be found at the S_1' pocket due to the insertion of 3 prolines (Pro-Pro-Pro) in the sequence prior to the hydrophobic cluster of isoleucines (297, 299 and 301).

5 Conclusions

This study has established some important differences between the statine and aminoalcohol transition-state analogues which indicate that statine is effectively a dipeptide whereas the shorter aminoalcohol analogue, despite the insertion of the mainchain methylene, behaves as a single amino acid. The interactions between the aminoalcohol inhibitor and endothiapepsin are essentially the same as those observed with reduced bond and hydroxyethylene inhibitors. The different conformation of the P_3' Phe relative to polar sidechains at this subsite indicates that this is yet another pocket where the substrate is allowed sufficient flexibility to optimise favourable interactions with the enzyme.

REFERENCES

1. Ondetti, M.A. & Cushman, D.W. (1980) Ann. Rev. Biochem. 51, 283-308.
2. Hofbauer, K.G. & Wood, J.M. (1985) Trends in Pharmacol. Sci. 6, 173-177.
3. Szelke, M., Leckie, B., Hallet, A., Jones, D.M., Sueiras, J., Atrash, B. & Lever, A.F. (1982) Nature 299, 555-557.
4. Boger, J., Lohr, N.S., Ulm, E.H., Poe, M., Blaine, E.H., Fanelli, G.M., Lin, T-Y., Payne, L.S., Schorn, T.W., Lamont, B.I., Vassil, T.C., Stabilito, I.I., Veber, D.F., Rich, D.H. & Bopari, A.S. (1983) Nature 303, 81-84.
5. Gelb, M.H., Svaren, J.P. & Abeles, R.H. (1985) Biochemistry 24, 1813-1817.
6. Szelke, M. (1985) in 'Aspartic proteinases and their inhibitors' (Kostka, V., ed.) pp. 421-441, de Gruyter, Berlin.
7. Dann, J.G., Stammers, D.K., Harris, C.J., Arrowsmith, R.J., Davies, D.E., Hardy, G.W. & Morton, J.A. (1985) Biochem. Biophys. Res. Commun. 134, 71-77.
8. Rich, D.H. & Sun, E.T.O. (1980) J. Med. Chem. 23, 27.
9. Sibanda, B.L., Blundell, T.L., Hobart, P.M., Fogliano, M., Bindra, J.S., Dominy, B.W. & Chirgwin, J.M. (1984) FEBS. Lett. 174, 102-111.
10. Foundling, S.I., Cooper, J., Watson, F.E., Cleasby, A., Pearl, L.H., Sibanda, B.L., Hemmings, A., Wood, S.P., Blundell, T.L., Valler, M.J., Norey, C.G., Kay, J., Boger, J., Dunn, B.M., Leckie, B.J., Jones, D.M., Atrash, B., Hallett, A. & Szelke, M. (1987) Nature 327, 349-352.
11. James, M.N.G., Sielecki, A.R., Salituro, F., Rich, D.H. & Hofmann, T.(1982) Proc.Natl.Acad.Sci.USA. 79, 6137-6142.
12. Bott, R., Subramanian, E. & Davies, D.R.(1982) Biochemistry 21, 6956-6962.
13. Pearl,L.H. & Blundell,T.L.(1984) FEBS.Lett. 174, 96-101.

Computing and Trial and Error in Chemotherapeutic Research

A. J. Everett

THE WELLCOME RESEARCH LABORATORIES, LANGLEY COURT,
BECKENHAM, KENT BR3 3BS, UK

Introduction

Two years ago it was felt that the reasons for our merging
the computer chemistry and QSAR groups might be interesting
enough to merit a contribution to this part of the conference
particularly because of the associated computer hardware and
software requirements.

I should make it clear that I am speaking on behalf of the
following group of ten colleagues, who are applying computing
techniques in one form or another to increasing the chance of
discovering medicines.

John Champness   Robert Glen   Alan Hill
Brian Hudson   Richard Hyde   John Lindon
David Livingstone Elizabeth Rahr   Sally Rose
George Tranter

In addition, Dr. C. Beddell is extending heuristic methods
to macromolecules, in particular, of course, proteins.

At a technical level any one of this team could give a
talk on this subject far better than I. However I suspect that
many of them would not wish to commit themselves to such a
simplified view of the discovery process. After all, they do
have to respect the conventional straightjacketed wisdom of the
medicinal chemical hierarchy.

Discovery and Design

Let us start with some semantics and take the word
'design'. In my opinion this is one of the most ill-used words
in the vocabulary of medicinal chemistry. Whatever vintage of

medicinal chemists we are, I would submit that we do not design medicines; we discover them. Of course it is convenient for fund raising purposes to assign a significance to computer-based systems in the discovery of medicines which far trans-cends their actual contribution. Synergism yes, but design no. Design, according to the Concise Oxford Dictionary, involves contriving, planning, purpose and intention. I would draw your attention to the last two. As scientists we are all skilled at rationalising results so I have little doubt that after I have finished speaking I will be told that there are a number of medicines on the market that have been designed, but I suspect that they will be supported by excellent post hoc explanation. It is just not true in my experience to suggest that we can aim at the cure and design a molecule to provide it. I challenge anyone to provide clear and unequivocal examples of this occurrence.

I feel it is about time that someone said in a semi-public sense, what is mostly said over a pint of beer between medicinal chemists. It is, that all the devices available to us from ab initio and molecular dynamics calculations through advanced statistical techniques like non-linear mapping, to receptor definition by X-ray methods, are all but single aspects of the process of an intellectual synergism which may, or may not, lead to an effective cure. To over-emphasise any one of these is more indicative of the worker's tunnel vision than of the intrinsic merits of the particular technique. We should all be wary of fashionable shibboleths. I do not know the total capital cost that can be attributed to elegant structuring of a number of macro-molecular species including enzymes. I do know that we, as a Company, spent 12 years concentrating on the crystallographic coordinates of dihydrofolate reductase as a basis for finding a successor to our anti-bacterial Trimethoprim. I also know that we did not succeed. Our thinking was dominated by the receptor. To hell with transport and distribution which turned out to be the decisive failure point.

Not just in our Company but in others the coordinates of the renin enzyme have been repeatedly offered as a way forward for the development of a powerful anti-hypertensive. You can be sure that if and when that anti-hypertensive arrives, it will be post hoc rationalised into a proposition in which graphical techniques will appear to be predominant. The working synthetic medicinal chemists will smile wryly and know the score. Of course, the truth is that the graphics and the computing will have acted as a nucleus around which the project was congealed. In other words, it has synergised, it has not actually led. Indeed, if it did, we would probably end up with the same situation as occurred with Trimethoprim: a

concentration on the physical receptor to the exclusion of all
other factors, in this instance stability via the oral route,
together with some rather nasty toxic effects.

A very much more up-to-date situation is that which
applies to the structuring of the reverse transcriptase
involved in the production of the AIDS virus. At present there
is a momentum which will culminate in the structuring of the
enzyme which has been produced and crystallised in our Virology
and Biochemistry Departments at Beckenham. Of course it is
sensible when seeking a cure to draw upon all the information
that we can get, but it is surely unwise to attribute to this
search for transcriptase architecture an overwhelming
significance in the process for finding suitable curative
agents. Apart from the time required to obtain this architecture, we need knowledge of the functionality of the active site
and its precise location. In the meantime the technique of
site-directed mutagenesis will almost certainly reveal indicators which will provide more valuable constraints. Indeed I
suspect that quite a lot of work based on our heuristic methods
will be in progress by the time the coordinates appear.

At this point you may be wondering how someone with a team
of eleven in the business of computer aided discovery of
medicines can be saying all this. I gave a partial explanation
in the beginning, but also I believe that the actual strategy
which we are deploying, and in fairness, others may be, is the
one which is most likely to lead to curative materials. It
does not put molecular graphics on a pedestal. On the contrary
it includes it as part of our overall heuristic strategy.
Incidentally, you will note that I go to some considerable
trouble to avoid the use of the word drug. I distinguish
between a drug which has a physiological action which may be
desirable or undesirable and a medicine, in which the desirable
features are predominant. I have been told that a drug is
something which when injected into a rat produces a paper. I
suspect that we would all agree on one thing - it is far easier
to achieve a drug than a medicine. Indeed I would argue that
all compounds are drugs, for all experience suggests that even
the components of those mixtures of compounds which we
designate as food have desirable and quite often undesirable
consequences.

Heurism

So what is our heuristic approach? Well you might say
that heuristic is merely another word for trial and error, but
in our experience we would like to suggest that it means that,
and more than that. It means a total research strategy
culminating in an expert system. It probably means one of the

biggest expert systems that one is likely to find. It highlights the fact that the pharmaceutical industry is taking a long time to produce simulation techniques similar to those which are already established in the aircraft and automobile industries.

Figure 1 encapsulates almost, but not all, of our thinking on this matter. I do not suppose for one moment that any single element in this is new. In fact I have hinted that the diagram itself may not be, but observation of others in the business leads me to think that at least it is not a re-statement of the obvious.

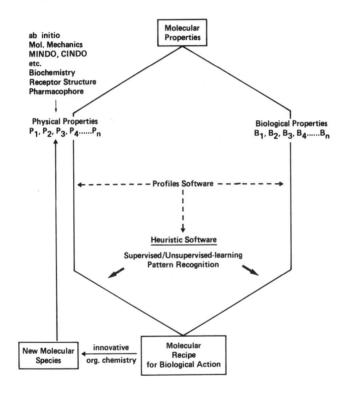

Figure 1 - The Heuristic Approach to the Discovery of Medicines

Before considering Figure 1, first let us now look at the origins of this heuristic approach. There are two branches of its ancestry - "traditional" QSAR and "traditional" computational chemistry.

    The early QSAR was typified by correlations between
Biological Activity and measured partition coefficients. The
limitations of such an approach are obvious and it is against
this background that we can appreciate the importance of Corwin
Hansch's work in the early 1960's. He introduced the concept
of QSAR based on predicted substituent constants. For the
environment of a substituted benzene ring, it was possible to
predict partition (π), electronic (σ) and steric (E_s or MR)
properties. Regression analysis was used to select the best
single or multiple correlation equation, and the activity of
proposed new molecules could be predicted to a level which was
hopefully better than random. However, we were effectively
working with a limited "fixed menu" of physicochemical
properties - albeit good ones.

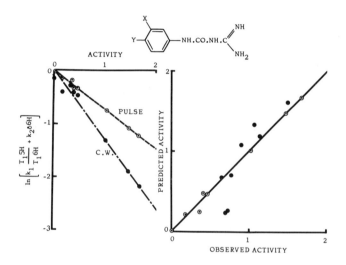

Figure 2 - The activity of Anti-malarials as a function of
proton chemical shifts and transverse relaxation times
  ⊙ Training set of molecules
  ● Test set of molecules

    A number of us were trying to get out of this straight-
jacket, and it is worth looking at Figure 2 which shows data
for some 16 amidino ureas intended as anti-malarials. By
painful examination of the n.m.r. data we were able to produce
this regression on 8 of the compounds. The regression so
generated was used to predict the remaining 8, and at that time
we were well satisfied with the result, but it was nothing more
than a physicochemical curiosity because the function that was
used to predict activity included the ratio of certain proton

relaxation times along with proton chemical shifts. The former was the major stumbling block because it lacked predictability. Only if the estimation of the anti-malarial activity by physicochemical means was cheaper and faster than the actual evaluation in vitro could this relationship be of any value. In short, it was smart but did not contain the essential element of predictability. However, it did indicate that pi, sigma and steric properties were not the end of the story.

Increasing contact with computational chemists, who were initially orientated towards the study of ligand-macromolecule interactions, led to an appetite for the properties which they could calculate in QSAR studies. In other words we were now faced with the option of going "a la carte". In fact why should the molecular properties essential to H_2 antagonism be those responsible for beta-blockade?

So what are the key features embedded in the plan in Figure 1?

(1)   It shows a continuous range of properties, both biological and physical, which we have arbitrarily separated.

(2)   At the present time and in general the number of physical properties far exceed the observed biological ones.

(3)   We can predict, using quantum mechanical methods and semi-empirical techniques, a very wide range of molecular properties. These can range from the electron density on a given atom or group of atoms through the dipole moment of the molecule, the moment of inertia, architectural dimensions etc. and may amount to as many as a 100.

If we accept that some of these properties are the significant ones in producing a biological response B_1 and toxic response B_2, the trick is to devise ways and means of seeking the key elements of P which optimise B_1 and reduce B_2. In other words some properties matter, some don't.

Property Selection

There are a number of techniques which can be used to approach this problem. They include hunch, linear regression, principal component analysis and non-linear mapping techniques. Along with these numerical methods we must be sensitive to the information available from biochemical studies and receptor structuring in so far as they can influence our property package. If we apply all of these techniques wisely we can, in favourable instances, arrive at a molecular recipe for the biological response.

Ideally this recipe should be representative of the molecular action rather than of individual atomic components. For example, in any given instance it may not be the electron density on a nitrogen which is the determinant in biological action, it may merely be part of a sub-set whose actions have been inflated by our manipulative manoeuvres. The true situation, which one day we will have achieved, is one where we are dealing in true molecular parameters, recognising that the whole molecule is involved in the medicinal action and that it is not just a vehicle for carrying around an activated heterocyclic nitrogen, for example. It may well be that the quantum mechanician may evolve a relationship which properly expresses this molecular concept, but one thing is for sure: whatever is evolved as a recipe should have the quality that it will permit regeneration of other molecules with similar characteristics. It need not do so very well, for if we now feed back these putative medicines into the beginning of the process again (Figure 1) and we use a non-linear map as the key method for property selection, we have an expert system where the process is only restricted by the computing power available to us and not by conceptual limitations.

By this time it will be quite clear that I have drifted a substantial way from the purely visual graphical displays which are so common now in medicinal chemical groups. Yes indeed they are needed but they are sub-sets of this whole process. Only in as much as they can affect our enlightened selection of properties have they a bearing on the discovery process, and by no stretch of imagination can the receptor, as defined by laboured X-ray analysis, be anything but a sub-set of this heuristic approach.

X-ray Methods

A word or two about X-ray methods applied to the structuring of enzymes. It has always seemed to me that this approach to medicines has a greater aesthetic than practical appeal. I suppose that this is borne out by the fact that relatively few pharmaceutical organisations have the hardware and staff for this purpose. The fact is that the lead time which it introduces is very long. First identify an enzyme with a syndrome, secondly isolate and crystallise the enzyme, thirdly structure it, fourthly identify the active site and fifthly set about building molecules that will fit this site. I doubt if this whole process could take less than two years even assuming the correlation between the enzyme and the illness is fully established. Furthermore, the dreaded transport and distribution problem is tacitly ignored. I do not want to be misunderstood. Of course there is room for this work, but it is of a long-term nature and probably still best resides in the academic world. But let it not be forgotten that in Wellcome we have pursued this course along with the best of them and we do not propose to stop, but I am quite confident that none of us believe that it is a single strategy around which the development of medicines should occur. It is, as I have said earlier, a sub-set of the complete discovery process.

The Nature of Properties

I have already proposed that a careful study of molecular properties with enlightened selection might lead to a molecular recipe for biological action. So a word or two about properties may be in order. Properties come in various guises, but the three things they must have in common are M, L, and T, and I suppose that at a fundamental level the molecular recipe is nothing more than an assembly of these dimensions in the right packets and proportions.

Properties also have three other characteristics. They can either be:

(a) Calculable - as one might obtain from various quantum mechanical programs like *ab initio* or MINDO or downmarket CINDO.

(b) Predictable - as for example the use of empirical tables to obtain ^{13}C chemical shifts or log P from the Pomona software.

(c) Guessable - where we can make a reasonable estimate of the property but with fairly large error limits. I think I would include pKa in this category.

I suppose we should not get obsessed by the veneer of respectability that very large number crunching can confer. Quantum mechanics will spew out a very large range of properties with numbers which relate to a "free space" molecule. Molecular mechanics will give us optimised geometry using our simple mechanical concepts and will fail quite miserably in certain situations; for example beta lactam derivatives. On the other hand, log P and pK_a and carbon 13 shift may have a far less rigorous basis for their estimation but will be derived from the real world rather than the clarified one of quantum calculations. In that sense we are led to distinguish between accuracy with a large variance, and precision, whose mean is miles away from the "true" parameters, with a small variance.

Incidentally, if indeed we are trying to take proper account of M, L and T, it is rather limiting to use a purely geometric or L factor dictated by conventional receptor structuring as the determinant for progress. Quantum mechanics and molecular mechanics with all of their faults at least combine M, L and T and no doubt we will evolve to the proper use of molecular dynamics in validating and seeking more realistic properties for the heuristic approach.

It will be evident that one of the themes which dominates at least my thinking is the rejection of architectural geometry as the primary constraint in seeking medicines and in its place, a recognition that it is the total molecule with its full range of properties which matters in achieving a given biological property. And here a new problem arises, for when is a property a P, or molecular property, and when is it a B, or biological property; and, one might add, when does a biological property become a medicinal property? My distinction between drug and medicine is becoming evident. To cite an example of this dilemma, we only have to consider the binding constant K_i for, say, a dihydrofolate reductase inhibitor, where already the use of programs such as AMBER or indeed ENERGY and even more recent advanced calculations, leave one to believe that K_i is not strictly a biological property; I think that Dr. Lee Kuyper and others in Burroughs Wellcome U.S. who have carried out much of this work would subscribe to the view that it is a predictable physical property. Maybe that is a little optimistic at the moment, but certainly it is the trend. In short, to aim at a renin inhibitor by seeking the biggest K_i is, in a way, putting the cart before the horse. K_i is a physical property which is not uniquely and directly related to the behaviour of the molecule as a medicine. However it is almost certainly related to its behaviour as a drug.

What should we do about K_i? Well, if we can predict it badly or even in the guessable category I would submit that it is on the left hand side of Figure 1 along with other molecular properties. This does affect the way we go about things for now we must seek a more relevant biological property which is more representative of the likely behaviour of the inhibitor as a medicine. This may imply that instead of doing scores of K_i measurements in a biochemical laboratory we do very many fewer in a pharmacological environment in animal models which are much closer to the hypertension which we wish to ameliorate.

In this sense our heuristic approach is ceasing to be just a paper exercise and is beginning to affect the way we should be going about things.

The world is ridden with inconsistency and I claim no exception, for one aspect of the heuristic approach which bothers me is an essentially geometric one. Chirality sums it up. I find it worrying that in a natural environment which is, in many instances, peculiarly sensitive to symmetry, we pay little more than lip service to this concept in our quest for medicines, at least in our laboratory. It is only now that we are beginning to be aware of the chiral insensitivity of our work. I have no smart answers, I merely wish to place on record our recognition of the problem and the need in the future to feed in properties which represent this aspect of the molecular characteristics of compounds.

Examples

In principle the Pharmaceutical Industry is shackled when it comes to the question of current and useful examples of their work. Mostly we are forced to present work which is, at best, delayed by patent considerations or, at worst, disguised so that we do not aid our competitors. I am no exception to this rule and, therefore, I apologise in advance for the next few figures where they have been purged of structure, the one thing in which we are interested. However, since I am evidently providing the light relief at the very end of a serious Symposium, you may not be unduly worried. The pictures which I have to show are ones which illustrate the nature of the heuristic process rather than the development of any particular medicine, although in fairness to my colleagues I think it should be clearly stated that the impact of their methods is very noticeable in certain areas of our work at Wellcome.

Let me first mention the non-linear map because that figures rather significantly in the heuristic process. For those of you who are familiar with the statistics of this process then I apologise for the simplicity of my explanation. For those of you who are not, then at worst I will have indicated an approach which you can follow up with people who really know.

If we start at the point where we recognise that the physicochemical and biological properties of molecules are related in some complex and unknown fashion, then given a training set of molecules with known activity, heuristic software employing pattern recognition techniques is capable of defining the relevant physicochemical properties. The

quest, of course, is a recipe of molecular properties which
are desirable for the required activity. There are those who
are worried that the recipe is limited by the chemical range
of the training set of molecules. At first sight the argument
seems unassailable, but in my view it becomes less and less
valid the further we go in excluding geometrical parameters to
the preference of more fundamental electrical ones.

The computer output consists of a map in two dimensions
where each point represents a molecule and the closeness of
one point to another is related to the similarity of the whole
package of properties of the one molecule to that of the
other. Those molecules similar in physicochemical properties
will lie close together in the map. If for example the
property package consists of eight parameters then because one
really needs eight axes to describe eight properties, the two
axes shown on this model have the complex combination of all
eight. They have no physical meaning; hence the term
non-linear map.

Before looking at a few results, I think it worth
mentioning that the mechanics by which computed and predicted
properties are introduced into a pattern recognition software
is through a software suite which we call PROFILES. This is
not the place to give a poor presention of this work, but it
needs to be known that Robert Glen and Sally Rose found it
necessary to generate this software in order to provide an
easy mechanism for applying the heuristic method (Figure 1).
For those who are interested there are more details in the
paper of June 1987 in the Journal of Molecular Graphics.

In essence it allows the specification of a data base,
the calculation, retrieval and storage of properties, the
calculation from structure files of additional properties, the
retrieval of properties from quantum mechanical programs,
calculation of additional properties from QC programs, as for
example the calculation of superdelocalisability from MO data,
the manual input of physical properties, for example pKa, the
definition and manual input of biological and toxicological
data specific to the database and, most importantly, the
creation of input files for the pattern recognition program
package ARTHUR and the data manipulation package RS1.

As an aside I should emphasise that although I am
referring frequently to non-linear maps and pattern
recognition techniques, our group employs a full range of
statistical methods including, of course, principal component
analysis. Indeed it is worth taking a glance at Figure 3
which is taken from Dr. Glen's paper, because this does show
grouping of compounds related to GABA and is able to

differentiate agonist, partial agonist and inactive using principal components.

GABA Mimetics

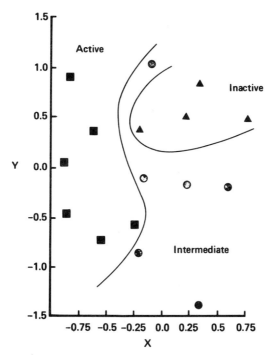

Figure 3 - Principal Component plot for fifteen GABA mimetics

In Figure 4 we have the results of a very early investigation which gave us initial encouragement to proceed further with non-linear maps. Here I can be compound specific because we were looking at a dead series of compounds which were, on the one hand, phenyl benzyl ethers and, on the other, the corresponding ring-closed series where the oxygen is part of a pyran ring system. They were part of an unsuccessful search for anti-rhinovirus medicines.

Figure 4 shows the result of plotting a non-linear map for these phenyl ethers and pyrans where all the properties were predicted nuclear magnetic resonance chemical shifts for the protons and carbons common to both series. That is, all ^{13}C atoms other than 7 and 8 and for protons the 2,5,9,14 and 15 atoms since these positions were invariably unsubstituted. The ^{13}C shifts were normalised from 0 to 1. These compounds

were tested for plaque reduction of Rhinovirus type 1B and
classified from log ED 50 data into high, medium and low
potency.

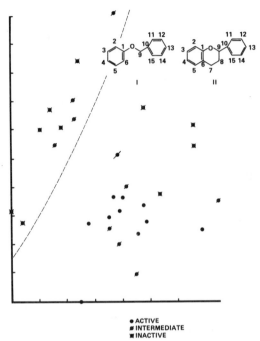

● ACTIVE
◪ INTERMEDIATE
✖ INACTIVE

Figure 4 – A non-linear map for the anti-rhinovirus activities
of phenyl ethers and the corresponding pyrans

        The n.m.r. people stated that there were no obvious
spectroscopic differences between these rather close series
which would allow one to distinguish one from the other.
However, the non-linear map succeeded in doing so. It can be
quite easily seen that the linear and the cyclic ethers are
completely separated, the pyrans all being below the dotted
line. Bearing in mind the sensitivity of the n.m.r. spectrum
to molecular structure, one might argue that this is to be
expected. But this is an aside, for the main purpose of the
exercise came when one marked in the activities of this
training set. I think you can see that when this is done
there is a very marked grouping of structural characteristics
which with hindsight tells us that the benzyl ethers were not
a good bet as antivirals.

        To summarise. From inspection it can be seen that:

(a)   the two chemical sub-series are effectively separated.
      and

(b)   the potency of any given compound can, to some extent, be
      deduced from its position on the map and/or the potency
      of its neighbours.   The performance of the map in
      defining that potency class to which the compound
      definitely does <u>not</u> belong is even better.

Another more recent example (Figure 5) shows the results
of applying the non-linear mapping technique to a number of
bicyclic amines.   Dr. Miller who originated these compounds
decided to apply the technique such that the mapping was the
result of applying the atomic charges on three of the atoms
together with the dipole vector in the Y direction and the
nucleophilic superdelocalisability on another atom, as well as
the X and Z principal ellipsoid axes;   a total of seven
parameters in all.

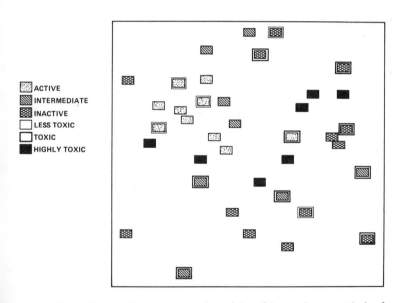

Figure 5 - A non-linear map for bicyclic amine antivirals
(plaque reduction influenza type A)

I think that the map clearly indicates that there is a
volume of multidimensional space which is a good target, but
my goodness, what a recipe!   This clearly highlights a success
and an intrinsic problem.   The problem is that of reversing
the process: from a recipe to a new series of molecules.

Most of these parameters are well understood by the physical
scientist. They are less well understood by others and at
present this makes the very process of predicting other active
compounds rather difficult. It highlights one of my original
points that the intelligibility of the recipe may indeed be a
significant constraint on the properties which are selected in
the heuristic process.

In Figure 6 we have a map from a training set of
molecules with known activities where the properties chosen
were the three principal ellipsoid axes, the orthogonal
components of the dipole moment, the total dipole moment and
the Van der Waals volume.

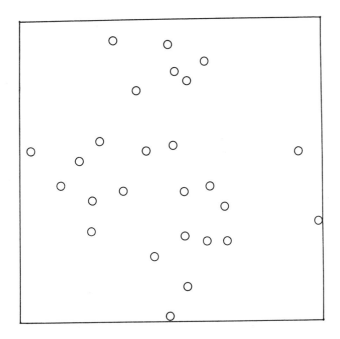

Figure 6 - A non-linear map for putative Aspartate Agonists &
Antagonists

If we now mark the classified biological activities of
all of the molecules on this map, (Figure 7) then the training
set takes on a more structured appearance. We see that we

have effectively grouped not just active and inactive, but agonist, antagonist and inactive.

A critical observer will notice that there are a circle and triangle in the wrong positions. For the sake of tidiness we should know that the circle is a partial agonist and the triangle derives from a compound which acts through an entirely different biochemical mechanism from the one which is involved in all the other compounds. In other words the molecule represented by the triangle probably does not act at the same receptor as the others. This is but one small example of the ability of this approach to reveal hetero- geneity of mechanism of biological action and highlights the need to recognise that this can be a limitation in applying the non-linear mapping technique.

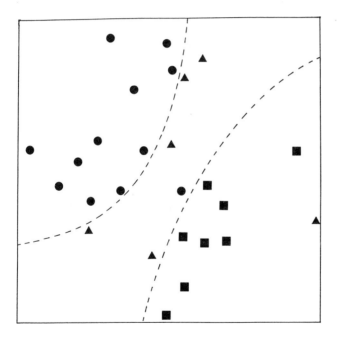

Figure 7 - A non-linear map showing the hyperspace correspond- ing to Agonism ■ , Antagonism ● and inactives ▲

The Future

So referring back to our original slide of the heuristic method, where are we going? What are the limiting factors in developing the approach? Is it realistic in the sense that simulation techniques are viable? In approaching the problem as being that of the generation of an expert system, where do our efforts have to be concentrated? What approaches are there to selecting the correct properties? Which ones relevant to biologial interaction should we be considering? How do we depart from the very considerable restrictions of the quantum mechanical approach which restricts us to molecular properties of single molecules rather than aggregates or solvates? How should properties be selected in respect of their usefulness in predicting new molecular species? The list is almost interminable but it is possible to focus on a few key factors.

One of the foremost, of course, is the selection mechanism for the properties such that they constitute the determinant factors in the biological action. One obvious way, albeit expensive in computing and rather short of subtlety, would be to take all possible combinations of the properties and use the maximising of grouping in the non-linear map as the figure of merit for the recipe. Another might be to recognise and to give weight to those properties which bear upon selected key functions. These might include a hydrogen-bonding parameter or parameters and in this context one can perhaps consider any drug or medicine as operating as a mixed donor and receptor of hydrogen bonds. In a more extreme sense, this would lead one to consider a peptide, for example, as an assembly of deactivated carbonyl receptors along with a corresponding number of activated NH donors. If one were to believe this, then these properties would determine the architecture rather than the other way round.

So one way forward for selecting properties is to orient them around interactions which we believe have a direct correlation with the biological process. Common sense will also dictate that where, for example, one is developing a molecule which is known to be intimately involved in a process which is mediated by electron transport, one would seek electrochemical properties to reflect this. Because of the correlation between free radical action and electron transport, one would then be led to seek ESR properties. When all these insights have been exhausted, one is still left with the arithmetic methods of restricting the choice of properties which we do already. For example, amongst others, we remove a property which is highly correlated with another.

So, by slanting our initial set of properties so that they more readily recognise some of our beliefs about interaction with receptors at the molecular level, laced with statistical techniques for removing redundancy, and with the liberal use of non-linear mapping techniques for further reduction, we have a way forward.

The next most fearsome problem is the development of logical techniques and the corresponding software to translate a set of essentially non-geometric recipe properties into suggested molecules which carry this recipe within them but are structurally different from the training set. As I said earlier, this itself is a restriction on the properties which we can usefully employ. As far as we are concerned this reverse process has a long way to go before we can see an expert system in sight.

To the cogniscenti, much of what I am saying will be seen as artificial intelligence applications, where the boundary between supervised and unsupervised learning is constantly being attacked. In my own mind I have little doubt that the broad proposition that I am putting forward will be welcomed by few and rejected by many. I will have no illusions that this will be an essentially emotional response.

Application of Molecular Dynamics in the Prediction of Accessible Conformations

D. J. Osguthorpe,* P. Dauber-Osguthorpe, R. B. Sessions,
P. K. C. Paul, and P. A. Burney

SCHOOL OF CHEMISTRY, UNIVERSITY OF BATH, CLAVERTON DOWN,
BATH, AVON BA2 7AY, UK

The use of theoretical chemistry techniques as an aid to drug design is rapidly becoming an integral part of any drug development program. Our current interest is in the application of the technique of molecular dynamics in the drug design field, particularly in the area of peptide hormones and in receptor-based drug design.

Molecular Dynamics Technique

A simulation of the dynamics of a molecular system ("molecular dynamics") is obtained by solving Newton's equations of motion for the system.

$$F_i = -\frac{\partial V}{\partial x_i} = m_i a_i \qquad i=1,n \qquad (1)$$

where F_i, x_i, m_i, and a_i are the force on atom i, the coordinates, mass and acceleration of atom i, and V is the energy of the system.

Given a potential energy function which describes the energy of the system as a function of the coordinates, the force on an atom is defined by the first derivative of the energy, V, with respect to the cartesian coordinates of that atom.

In molecular dynamics, as in energy minimisation[1] or normal mode analysis,[2] the potential energy of the molecule, V, is represented as an analytical function of all internal degrees of freedom and interatomic distances of the system as in equation (2).

$$V = \Sigma\{D_b[1 - e^{-\alpha(b-b_0)}]^2 - D_b\} + 1/2\ \Sigma H_\theta(\theta - \theta_0)^2 \qquad (2)$$
$$+\ 1/2\ \Sigma H_\phi(1 + s\cos n\phi) + 1/2\ \Sigma H_\chi \chi^2$$
$$+\ \Sigma\Sigma F_{bb'}(b - b_0)(b' - b_0')$$
$$+\ \Sigma\Sigma F_{\theta\theta'}(\theta - \theta_0)(\theta' - \theta_0') + \Sigma\Sigma F_{b\theta}(b - b_0)(\theta - \theta_0)$$
$$+\ \Sigma F_{\phi\theta\theta'}\cos\phi(\theta - \theta_0)(\theta' - \theta_0') + \Sigma\Sigma F_{\chi\chi'}\chi\chi'$$
$$+\ \Sigma\epsilon[2(r^*/r)^{12} - 3(r^*/r)^6] + \Sigma q_i q_j/r$$

This representation is known as a Valence Force Field, and reflects the energy necessary to stretch bonds (b), distort angles (θ) from their unstrained geometries, and twist torsion angles (ϕ). In addition, as we know from vibrational spectroscopy and normal mode analysis, these internal deformations are coupled, and this is represented by the cross terms (terms containing two internals, e.g. b and b', or b and θ). Finally, the nonbonded (or Lennard-Jones) and coulomb interactions, representing steric repulsions, dispersion or attractive forces, and electrostatic interactions are given by the last three terms.

Similar force fields are the basis of all current programs that perform molecular mechanics, such as MM2,[3] CHARMM,[4] AMBER,[5] GROMOS,[6] and the program in use at Bath, DISCOVER.[7] They differ somewhat in the terms included in the energy expression, the functional form and most of all in the values used for the parameters. The accuracy of any calculation with such a force field depends on how well the functional form and potential parameters represent the actual energy surface of the molecule.

The parameters we are using for the Valence Force Field were determined by fitting a wide range of experimental data including crystal structures (unit cell vectors and orientation of the asymmetric unit), sublimation energies, molecular dipole moments, vibrational spectra and strain energies of small organic compounds. *Ab initio* molecular orbital calculations have also been used in conjunction with experimental data to give information on charge distributions (used to derive partial atomic charges), energy barriers and coupling terms.[8]

Having specified the potential, we define the initial conditions of the molecular system, i.e. we select a set of initial coordinates and velocities for each of the atoms. Once the initial conditions are given, Newton's equations of motion are integrated forward in time. This is done in practice by calculating the forces, $\vec{F}_i$, from the analytical derivatives of the energy, V, and the acceleration, $\vec{a}_i$, from Newton's law, as given in equation 1. We then take a small time step, Δt, of 10^{-15} sec, and by applying the acceleration, $\vec{a}_i(t)$, over this time period, we update the velocity, and position of each atom in the system, using a Gear,[9] Verlet,[10] or leap frog[6] algorithm. The forces and accelerations are then calculated at the new position and the whole procedure repeated. From this integration of Newton's equation, we obtain the trajectory of each of the atoms as a function of time. In this way, we can follow the concerted motions of the atoms in the molecule as they move in response to the forces exerted on them by the other atoms in the molecule and maintain an average velocity in accord with the temperature of the system. This trajectory completely describes both the static and dynamic properties of the system in the classical limit. We obtain statistical thermodynamic, spectral and structural properties as appropriate time averaged quantities, while dynamic properties such as structural fluctuations or conformational transitions may be monitored directly by analysis of the trajectory, as we shall see below, or by viewing the conformational motion on an interactive graphics system or from a movie of the same.

Applications of Molecular Dynamics

Molecular dynamics applications fall into two basic classes, the "true" simulation of experimental phenomena and conformation searching.

Simulation of Experimental Results

Inherent to minimisation techniques is the limitation that disordered systems can not be simulated at all. In particular, the solvent can not be handled properly without taking into account the mobility of the solvent molecule. Furthermore, the dynamic nature of the molecule itself is not reflected in a minimisation procedure. Thus, when computational results of an isolated minimised molecule are compared with solution experiments, like is not being compared with like.

The first application of molecular dynamics is thus the "true" simulation of experimental systems, where the mobility of the atoms is taken into account explicitly. The resulting molecular dynamics trajectory contains all the information necessary to compute thermodynamic properties as well as dynamic properties. Thus, calculated properties correspond directly to the experimental ones, and the values should agree to within the limits of accuracy of the force field and the experiment. This provides an important tool for testing the validity of the methodology and the force field. For example, in a recent study a molecular dynamics simulation of a peptide, using periodic boundary conditions to represent the crystal, was carried out. From the simulation of peptide crystal, the time average structure was computed and compared to the X-ray structure.[11] In addition to the ability to reproduce experimental results, this simulation also provided insight to the effect of the crystal forces on the dynamics of the peptide molecule. A simulation of a protein crystal, including solvent and ions, provided the ability to study the instantaneous and average water structure and monitor the ion migration with time, properties which can not be obtained in any other way.[12]

In the other class of molecular dynamics applications, the ability of dynamics to overcome potential barriers in an energy directed manner is utilised for conformational searching. In these applications the exact trajectory produced by the simulation is of less importance.

Conformation Searching with NOE Distance Constraints

Recently, the conformational searching capabilities of dynamics has been applied to the problem of determining a 3-dimensional structure of proteins and peptides from NMR data.[13, 14] In order to "direct" the dynamics trajectory, additional terms have been included in the force field which constrain distances between atoms according to NOE data from NMR, for example,

$$V_{NOE} = K_{NOE}(D_{ij} - D_{NOE})^2$$

where D_{ij} is the current distance between atoms i and j, D_{NOE} is the corresponding experimental value, from NOE data, and K_{NOE} is an arbitrary "potential parameter" determining the stringency of this constraint.

Because of these additional terms, trajectories from such studies have no physical significance and "dynamic" events occurring in such trajectories cannot be related to experimental data.

Accessible Conformation Searching and Molecular Design

The other area conformational searching has been applied to is molecular design, particularly for peptide hormones. Peptide hormones are currently of great interest as they control many of the homeostatic mechanisms of animals and humans. By modulating the behaviour of peptide hormones with agonists or antagonists, it is possible to control many medical conditions. There is currently a lot of interest in the pharmaceutical industry in designing drugs by mimicking the peptide hormone controlling the faulty or diseased biological mechanism. Since these hormones are highly flexible, an integral and major part of designing the drug is to find out what is the "active conformation" of the hormone, by investigating the energetically accessible conformations of the hormone and its analogues.

Molecular dynamics simulations of the peptide hormone Vasopressin[15] have shown that this technique does generate different conformations of the peptide hormone. This has lead to the development of procedures which attempt to determine the binding and active conformations of peptide hormones using molecular dynamics.[16]

These procedures were developed based on the idea that conformational recognition is the basis of receptor-ligand interactions. Initially the receptor recognises a "binding" conformation, which the ligand must be able to adopt to bind to the receptor. The receptor may recognise this conformation by looking for the correct positioning of certain functional groups, the binding groups. This means that the ligand has to be able to adopt a certain conformation, one which has the functional groups positioned correctly. Following binding, the response is generated by either a conformational change occurring in the ligand-receptor complex (which involves the ligand) or the correct positioning of certain functional groups (the active groups). Thus, agonists are capable of undergoing this conformational change or they have the active functional groups, whereas antagonists do not.

Therefore, if we find the accessible conformations of a peptide hormone and those of an antagonist, and then perform a structural cross comparison of these two sets of structures, conformations that both the peptide and antagonist adopt are putative binding conformations whereas conformations that agonists adopt but antagonists do not are putative conformations necessary for activity.

We have been investigating this procedure on a number of peptide hormones at Bath, in particular a newly discovered peptide hormone melanin concentrating hormone (MCH) and cholecystokinin (CCK)

Accessible Conformations of MCH and MCH Fragments.

Melanin Concentrating Hormone (MCH) is a peptide hormone that was discovered at Bath University.[17] It has the opposite effect to Melanin Stimulating Hormone (MSH) as it causes the melanin granules to concentrate at the centre of the cells, hence causing the fish skin to appear paler, but it is not a simple antagonist of MSH. Although it has been shown to be present in higher animals, including man, its function in these animals is, as of yet, unknown.

MCH is an oligopeptide of 17 residues with the sequence:

Asp-Thr-Met-Arg-Cys-Met-Val-Gly-Arg-Val-Tyr-Arg-Pro-Cys-Trp-Glu-Val

A disulphide bridge between Cys^5 and Cys^{14} forms an intramolecular ring of 10 residues. This hormone is undergoing intensive study at Bath, with both synthetic and biological studies being performed, in addition to the computational studies, to attempt to determine the activity requirements of MCH.

The synthesis of MCH and of analogue fragments has been carried out by solid phase synthesis using the FMOC continuous flow method of Atherton and Sheppard[18] and pentafluorophenyl esters. The peptides have been purified generally by a combination of gel filtration chromatography and preparative reverse phase HPLC.[19]

Preliminary studies of the biological activity of the analogue fragments indicates that the disulphide bridge is essential for activity. In addition the C-terminal tail (residues 15-17) seems to be necessary for the efficient functioning of the hormone. The synthetic fragment, MCH_{5-14} is currently under NMR investigation, to determine the solution structure. Assignment of the backbone protons has been completed, using a combination of 1D decoupling and 2D-COSY experiments. NOE and NOESY experiment are currently underway to determine sequence specific assignments and distances. Additionally, the conformational homogeneity is being investigated using the criteria of Kessler. Hopefully, the NOE data can then be used in constrained dynamics simulations to determine the solution conformation of the peptide.

Concurrently with these studies, we have been using the computational techniques described above to get an indication of possible active conformations. We have performed a total of 150 picoseconds of dynamics so far on MCH and 2 fragments, the cyclic MCH_{5-14} ring and the linear MCH_{5-14} fragment. It is anticipated that such an approach would give a clue to the importance of the cyclic ring and the tail pieces in determining the conformations and consequent activity of MCH.

In order to analyse the results of the dynamics in conformational terms, we draw time profiles or trajectories of main chain torsion angles, which show the different conformations accessed and the transitions occurring between them. In the case of MCH and the two fragments under consideration it is found that the largest conformational changes occur in the Gly-Arg region. Figure 1 shows the trajectories of ψ_{Gly} and ϕ_{Arg} of the full MCH molecule. This shows a frequent conformational transition of the Gly residue from the α-helical to the C_7 equatorial. This in turn affects the conformation of its immediate neighbour Arg, and it

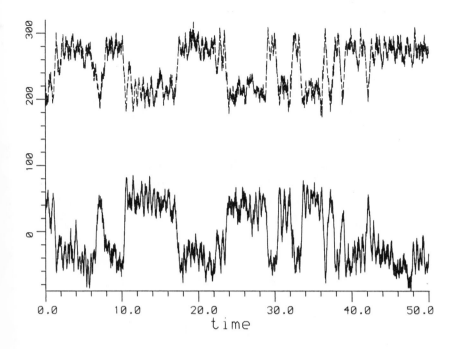

Figure 1   The trajectories of ϕ_{Gly} (full line) and ψ_{Arg} (dashed line) over a time period of 50 picoseconds. (Angles in degrees)

can be seen quite clearly that the conformational changes of ψ_{Gly} and ϕ_{Arg} are in tandem. The Gly residue is known to be quite flexible and therefore it is not surprising that this part of the MCH system shows the greatest mobility. In terms of conformational parameters and backbone structural features the (ϕ,ψ) of Gly and Arg move from around (-80,40) and (-150,100) to (-70,-40) and (-80,100) respectively. This can be construed as being due to the competition between the formation of a γ turn at Gly and a type II β turn across the adjacent Arg-Val junction. Surprisingly the two Val residues on either side of the Gly and Arg residues do not show any conformational changes. These two major conformations of MCH are shown in figure 2 (a) and (b).

(a)

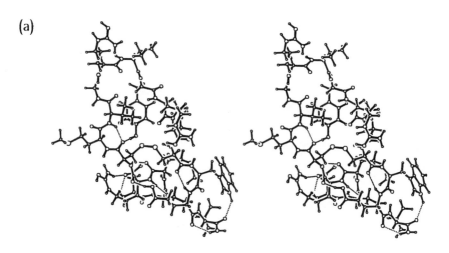

(b)

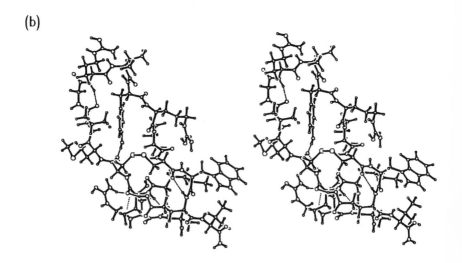

Figure 2.   The "snapshots" of some of the MCH conformations
                accessed during dynamics. Minimised conformations at (a)
                10 ps, (b) 12 ps, (c) 20 ps and (d) 40 ps

(c)

(d)

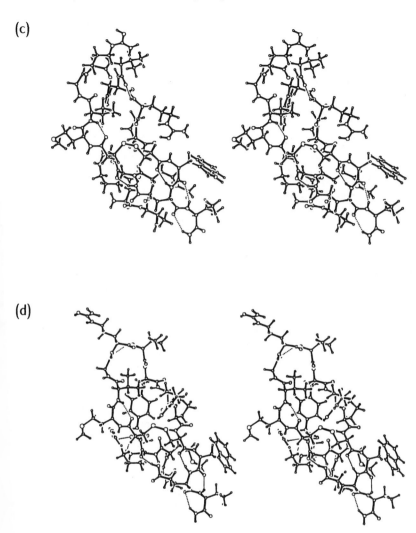

Figure 2.   continued.

Apart from these backbone conformational changes it is found that the rest of the ring is quite rigid. Other localised conformational changes in the MCH molecule include stretching or folding of side chains, formation of γ turns across Met[6], Pro[13] and sometimes across Arg[9], some of which are interrelated.

The next stage of the molecular design procedure is to determine the local minima being passed through. This is done by taking the instantaneous coordinates every picosecond and minimising each of these structures. Figure 2 (a)-(d) shows 4 selected conformational minima of MCH generated by this procedure which demonstrates the different conformations that have been found. One of the major features of these conformations is an internal cross-ring hydrogen bond from the Tyr[11] side chain hydroxyl to the backbone carbonyl of Cys[5] which may, to a certain extent, be responsible for the rigidity of the ring. We are currently starting dynamics from different initial structures to determine the importance of the presence of this hydrogen bond.

At this stage of the investigation, we cannot determine which conformations may be the binding or active conformations, although we may use the conformational ideas from the trajectory to direct the synthesis effort. For example, replacing the Gly with a D-amino acid residue, will constrain the flexibility of the peptide in this region, and enhance the preference of the α-helical over the C_7 equatorial conformation. Additional information is needed before we can make a more specific prediction of the active conformation. This emphasises the close relationship that is necessary in research of this kind between the synthetic, the biological and the theoretical studies to arrive at putative binding or active conformations.

One difficulty with accessible conformation searching by molecular dynamics is that it may be possible to miss some low energy conformations. For example, if two sets of conformations are separated by a high energy barrier between the two sets but only low energy barriers between conformations in each set (the "ergodic" problem), a dynamics trajectory starting in either set may never cross to the other set. We are currently addressing this problem for MCH by a complete generation of all possible ring backbone structures. A ring conformation is generated by taking each residue in turn and assigning one of a set of predefined values to the ϕ, ψ torsion angles. We have chosen a set of (ϕ, ψ) angles which span the low energy minima of the ϕ, ψ map, and can lead to all major secondary structures, including the α-helices, β-sheet, the β and γ turns and extended chains. This resulted in a set of 9 (ϕ, ψ) angles for residues 2-9 of the ring, seven angles for the Cys^5 ψ and four for the Cys^{14} ϕ. A total of $\approx 5 \times 10^8$ conformations have been considered. During the generation, we reject all conformations in which the distance between the C_β's of the Cys residues is not in the range 3.8-4.1Å, that is, it is not possible to cyclise the ring. This range takes into consideration that we are using a rigid geometry with specific ϕ, ψ angles and allows for possible ring structures where the ϕ, ψ angles could deviate from the table values. Additionally, hard sphere contact checking was used to reject further structures, using Van der Waal's radii of 1.4Å. The ring is then cyclised by an algorithm which adjusts the position of Cys^{14}-C_β such that the C_β-C_β distance is exactly 4.0Å. A pre-generated rigid geometry disulphide bridge in which the C_β's are 4.0Å apart is fitted onto the generated chain's C_β's and rotated until the geometry around the C_β's is tetrahedral. Two disulphide bridges are used, one with a torsion angle of 90° and one with -90°. Depending on the orientation of the C_α-C_β bonds this is not always possible for all generated rings. A final contact check is performed which leads to a further rejection of generated conformations. Each value of Cys^5 ψ leads to $\approx 40,000$ rings with C_β's within 3.8 to 4.1Å. Cyclisation of these conformations and further contact checking lead to a final set of $\approx 2,500$ conformations. This has been done for 7 Cys^5 ψ values for a total of $\approx 17,500$ rings.

Analysis of these conformations is currently underway.[20] Preliminary investigations show that around 30% of the generated conformations have one β turn, and more than 80% have at least one hydrogen bonded γ turn. Further, a high incidence for the starting positions (i+1 position) of β turns is found for Met^6, Gly^8 and Val^{10}. An interesting result is that the Tyr^{11}-Arg^{12} and the Arg^{12}-Pro^{13} junctions are not involved in any β

turn conformation. This is because of the conformational restrictions imposed by the presence of the Pro residue and the cyclisation of the ring. In particular a Tyr[11]-Arg[12] β type turn can be ruled out because of the short contacts between the N-C$_\delta$ of the Pro[13] residue and the C-O group of the Val[10] residue. Thus the Pro residue, which is known to be involved in β turns[21] in peptides, actually prevents the formation of any β turns in its proximity in MCH. However, the analysis shows that the propensity for γ turns to occur is the highest for Pro when compared with the other residues. Around 30% of all the equatorial γ turns found occur in the Pro residue while each of the other residues are found in about 10% of these turns, except for Arg[12] which, owing to the conformational restriction imposed by the adjacent Pro residue, is not found in any γ turn conformation. Further analysis will be used to determine how well the dynamics trajectory has sampled these conformations and selected generated conformations will be used as the starting structures for additional dynamics trajectories.

As additional data becomes available on the activity of the synthetic fragments of MCH, additional conformational searches will be performed which will hopefully lead to a putative active conformation for MCH.

Accessible Conformations of CCK and Some of its Analogues

We are studying CCK and its analogues and as activity data for the analogues is available, a complete study based on the molecular design techniques described above is possible.

Cholecystokinin (CCK) is a neuropeptide hormone present in both the gastrointestinal tract and nervous system.[22] In the gut it is present mainly as a linear 33 amino acid peptide (CCK33) and its functions include stimulation of both pancreatic enzyme secretion[23] and gall bladder contraction.[24] However, in the peripheral and central nervous system the majority of CCK is in the form of the C-terminal sulphated octapeptide fragment CCK8, which has been shown to exhibit all the hormonal properties of CCK33.[25] The function of CCK in the central nervous system has been linked to the regulation of appetite, analgesia,[26] and schizophrenia.[27]

The non-sulphated 8 residue CCK (CCK8NS), with the sequence:

Asp Tyr Met Gly Trp Met Asp Phe NH$_2$

has been shown to have similar receptor binding as CCK8[28] and, as the sulphate group does not add any major conformational degrees of freedom into the peptide chain nor the side chains, the accessible conformations for CCK8NS should be close to those for CCK8.

The first part of the design procedure is to perform an accessible conformation search and we have currently performed a 100 picosecond molecular dynamics simulation of CCK8NS. Every picosecond along this trajectory, the instantaneous structures were energy minimised. Out of the 100 minimised conformations 80 were found to be unique, i.e. by superimposing each structure on all others, the root mean square (RMS) deviation between all heavy atoms is larger than 0.01 Å

The second stage of the design process is to perform an accessible conformation search of an analogue of the hormone, and preferably as constrained an analogue as possible. We are studying the cyclic analogue

Boc-D-Glu-Tyr(SO$_3$H)-NLeu-D-Lys-Trp-NLeu-Asp-Phe-NH$_2$

which is cyclised through an amide bond of the side chains of D-Glu and D-Lys. A preliminary dynamics simulation has been carried out on the simplified molecule

D-Glu-Tyr-NLeu-D-Lys-Trp-NLeu-Asp-Phe-NH$_2$

for 25 picoseconds. Again, structures were minimised every picosecond to yield 25 conformational minima.

The final stage of the molecular design procedure is to perform a structural cross-comparison of the accessible conformations of the native compound with those of the analogue. The first cross-comparison was done by fitting the backbone atoms of the structures.

We are currently in the process of performing the full analysis of these two sets of conformations.[29] An important part of this analysis is to consider a wide range of "similarity indicators". This is necessary as we do not know if the position of the backbone atoms, which is what the fit of the backbone atoms gives us information about, is essential to the definition of active or binding conformations. Thus comparisons are carried out using properties such as torsion angle values, location and relative orientation of side chain atoms, hydrogen bond networks, position of electrostatic charges, etc. Figure 3 shows a fit of a conformation of the native compound with a conformation of the cyclic analogue based on the residues with aromatic sidechains.

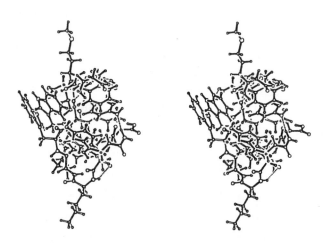

Figure 3   Superposition of a conformation of CCK8NS with a conformation of the cyclic analogue based on the aromatic sidechains

Accessible Conformation Searching with a Known Receptor Structure

We are also using accessible conformation searching to aid us in the molecular design of compounds when the structure of the receptor is known. In particular, we are applying this to the design of new inhibitors to Phospholipase-A_2 (PLA$_2$) by investigating the binding of known ligands to PLA$_2$, using the crystal structure of extra-cellular PLA$_2$ determined by Drenth *et. al.* [30]

PLA$_2$ hydrolyses phospholipids at the 2-acyl ester linkage, and is therefore responsible for liberating arachidonic acid from the phospholipid membrane pool. Arachidonic acid is further metabolised and leads to inflammatory mediators such as prostaglandins, prostacyclins and leukotrienes, and thus plays a key role in the inflammatory response. Control of this enzyme is likely to be of therapeutic value in the treatment of diseases such as rheumatoid arthritis and atherosclerosis.

We have performed a preliminary minimisation of the X-ray structure to provide a starting structure for the modelling. This is necessary as the X-ray structure is a time-average structure, not a minimum structure, and a minimum energy structure is required in order to make an energetic analysis of bound ligands consistent. We included 288 waters in the minimisation, these are the X-ray waters and further waters generated to surround charged groups to a distance of 3.5Å and the active site to a distance of 10Å. Bridging X-ray waters which hydrogen bond to at least two protein residues were also included, since they are essential for maintaining the integrity of the protein structure. The structure was minimised to an average derivative of 0.02 kcal mol^{-1} and figure 4 shows the difference between the X-ray and minimised structures. The RMS deviation between all heavy atoms of the backbone of the α-helical and β-sheet secondary structures is 0.862Å.

For the first step of investigating ligand binding to PLA$_2$, we decided to dock a substrate into the active site to gain a better understanding of how the enzyme binds compounds. This information can only be obtained from theoretical studies since substrates are not usually found in X-ray structures as the enzymes are generally still active and thus the substrate would react over the course of the X-ray data collection making determination of the substrate density impossible. The active site domain of PLA$_2$ has been identified by a number of experiments.[31, 32] In the case of the PLA$_2$ X-ray structure nothing was found in the active site except for a solvent molecule, a pentanediol, as the solvent of crystallisation was a 50% mixture of this diol with water. We extracted from the Cambridge Crystallographic database a phospholipid structure [33] in a

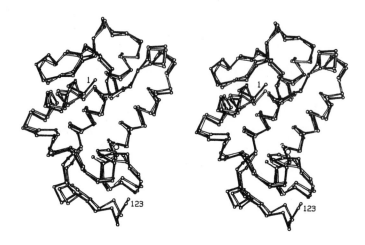

Figure 4   C-α trace of PLA$_2$ (open bonds) and minimised (filled bonds) structures

conformation most similar to that usually found in solution.[34] This substrate was initially visually docked into the active site of the minimised protein in a fashion consistent with the spatial restrictions imposed by the mechanism proposed by Drenth.[35] This involves the His[48] residue in the active site, Asp[99], the calcium ion and an active site water. In the first stage, a water hydrogen bonded to the His[48] acts as a nucleophile and attacks the ester link to form the charged tetrahedral intermediate which is stabilised by the Ca^{2+}. This complex collapses to products, and finally proton transfer to His[48] completes the catalytic cycle.

In order to attain a satisfactory docking, it was necessary to make some modifications not only to the substrate conformation but also to the protein structure. The first problem was residue Tyr[69]. In the X-ray structure the side chain of this residue is completely blocking the entrance to the active site and it was necessary to rotate the side chain torsion angles to position it away from the active site. Then, although the substrate could be docked, only a reasonable fit could be achieved, with holes in the space between the phospholipid and the protein. On investigation, it turned out that the side chain of Asp[49] in the X-ray structure is coordinated to the Ca^{2+}, however, if we rotate the side chain torsion angles of this residue it can be moved away from the Ca^{2+} and replaced by the PO_4^- of the phospholipid. This leads to a much better fit of the rest of the substrate in the active site. Additional evidence supporting such a binding is that there is a class of PLA_2's in which the Asp[49] is replaced by Lys. These enzymes are still active and, moreover, the Ca^{2+} can be shown to be bound only after the phospholipid.[36] This complex was energy minimised and in the resulting structure the phospholipid is snugly bound in the active site with the catalytic water in the correct position just above and behind the carbonyl carbon of the cleaved ester link, see figure 5.

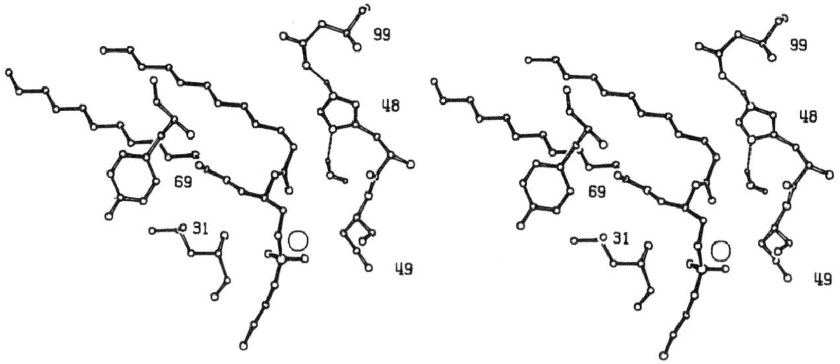

Figure 5   Phospholipid docked into PLA_2 active site (Minimised complex)

We are currently docking other known inhibitors to this enzyme, in particular a transition state analogue [37] and the fluorescence probe 1,8-anilinonaphthalene sulphonic acid (ANS),[38] see figures 6 (a) and (b). In particular, the *in vacuo* minimised structure of ANS fits snugly into the active site.[39]

What this study has brought out so far is the importance of taking into account that the receptor as well as the ligand can be distorted or

(a)

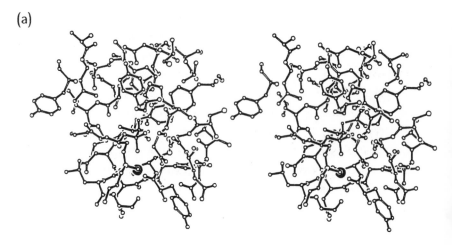

(b)

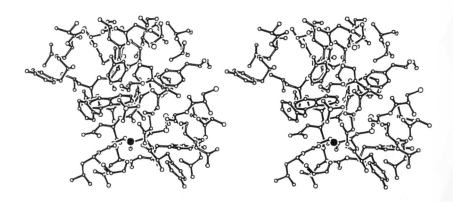

Figure 6   Inhibitors docked into the PLA₂ active site
              (a) a hydrated fluoroketone phospholipid analogue
              (b) 1,8-anilinonapthalene sulphonic acid (ANS)

rearranged on binding. In order to avoid the pitfalls of simple rigid docking, we are currently using molecular dynamics of the phospholipase-inhibitor complexes to explore the local accessible conformations of the active site. We are also examining the dynamic nature of the protein - ligand binding procedure by investigating the low frequency normal modes of the protein. These normal modes are the natural motions of the protein and may be important in substrate binding as well as in product release. For example, these modes may suggest low energy distortions of the protein which lead to easy access of the ligand to the active site, or better ligand - protein interaction.

Summary

Molecular dynamics is a powerful tool in the molecular modellers kit which can aid in the understanding of the conformational properties of drug molecules and thus aid in the molecular design component of a drug design strategy.

References

O. Ermer, *Structure and Bonding*, **27**, 161, Berlin (1976).

E. B. Wilson, J. C. Decius, and P. C. Cross, *Molecular Vibrations*, McGraw Hill, New York (1955).

U. Burkert and N.L. Allinger, in *Molecular Mechanics*, American Chemical Society, Washington, D.C. (1982).

B. R. Brooks, R. E. Bruccoleri, B. D. Olafson, D. J. States, S. Swaminathan, and M. Karplus, *J. Comp. Chem.*, **4**, 187 (1983).

S. J. Weiner, P. A. Kollman, D. A. Case, U. C. Singh, C. Ghio, G. Alagona, S. Profeta, Jr., and P. Weiner, *J. Am. Chem. Soc.*, **106**, 765-784 (1984).

H.J.C. Berendsen, J.P.M. Postma, W.F. van Gunsteren, A. DiNola, and J.R. Haak, *J. Chem. Phys.*, **81**, 3684-3690 (1984).

Available from Biosym Technologies, San Diego, USA.

P. Dauber-Osguthorpe, D. J. Osguthorpe, J. Wolff, and A. T. Hagler. in preparation.

9. C. W. Gear, *Numerical Initial Value Problems in Ordinary Differential Equations*, Prentice Hall, Englewood Cliffs, New Jersey (1971).

10. L. Verlet, *Phys. Rev.*, **159**, 98 (1967).

11. D. Kitson and A.T. Hagler. private communication

12. P. Avbelj, D. Kitson, J. Moult, A.T. Hagler, A. Sielecki, and M.N. James. private communication

13. G.M. Clore, A.M. Gronenborn, A.T. Brunger, and M. Karplus, *J. Mol. Biol.*, **186**, 435-455 (1985).

14. R. Kaptein, E.R.P. Zuiderweg, R.M. Scheek, R. Bolens, and W.F. van Gunsteren, *J. Mol. Biol.*, **182**, 179-182 (1985).

15. A. T. Hagler, D. J. Osguthorpe, P. Dauber-Osguthorpe, and J. C. Hempel, *Science*, **227**, 1309 (1985).

16. E.L. Baniak, L.M. Gierasch, A.T. Hagler, J. Rivier, T. Solmajer, and R.S. Struthers, in *Program of the Ninth American Peptide Symposiumi, Toronto, Canada*, p. 125 (1985).

17. B.I. Baker and T. Rance, *Gen. Comp. Endocrinol.*, **37**, 64-73 (1979).

18. E. Atherton, D.L.J. Clive, and R.C. Sheppard, *J. Am. Chem. Soc.*, **97**, 6584 (1975).

19. M.M. Campbell, D. Brown and P. White. to be published.

20. P.K.C. Paul, M.M. Campbell, and D.J. Osguthorpe. in preparation

21. G.D. Rose, L.M. Gierasch, and J.A. Smith, *Adv. Protein Chem.*, **37**, 1 (1985).

22. J.J. Vanderhaeghen, J.C. Signeau, and W. Gepts, *Nature*, **257**, 604-605 (1975).

23. A.A. Harper and H.S. Raper, *J. Physiol*, **102**, 115-125 (1943).

24. A.C. Ivy and E. Oldberg, *Am. J. Physiol.*, **86**, 599-613 (1928).

25. M.A. Ondetti, B. Rubin, S.L. Engel, J. Pluscec, and J.T. Sheeham, *Am. J. Dig. Dis.*, **15**, 149-156 (1970).

26. G. Zetler, *Psychopharmacology Bulletin*, **19**, 347-351 (1983).

27. N.P.V. Nair, S. Lal, and D.M. Bloom, *Prog. Brain Res. (Psychiatri. Disorder; Neurotransm. Neuropept.)*, **65**, 237-258 (1986).

28. P. Gaudreau, R. Quiron, S. St Pierre, and C.B. Pert, *Eur. J. Pharmacol.*, **87**, 173-174 (1983).

29. P.A. Burney and D.J. Osguthorpe. in preparation
30. B.W. Dijkstra, K.H. Kalk, and J. Drenth, *J. Mol. Biol.*, **147**, 97 (1981).
31. H.M. Verheij, J.J. Volwerk, E.H.J.M. Jansen, W.C. Puyk, B.W. Dijkstra, J. Drenth, and G.H. de Haas, *Biochemistry*, **19**, 743-750 (1980).
32. J.J. Volwerk and G.H. de Haas, *Lipid Protein Interactions*, **1**, 69-149 (1982).
33. M. Elder, P. Hitchcock, R. Mason, and G.G. Shipley, *Proc. R. Soc. Lond. A*, **354**, 157-170 (1977).
34. H. Hauser, W. Guyer, I. Pascher, P. Skrabal, and S. Sundell, *Biochemistry*, **19**, 366-373 (1980).
35. B.W. Dijkstra, J. Drenth, and K.H. Kalk, *Nature*, **289**, 604-606 (1981).
36. J.M. Maraganore and R.L. Heinrikson, *Biochem. Biophys. Res. Commun.*, **131**, 129-138 (1985).
37. M.H. Gelb, *J. Am. Chem. Soc.*, **108**, 3146-7 (1986).
38. M.G. van Oort, R. Dijkman, J.D.R. Hille, and G.H. de Haas, *Biochemistry*, **24**, 7987-7993 (1985).
39. R.B. Sessions, M.M. Campbell, and D.J. Osguthorpe. in preparation

19